TECHNICAL WRITING

Process and Product

TECHNICAL WRITING

Process and Product

Charles R. Stratton
University of Idaho

Holt, Rinehart and Winston
New York Chicago San Francisco Philadelphia
Montreal Toronto London Sydney
Tokyo Mexico City Rio de Janeiro Madrid

An earlier version of some of the material in Chapter 1 appeared in "Technical Writing: What It Is and What It Isn't," *Journal of Technical Writing and Communication*, 9:1 (1979), and of some of the material in Chapter 3 in "Needs Assessment for Communication System Design" in the same journal, 6:2 (1976). An earlier version of the material in Chapter 22 appeared in "Analyzing Technical Style," *Technical Communication*, 26:3 (1979).

Library of Congress Cataloging in Publication Data

Stratton, Charles R.
Technical writing.

Includes index.
1. Technical writing. I. Title.
T11.S718 1984 808'.066'6 83–12672

ISBN 0-03-056733-5

Address correspondence to:
383 Madison Avenue
New York, N.Y. 10017

Printed in the United States of America
Published simultaneously in Canada

4 5 6 7 090 9 8 7 6 5 4 3 2

CBS COLLEGE PUBLISHING
Holt, Rinehart and Winston
The Dryden Press
Saunders College Publishing

□ Preface □

I have written this book for students. I have used an easygoing writing style that students enjoy—at least the 400 or so students who have used preliminary drafts have enjoyed reading it. They have found the material in it easy to read, easy to understand, and easy to remember. The students I have aimed at are those majoring in a variety of scientific and technical disciplines, students who, although they may not have great love of writing, realize the importance of writing to their professional development. In general, I have oriented the book away from school writing and toward on-the-job writing, although I do pause from time to time and offer "Success Strategies for Student Writers" in recognition of the fact that the book is, at base, a text to be used in a technical writing class.

Throughout the book, I stress the *process* of writing and revising, rather than dwelling on the various *products* of writing. I truly believe that the most valuable thing students can take from a writing class is knowing how to write rather than knowing what to write or what their writing should look like when they're through. In another departure from mainstream technical writing texts, I acknowledge the legitimate place that adequate writing has in the world of work. Not everything must be excellent. (What is an excellent requisition, anyhow? Or an excellent telephone memo?)

The book has four parts. In Part One—Preliminaries, I discuss some of the differences between college writing and professional writing and offer my views on some of the kinds of writing that is done in the world of work and on the different kinds of writing that different technical and professional people are asked to do in business, industry, and government agencies. I go on to discuss adequacy and excellence in technical writing, offering characteristics of each and inviting students to gather samples of each from their own disciplines. Finally, I discuss the importance of analyzing communication contexts in terms of audiences, purposes, and types of information. The thrust of Part One is to help students move away from writing writing and toward communicating information.

Part Two—Basic Writing Techniques gets student writers on the road to adequacy, with a bit of standard formats; an overview of data-gathering, initial draft writing, and revisions; procedures for handling manuscripts, editing for correctness, developing graphics and illustrations, and handling documentation in reports. In the six chapters of Part Two, I also introduce some terminology and concepts that will prove useful in later years as young professionals begin to interact with editors, graphic artists, typesetters, and printers.

Part Three—Types of Written Communication, as its name implies, gets into the nuts and bolts of correspondence, documents, and publications. The movement in Part Three is from the common and routine (letters, memos, and minor reports) to the less frequent but more demanding (management decision-making documents and technical articles)—a movement from the demands of adequacy to the demands of excellence, as it were.

Finally, in Part Four—Revising for Excellence, I present some new ideas (at least new for most readers) on style in technical writing and some suggestions on improving writing style to increase the likelihood that the end product will be judged excellent. I am convinced that style is the essence of excellence, and I am equally convinced that habits of effective style can be taught and learned in a systematic fashion.

My debts are many in this undertaking. Let me acknowledge a few of them. I am grateful to my colleagues in the English Department at the University of Idaho who read and revised early drafts and served as valuable sounding boards—especially Doug Adams, Richard Dozier, Dick Hannaford, Teo Sipahigil, and Leo Storm. I am grateful as well to my colleagues elsewhere who reviewed intermediate and final drafts and offered many suggestions for improvement—especially Dave Carson (Rensselaer Polytechnic Institute), Margot Haberhern (Florida Institute of Technology), Tom Pearsall (University of Minnesota), and Jim Souther (University of Washington).

I wish to offer special thanks to Janyce Paulsen who keyboarded the entire manuscript many times and to the two finest English editors anywhere (they happen to be at Holt)—Susan Katz and Nedah Abbott. All three have been special people in my life. Let me acknowledge, also, the invaluable assistance of Frank Habermann (Creative Director, The Trane Company), Lynn Swigart (Service Publications Manager, Caterpillar Tractor Co.), Jack Hammon (President, Interwest Engineering), and H. L. Kirk (Senior Project Editor, Holt, Rinehart and Winston).

Finally, and most importantly, I wish to thank the countless students who labored in my classes with preliminary drafts of the book. It is they, more than anyone, who have made the book what it is because they represent the students everywhere that I have written the book for.

C. R. S.

Moscow, Idaho
August 1983

□ Contents □

PART TWO

BASIC WRITING TECHNIQUES

TECHNICAL WRITING

Process and Product

☐ ONE ☐

Preliminaries

It is customary in a textbook such as this to talk about technical communication in general terms a bit before settling down to talk about specific techniques and specific forms. It is also useful. For most of you, the three chapters of Part One will present some rather new—and perhaps rather disturbing—concepts. For a dozen years or more you have taken classes in English or language arts. You have been exposed to a great wealth of knowledge about writing and reading and about grammar and usage, but most of this exposure has been oriented toward communication skills that serve you in your role as student. You have learned how to write for your classes—both English classes and other classes. But now you have to start thinking about communicating as a professional—as a technical expert. This is the primary goal of Part One: to help you begin to make the transition from student writer to technical writer—or technical person who writes.

Chapter 1, "Writing in Business, Industry, and Government," opens with a discussion of college writing versus professional writing and then goes on to talk about who writes what in the world of work. As you study this chapter, project ahead in your own professional life and try to predict kinds of writing you will be asked to do when you get on the job. Chapter 2, "Adequacy and Excellence," suggests that there is a place in the real world for adequate writing as well as excellent writing and then outlines some of the characteristics of each. One thing you should note from this chapter is that excellent writing requires a lot more effort on your part than does adequate writing. Finally, Chapter 3, "Analyzing the Communication Context," discusses audience, purpose, and type of information and the ways in which these factors of the communication context affect what you write and how you write it. Chapter 3 also introduces a device you will find in many of the remaining chapters of the book: "Success Strategies for Student Writers." As a whole, *Technical Writing: Process and Product* addresses you in your role as a developing professional and presents information, concepts, and skills that you will find most useful in on-the-job communication contexts. Of course, no matter how realistic you attempt to make it, writing in the classroom is different from writing on the job, hence the "Success Strategies." In them, I address you in your role as student writer and offer suggestions designed to make your task as a student a little easier or to discuss aspects of writing and revising that are peculiar to classroom writing and have little application in the world of work.

Part One, then, covers preliminaries. But these preliminary concepts form the foundation for the entire book. Moreover, these concepts will be new for most of you. I would suggest reading these first three chapters with some care and then returning to them after you complete *each* of the remaining major sections of the book. If you do this, you will find that not only does each major section shed new light on the preliminary concepts but also that a rereading of the first three chapters will shed valuable light on each major section.

1

Writing in Business, Industry, and Government

Technical writing is basically the kind of job-related writing you will do in your profession or discipline. Whenever you are dealing with things pertaining to a particular art, science, or trade, you are dealing with things technical. As an English adjective, the word *technical* is derived through French from the Greek *technikós*—"technique" or "technic." It is commonly assumed that for writing to be "technical" it must be filled with symbols, equations, and acronyms, and be generally obtuse if not downright unreadable. This is not the case. Technical writing is a pragmatic sort of writing designed to generate some observable response on the part of its readers. It is writing designed to get people to do things. Technical writing is an act of communication rather than an act of self-expression. The information you present is more important than your attitude toward it. Technical writing, then, is simply the writing you as a person in a particular field are going to be doing when you are on the job.

College Writing and Professional Writing

As a college student, you have studied composition and you have done quite a bit of writing for classes. This is fine, for after all technical writing is *writing*. All that you know about organization and development, about unity and coherence, about words, sentences, and paragraphs is germane to technical writing. Technical writing—like any writing—demands care with spelling, punctuation, grammar, and usage. But the writing you do as a professional will have a wholly different orientation from the writing you have done as a college student. The mechanics of the two types of writing are similar, but the audiences, purposes, and techniques are vastly different.

The essay you write for a final examination typically has an audience of one—your instructor—who knows more about your subject than you do and is less interested in what you say than in how well you say it. Much the same can be said for your lab reports, your term papers, and those essays you wrote for freshman composition ("What I Did Last Summer," "My Views on Genetic Engineering," "My First Love Affair"). The reports and proposals you write on the job will typically have large and diverse audiences who lack your special and immediate knowledge of the material at hand but need to gain that knowledge in order to pursue their own job activities.

Your instructors read your college essays and term papers primarily to assess your performance and to get an idea of your grasp of the material you cover. Their professional lives do not change measurably as a result of having read what you write. Your purpose in writing college papers is not to inform but to impress. You are concerned with what *you* get out of your writing and may care not a whit what your instructor gets out of it. Your professional writing, on the other hand, must be instrumental. The readers of your reports, instructions, and proposals will read what you write because they need the information you possess to make decisions, perform their jobs, and take actions based on your recommendations. Your purpose in writing becomes theirs; you write to communicate the information they need, and your chief goal is to provide that information as clearly, concisely, and completely as you can.

Finally, techniques. Your college essay was neatly handwritten on lined tablet paper. It had no headings or subheadings, no illustrations, no title page, and no appendixes. You generated only a single original copy and turned it in with comfortable finality, knowing that—good or bad—you would never touch it again. All this will be different with your professional writing. You will generate rough drafts that will then be typed, typeset, keyboarded into a computer, or what have you. Nearly everything you write will include drawings, photos, tables, or charts. Headings, subheads, tables of contents, and indexes will be the rule rather than the exception. Dozens or even hundreds of copies of your reports and manuals will be printed, and your writing will come back to you months—even years—later for revision and updating.

□ Communication versus Self-Expression

In college you write primarily as an act of self-expression. On the job you will write primarily as an act of communication. In college your writing is primarily creative and expository. On the job your writing will be primarily technical, or instrumental.

Technical writing is not creative writing. Although as a technical writer you must be creative, a report is different from a poem. When you write an ode or lyric, you are centering your attention on the relationship between yourself and your *subject*. The reader is present but out of focus. You do your thing with your subject and hope your reader likes it. (Or maybe you don't care whether your reader likes it or not.)

Technical writing is not expository writing. Although as a technical writer you will use description, narration, and argumentation, a report is not an essay. When you write an essay, you are centering on the relationship between yourself and your *reader*. The subject is there, but your principal concern is to bring your reader to an appreciation of your point of view about the subject. The stress is on the relationship between writer and reader, and the subject fades somewhat into the background.

We can view these relationships graphically by regarding the writer, reader, and subject of a writing situation as the vertices of an equilateral triangle (see Figure 1–1). Each mode of writing—creative, expository, and technical—has its principal domain along one of the sides of the triangle. Creative writing, as represented by the poem, highlights the link between writer and subject (see Figure 1–2), while expository writing, as represented by the essay, highlights the link between writer and reader (see Figure 1–3).

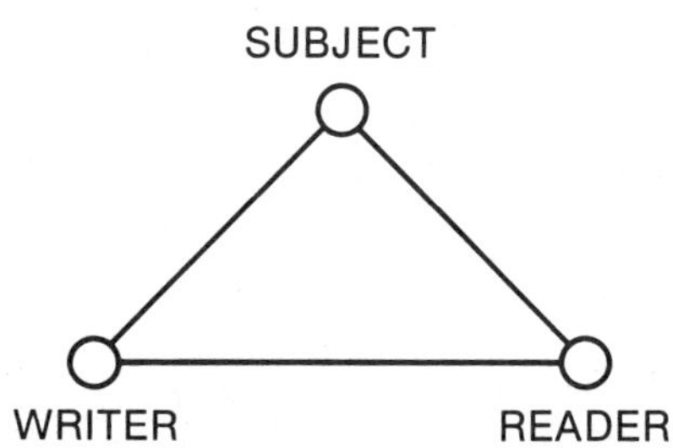

Figure 1–1 The Communication Triangle

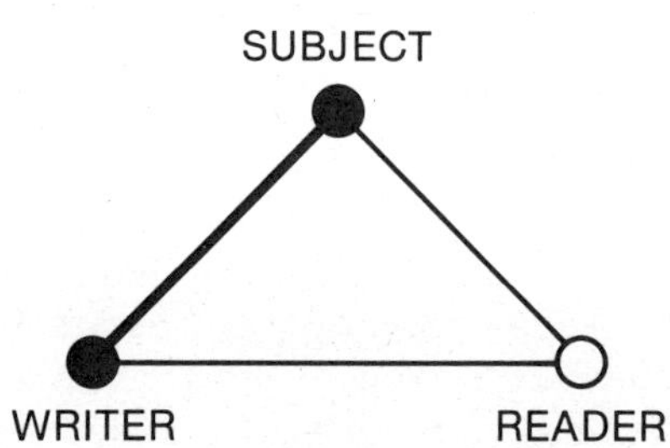

Figure 1–2 Creative Writing Focus

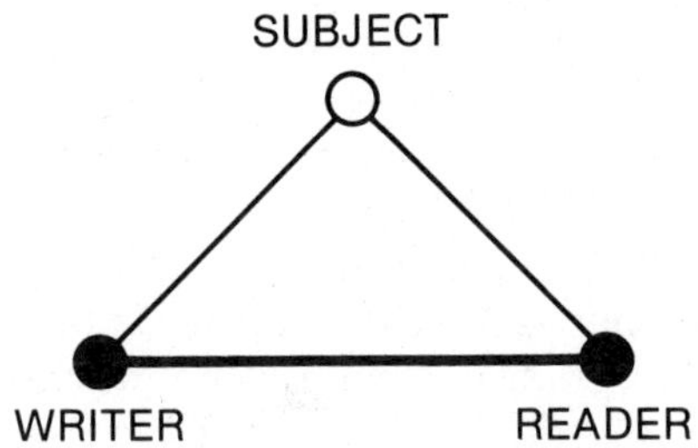

Figure 1–3 Expository Writing Focus

What of technical writing, then? When you write a report, you as writer must fade into the background. The focus must be (see Figure 1–4) on the relationship between *subject* and *reader*. As a technical writer you must be willing to efface yourself so that the light in your writing shines fully on the relationship between information and audience. You studied literature to learn how the writer approaches the subject. You studied composition to learn how you as a writer should approach your audiences. You study technical writing to learn how to help your audiences approach your subject. In all three modes of writing we are concerned with the interactions of writer, subject, and reader, but in each we bring a particular pair of these to the foreground and move the third to the background. The thing that makes technical writing unique (and somewhat intimidating) is that with technical writing it is you as writer who must take a secondary role.

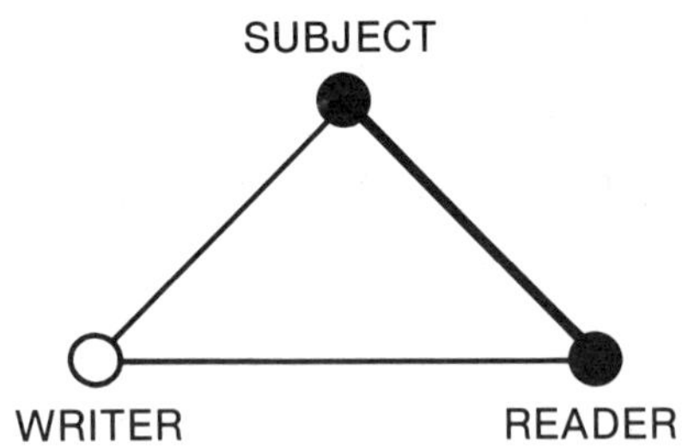

Figure 1–4 Technical Writing Focus

In fact, these differences in the roles of self and audience account for most of the distinctions between technical writing and other kinds of writing. You begin an essay or a story with an awareness of what you want to do and generally how you want to do it. Through your writing you express your observations, your ideas, your feelings, your personality, your insight, your ability to perceive. Your goal is to make contact with a reader's mind and feelings and carry the reader along with a developing idea of your own. The successful essay or story starts with a clear picture of your position vis-à-vis a subject and then draws your audience toward that position.

You must begin a report or proposal, however, with a clear picture of who will read and why. You must constantly keep your intended readers in mind, being considerate of their degrees of knowledge and understanding and constantly keeping before you the effect you want to have on your readers or the definite response you want to elicit from them. What looks or sounds good to you is not enough; what you write must look and sound good to your readers. The successful report starts with a clear understanding of where the reader is and then carries the subject matter to the audience.

□ The Report as an Instrument of Change

People don't write reports just for the heck of it. It's not likely that your boss will ever say "Since things are slow this afternoon, why don't you write a report?" People don't read reports just to pass the time of day, either. No one I know likes to curl up by the fire with a glass of warm milk and read a good technical report. Yet lots of reports get written and read. Why? The reason is simple: people read technical reports to gain information that helps them in their jobs. No other reason. If a report contains no new information (or if the information cannot readily be extracted from the report) or if the information has no utility for someone in the performance of his or her job, the report will not be read. In order to be useful, a report must make a difference.

A manager reads your recommendation report or feasibility analysis or proposal because he or she must make a management decision that affects the future of the firm or organization. If your report helps in the making of that decision, it's a good report—it makes a difference. An equipment operator reads your instructions because he has to run a piece of machinery. If they help him run the machinery more safely and more effectively, they're good instructions—they make a difference. The same is true for your technical articles, your state-of-the-art reports, your environmental impact statements, your specifications. The document is read only because someone has a job-related technical problem and needs your information to help solve it. Your document is a good one only to the extent that it helps the person reading it bring about some change, some improvement, in his or her firm or organization or professional life.

So to be effective in the technical writing you do, you must be able to specify the effects—the changes—your document will have and then channel all your writing energies to help achieve your desired effects: to help cause your desired changes. The document that fails to have an effect on its readers is at best useless. At worst, it is a waste of time and money for both writer and readers. None of the writing you do on the job will be judged mainly on grammar and spelling and punctuation, mainly on how well you express your ideas, or mainly on how well the reader likes it. It will be judged on how effective it is as an instrument of change.

□ Major Types of On-the-Job Writing

It might be a good idea here to look at some of the major sorts of writing you as a technical person will be asked to produce when you get on the job. For openers, we can identify three main categories of on-the-job writing:

- □ **Correspondence**—the letters and memos that flow between individuals in organizations, either to provide a written record of conversations or to serve as brief exchanges of information on limited topics.
- □ **External publications**—the formal pieces of writing that go outside the organization and serve, among other things, to project the organization's public image.
- □ **Internal documents**—the reports, proposals, plans, and procedures that stay inside the organization and keep the organizational wheels turning (or *from* turning, some would say).

As we consider each of these categories, you should try to get a feel for the relative importance of each in terms of your particular postgraduate plans. I can make general observations here, but only you can translate such general observations into specific projections that correlate with the kind of work you intend to do after you finish college. Here are some questions to keep in mind as you assess the importance of various categories and subcategories of on-the-job writing:

- □ **How often will I be asked to write something like this?** We can express frequency in terms of instances per week or per month you will have to turn out a particular piece of writing.
- □ **How much will I be asked to write?** We can express quantity per instance either in terms of number of words or number of pages. For most purposes, number of pages is sufficient.
- □ **How many people will read what I write?** We can express total readership in terms of the number of copies of the piece of writing that will be reproduced and distributed.

The importance to you, then, of a particular type of writing has to do with the number of times you will write such a piece, the number of pages you will write each time, and the number of people who will read each one.

Correspondence

On the whole in business, industry, and government agencies, letters and memos account for the overwhelming majority of *instances* of technical writing. Chances are that you will write more letters and memos than anything else. However, your letters and memos will be short—most of them less than a page—and most of them will be read by only a few people—three or four, on the average. For most technical people in most jobs, the correspondence they write is pretty much routine and not terribly demanding.

External Publications

Here are the major categories of external publication generated by industrial firms, major businesses, and government agencies:

- Sales, marketing, and promotional literature
- Manuals, handbooks, and parts books
- Reports of research information
- Corporate financial reports

Sales, marketing, and promotional literature includes catalogs, price sheets, advertising, bulletins, brochures, and the like. The high-volume pieces in this category are the sales catalogs, which run thirty to forty pages or more and are printed in quantities of tens of thousands.

Manuals, handbooks, and parts books make up a category we can call product literature; that is, documents and publications that accompany a product after it is sold. Manuals and handbooks run from thirty to 300 or more pages and are often printed in quantities of five to ten thousand.

Reports of research information include reports on contracted research, technical bulletins sent to clients and customers, conference papers, and technical journal articles. A researcher might generate something on the order of 200 to 300 pages of research reports each year, with each report being reproduced by the dozens (for contracted research) to the hundreds (for technical bulletins) or into the thousands (for journal articles).

Corporate financial reports include such things as annual reports, quarterly and semiannual reports, stock prospectuses, financial statements, and auditor's reports. These might run dozens of pages and hundreds of copies, except for stockholder reports, which are distributed by the tens of thousands.

Internal Documents

The principal categories of internal writing for businesses, industrial firms, and government agencies are:

- □ Routine project reports
- □ Management decision-making documents
- □ Personnel documents and policy and procedure manuals
- □ Archival data reports

Routine project reports include progress and status reports, periodic activities reports, inventories, financial reports, sales reports, travel reports, and other documents that serve mainly to keep top management informed about what's happening with company business and who's doing what. Collectively, they constitute the greatest bulk of internal writing and provide the grist for the administrative mill.

Management decision-making documents are the most important of the various kinds of internal writing you will be asked to do. These are the proposals, recommendation reports and feasibility analyses, justifications, and long-range planning papers that enable executives and administrators to make informed and sound decisions with respect to the company's or organization's major business.

Personnel documents and policy and procedure manuals include job descriptions, work assignments, personnel evaluations, organizational charts, and the "rules and regulations" of a firm or agency. Collectively, they constitute the major vehicle the leaders of an organization use to communicate formally with the employees. All but the smallest of organizations use personnel, policy, and procedure documents, and they can run to scores, if not hundreds, of pages collectively. Distribution depends, of course, on the size of the organization, but you can figure on at least one per employee.

Archival data reports represent a rather ill-defined catch-all group of documents. What I have in mind here are all the lists, tables, reports, and papers that serve no immediate informational or decision-making function but nevertheless contain data and information that might prove useful at some later date. These tend to be written, submitted, and filed rather than read.

□ The People Who Write

On-the-job technical writing is limited pretty much to management-level employees; the hourly employees—assembly-line workers, custodial staff, typists, file clerks, and the like—do not contribute significantly to the volume of written technical information. Among management-level, or salaried, employees, we can identify five more-or-less discrete groups:

- □ Supervisors, managers, and executives
- □ Technical experts
- □ Researchers
- □ Technicians
- □ Professional writers and editors

These groups of employees are characterized both by levels of education and experience and by job title and description. None of the groups, actually, is water-tight; individual employees may hold membership in more than one group and others may be in transition from one group to another. The borderlines tend to blur. But if we focus on the centers of the groups rather than on the edges, we can see some distinct classes of employees emerging, classes that will help us come to grips with the question of who writes what in the world of work.

Supervisors, Managers, and Executives

The administrators within a firm or agency are the line officers, the people who don't so much do things as see that things get done. They range all the way from foremen and straw bosses to presidents and chairmen of boards. Administrators come from the ranks of technical experts, researchers, or technicians, for the most part, and most hold bachelor's or master's degrees—either in business administration or related fields, in liberal arts, or in science or technology.

Technical Experts

The technical experts are the specialists in an organization. In business, industry, and government we find technical experts serving as accountants, economists, systems analysts, design engineers, sales representatives, service engineers, corporate lawyers, purchasing agents, personnel officers, statisticians, recruiters, laboratory scientists, librarians, and administrative assistants. Technical experts typically have bachelor's or master's degrees in their field of expertise and five or more years of experience in their fields. Education and experience isn't the whole story, though—job title and description are important, too. Many who qualify as technical experts will be working either as researchers or as administrators.

Researchers

Researchers are often classified as technical experts (for indeed they are), but here we will regard them as a separate group because the writing they do is significantly different from that of other technical experts. Researchers are involved primarily in the business of generating new information: new theories, new analyses and concepts, ideas about new ways of doing things, and ideas about new things to do. Many (perhaps most) researchers are employed by colleges and universities or by major research laboratories such as Los Alamos Scientific Laboratory, Battelle Memorial Institute, or Bell Laboratories. On the other hand, most firms except the very smallest have a research or research and development group somewhere within them. Researchers typically hold master's or Ph.D. degrees in science (including social science) or engineering and have ten or more years of research experience behind them.

Technicians

The technicians are the backbone of any organization. They are the people in the management ranks that get things done. Collectively, they constitute the largest of the five groups of management-level employees. Their ranks include the draftsmen and designers, the computer programmers, the senior secretaries, the lab technicians, the bookkeepers, the engineer's aides, the equipment operators, the repairmen, and such skilled craftsmen as toolmakers, patternmakers, electricians, welders, and steam fitters. Technicians' educational backgrounds include high school diplomas, associate degrees, bachelor's degrees, and special training certifications. The holders of lower degrees and certificates typically have extensive on-the-job experience, while the college graduates who function as technicians have sound theoretical training but typically lack experience.

Professional Writers and Editors

The professional writers and editors we consider here are those who are employed in business, industry, and government—I don't include journalists, advertising copywriters, and free-lance writers, who work in what we might call the communications industry. These people are all professionals, to be sure, but the group I want to call your attention to are the professional communicators who work *outside* the communications industry itself: who are employed by firms and organizations whose primary activity is producing and distributing goods and services. Job titles include technical writer, technical editor, documentation specialist, information officer, editorial assistant, copywriter, and publications engineer. Some professional writers and editors have bachelor's or master's degrees in technical writing, science writing, or technical communications. Others have BAs or MAs in journalism or English. Still others have degrees in science or engineering. What they have in common is that they enjoy writing and they are very good at it.

□ Who Writes What?

As a last step in this brief survey of writing in business, industry, and government, let us look at who writes what (or what is written by whom, if you prefer).

Correspondence

Everybody writes letters. But out of every hundred business letters written, eighty are written by supervisors, managers, and executives. The rest are spread evenly among technical experts, researchers, technicians, and professional writers and editors. The same is true for memos.

Sales, Marketing, and Promotional Literature

Technical writing in this category is handled almost exclusively by the professional writers and editors. Technical experts and supervisors and managers review and approve marketing pieces, but rarely (and only in the smallest of firms and organizations) are they involved directly in the writing and editing.

Manuals, Handbooks, and Parts Books

Once again, these forms of external publications are written and edited by the professional writers and editors in all but the smallest of firms and organizations. In some organizations, technical experts write the initial drafts of manuals, instructions, and handbooks, but in others the technical experts only provide basic technical data for the writers and editors. The technical experts—and administrators as well—review product literature for technical accuracy and general suitability, but for the most part they don't get involved in the mainline writing, editing, and publishing of such documents.

Reports of Research Information

Research reports, conference papers, and journals are generally written by the research scientists themselves. They will be reviewed by supervisors and managers in the research divisions, and they will be edited—to a greater or lesser degree—by a technical or scientific editor if the group has one. Most research groups of any size have one.

Corporate Financial Documents

Corporate financial documents are usually written by the technical experts in the appropriate fields—accounting, economics, and law. Those that have a promotional function to them—notably the annual report—will be edited and perhaps rewritten by advertising copywriters or public relations personnel.

Routine Project Reports

As with correspondence, everybody files a monthly time sheet and activities report. Similarly, just about everybody in the organization has an occasion to write a progress report, status report, travel report, or some other routine document. Once we get past the lowest level of routine project report, however, the administrators take over. Supervisors, managers, and executives generate something over half the total manuscript pages of routine project reports.

Management Decision-Making Documents

Documents used by management for decision-making purposes are written for the most part by technical experts. (If the managers had the expertise, they wouldn't need the documents.) Occasionally, we find middle-management administrators preparing recommendation reports and feasibility analyses for higher administrators, but mostly it's the technical experts who do so. The number as well as the importance of documents of this type depends largely on the size of an organization. The family farmer, the corner druggist, and the service-station owner don't write recommendation reports to themselves—they just make up their minds what to do and do it. On the other hand, corporations and government agencies are just too large and cumbersome to rely on personal judgments and oral opinions. They need planning documents, scoping statements, feasibility studies, recommendation reports, and a host of others to conduct their day-to-day business.

Personnel Documents and Policy and Procedure Manuals

Job descriptions, personnel evaluations, and the like are written for the most part by supervisors and managers unless the firm or organization is large enough to have personnel experts. Similarly, policy and procedure manuals are written by administrators or by personnel experts if they are available. Often, though, professional writers and editors get into the act—especially with the larger or more important pieces of writing.

Archival Data Reports

Data reports, like correspondence and project reports, are written by just about everyone in the firm or organization, but they tend to cluster around the people who generate data—technicians and technical experts. That portion of the researcher's reports that stays inside the organization could be classified as *data* reports rather than *research* reports, although the distinction is somewhat tenuous.

□ Conclusions and Implications

What—if anything—does all this tell us about the need for skills in written communication on the part of various segments of the work force and (by extension) the demands placed on the college technical writing courses? I see three major conclusions.

- □ We find widely divergent demands placed on different segments of the salaried work force—both in the amount of writing they do and in the nature of the documents and publications they are asked to write.
- □ A given salaried worker or management-level employee is faced with different demands as he or she progresses through different stages of career development.
- □ In assessing the importance of writing skills for professional people who write, we have to consider three factors:
  1. The number of occasions on which an individual has to write something.
  2. The total number of manuscript pages an individual writes.
  3. The total number of pages reproduced and distributed (hence the total readership) for the writing of a given individual.

People with scientific and technical training typically spend their first three to five years working as technicians. Then they work as technical experts for ten to fifteen years before moving into supervisory and management ranks. Some, of course, spend their entire careers as technicians and technical experts; others move much more rapidly into and through supervisory and management levels; still others—probably most—find their level in the junior- and middle-executive ranks, retiring as group supervisors or department managers.

As technicians, they don't have many occasions to write and they don't generate many manuscript pages when they do. As technical experts, they both write on more occasions and generate more pages. It is as supervisors and managers, however, that they have many more occasions to write and generate significantly many more manuscript pages in a given week or month or year. Moreover, the pages they do write are reproduced and distributed in significantly more copies, so their total readership grows exponentially. Even as supervisors and managers, though, they don't write the many-page, high-volume pieces for external consumption—the sales catalogs, product brochures, operation and maintenance manuals, user handbooks, and corporate annual reports. These remain in the hands of the professional writers and editors.

Writing about Writing

1. Select a recent school paper you have written—a term paper or lab report or essay. Describe in some detail what changes you would have to make to make it suitable for publication in a nonacademic setting.
2. Write a poem about birth. Write an essay about some aspect of birth. Write a technical description of birth. Then describe the differences—both in the pieces of writing themselves and in your thought processes and writing procedures as author.
3. Proceed as in Assignment 2 for some other major milestone of human

life such as first day of school, commencement, marriage, birth of a child, retirement, death.

4. Write an essay on the concept of "the report as an instrument of change." Aim the essay at fellow students who have not taken or are not taking a course in technical writing.
5. List and briefly describe the varieties of external publications and internal documents produced by a state or federal agency you are familiar with.
6. List and briefly describe the varieties of external publications and internal documents produced by a business or industrial firm you are familiar with.

2

Adequacy and Excellence

Quality in writing, or effectiveness in written communication, is a continuous variable that ranges (so we presume) from zero to one hundred, from horrible to magnificent, from useless to maximally effective. On this continuous scale of writing quality, however, there are two significant points: (1) that point below which the writing is unacceptable and (2) that above which the writing is excellent. These points are so significant that they virtually reduce the continuum of quality into three distinct categories: excellent writing, adequate writing, and unacceptable writing (see Figure 2–1, page 18). This division is particularly characteristic of the kind of writing we are discussing here—what we call instrumental writing, professional writing, writing as an act of communication; in short, technical writing. In fact, we can define these three categories of technical writing operationally:

- **Excellent technical writing**—the kind of writing that will gain points for you in on-the-job situations; the kind of writing that will lead to raises, promotions, and other recognition of the excellence of your work; the kind of writing that will consistently get your recommendations accepted, your proposals funded, and your reports praised.
- **Adequate technical writing**—the kind of writing that will keep from embarrassing you in on-the-job situations; the kind of writing that will be accepted and passed on, largely without com-

ment; the kind of writing that won't hurt you but certainly won't help you.

- **Unacceptable technical writing**—the kind of writing that you'll get back to do over again until it's right; the kind of writing that will clamp a ceiling on all your other technical skills; the kind of writing that will be a constant impediment to your professional growth.

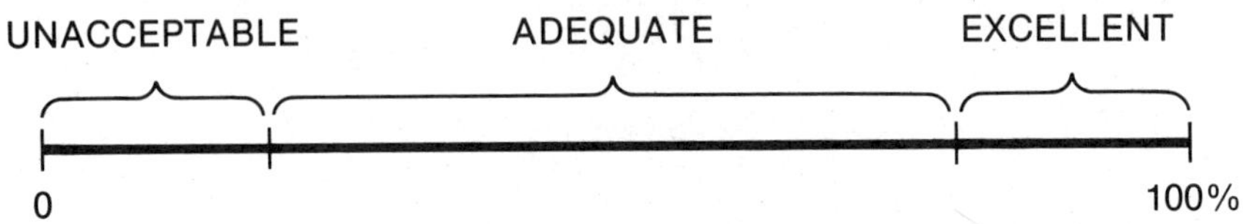

Figure 2–1 The Continuum of Writing Quality

How do these three categories of technical writing distribute? When we consider all of the on-the-job technical writing that is produced, what percentage of it is judged to be excellent? What percentage of it is judged to be adequate? What percentage of it is judged unacceptable? In a recent survey I conducted of some 200 editors and publications managers, managers estimated that about 20 percent of the writing that crossed their desks was unacceptable, slightly less than 15 percent was excellent, and the remaining 65 percent was adequate but less than excellent. Another informal survey of teachers of technical writing showed that about 22 percent of the students enrolled in such courses either fail or withdraw, about 18 percent earn As, and the remaining 60 percent earn Bs, Cs, or Ds. For convenience, let's put the cutoff points at 20 percent and 80 percent, as shown in Figure 2–2.

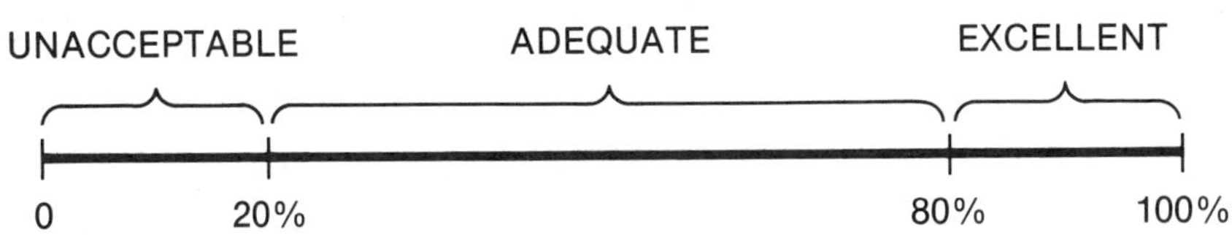

Figure 2–2 Cutoff Points for Adequacy and Excellence

As a student writer you thus have two goals. First, you must acquire (or demonstrate) the skills necessary to generate consistently acceptable or adequate technical writing. Second, you must strive to acquire the skills necessary to generate excellent technical writing so you will be able to do so on those occasions that demand excellence. Three key questions you need to ask are:

- How do I recognize adequate technical writing when I see it?
- How do I tell the difference between adequate technical writing and excellent technical writing?

- In what circumstances is adequate technical writing good enough, and when is excellence to be expected?

Criteria for Adequacy

How good is good enough? We can start by answering the question negatively. If a report fails to solve the technical problem it is supposed to solve, it is unacceptable. If a report is inaccurate, it is unacceptable. If a report is incomplete, it is unacceptable. If a report is so poorly organized that its readers can't use it, it is unacceptable. And finally, if a report has so many minor errors in it that it is an embarrassment to the firm or organization publishing it, it is unacceptable. Let's convert these to positive attributes. For a report or other piece of technical writing to be judged adequate, it must be:

- **Useful.** It must be directed toward a real organizational problem and it must contribute toward the solution of that problem.
- **Accurate.** All of the information in it must be sound, reliable, and valid. It must tell the truth and nothing but the truth.
- **Complete.** It must leave out no pertinent data or steps; it must touch all the bases. It must tell the whole truth.
- **Orderly.** It must follow the accepted conventions of organization. The information in it must be readily retrievable by the people who need it.
- **Correct.** It must be essentially free from the common errors of grammar, writing mechanics, and usage that are discussed in freshman composition courses.

Notice that each of these attributes or criteria for adequacy is pretty much an objective quantity, a variable whose value is determined largely by external standards. There's not much room for opinion. As a writer, you don't have much choice about accuracy, completeness, or correctness; the standards and guidelines are firm, and your job is to learn the standards and apply them as thoroughly as you can. (Those aspects of your writing where you *can* exercise choice we discuss under the heading of Style; see Part Four, "Revising for Excellence"). Also, keep in mind that trade-offs don't count when we're talking about the criteria for adequacy. Failure to meet minimum standards in any of these five areas is a fatal error—one that leads to unacceptability. You can't compensate for inaccurate information with superior organization. You can't compensate for poor organization with meticulous grammar and spelling. If your report is useless it is unacceptable no matter how accurate, complete, and well-organized it is. And an incomplete report is worse than useless; would you want to fly with an airline pilot who only knows seven-eighths of the route?

□ Levels of Adequacy

Minimal adequacy, really, is a modest achievement. If 20 percent of what is written is unacceptable, a report that would score 21 on a scale of zero to 100 is minimally adequate. Moreover, the range of adequacy is a wide one. If you start with a minimally adequate report (21 percent) and make it twice as good (42 percent), it is still adequate. If you make it three times as good (63 percent), it is still only adequate. In order to achieve excellence, you must make a fourfold increase (84 percent) over a minimally adequate report. It makes some sense, then, to talk about levels of adequacy and to try to distinguish between barely acceptable writing on the one hand and better-than-average, or pretty darn good, writing on the other. For some communication contexts, good enough is good enough. Other contexts (see below) demand nothing less than excellence. Yet in the real world—just as in the world of academe—there is a place for B-level work. Here are some characteristics of above-average (but less than excellent) writing:

- □ **It is adequate in every respect.** It meets the minimum standards of utility, accuracy, completeness, organization, and correctness.
- □ **It is more than adequate in some respects.** It excels in at least some areas of utility, accuracy, completeness, organization, or correctness.
- □ **It is brief.** It focuses closely on the problem it addresses and avoids telling readers what they already know or what they don't need to know.
- □ **It is smooth.** It is free from the choppiness of short sentences and from excessively awkward constructions.
- □ **It is direct.** Above-average writing is verb-oriented rather than noun-oriented; it favors agent-as-subject clauses; it avoids hedging; and it avoids wordy, roundabout constructions.

Notice that these characteristics of above-average writing are rather more scalar than polar. The criteria for minimal adequacy are two-valued, for the most part. A report is either accurate or inaccurate, complete or incomplete, correct or incorrect. As we move up the scale of quality into the 50-to-80-percent range, though, we begin to encounter sliding values. How direct is direct enough? How smooth? How brief? Furthermore, the values here are quite a bit more subjective and susceptible to individual judgment and assessment. But such is the nature of writing, and technical writing is no exception. The levels of adequacy merge with one another as we move from 20 to 40 to 60 on the scale of quality, and the distinctions begin to blur. Lights do not flash nor bells ring when a report improves from minimally adequate to above-average.

Obviously, the importance of a particular type of writing and a particular communication context come into play here. All writing, of course, must be minimally adequate. If yours isn't, you will get it back to do over again. For the kind of writing you do frequently, and certainly for writing that has a large readership, you will want to generate products that are more than minimally acceptable. Similarly, the more important (in terms of frequency, length, and readership) a piece of writing is, the better it ought to be. At the same time, it doesn't pay to get hung up on levels of adequacy; there are more similarities than differences between low-adequate and high-adequate writing.

□ Checklist for Adequacy

Writing at a level of adequacy is, in fact, so much a mechanical process that you can profitably use a checklist to make sure you have touched all the bases. As you discover the minimum writing requirements in your firm or organization, write them down. Then each time you write a report, go over your checklist item by item to make sure you haven't overlooked something that will mean the automatic rejection of your report. Don't be ashamed or embarrassed about using a checklist—airline pilots, among other professionals, use checklists all the time. Here is the checklist I ask students in my technical writing classes to use. The particular requirements in your class may be somewhat different, but this checklist will get you headed in the right direction.

Checklist for Adequacy

FORMAT

_____ Assignment cover sheet is present and properly formated.
_____ Design criteria are specified (see Chapter 3).
_____ Elements of formal reports are present and properly located (see Chapter 4).
_____ Plain, opaque white paper is used.
_____ Top, bottom, and side margins are proper on every page.
_____ Page numbers are present and proper (see Chapter 6).
_____ Headings and subheads are present and proper (see Chapter 6).
_____ The paper has sufficient figures and/or tables (see Chapter 8).
_____ Captions, callouts, and references are proper (see Chapter 8).
_____ Bibliographic references are proper (see Chapter 9).
_____ ____
_____ ____

ORGANIZATION (See appropriate chapters in Part Four)

_____ All necessary sections are present.
_____ Sections and subsections are in the proper order.

- ____ Sections and subsections are properly balanced (see Chapter 5).
- ____ Sections and subsections are organized internally (see Chapter 5).
- ____ The paper follows the general outline for the type of writing it is.
- ____ ______________________________
- ____ ______________________________

USAGE AND MECHANICS (See Chapter 7)

- ____ Each page is free from spelling and typographical errors.
- ____ Each page is free from problems with major punctuation.
- ____ Each page is free from problems with minor punctuation.
- ____ Abbreviations and symbols follow conventions.
- ____ Capitalization follows conventions.
- ____ Underlining and italics follow conventions.
- ____ All verbs agree with their corresponding subjects.
- ____ All pronouns agree with their corresponding antecedents.
- ____ Headings, subheads, and listed items are grammatically parallel.
- ____ No false parallelism exists.
- ____ The paper is free from dangling modifiers.
- ____ The paper is free from sentence fragments.
- ____ ______________________________
- ____ ______________________________

BASIC WRITING STYLE

- ____ Paper addresses a real(istic) problem.
- ____ Paper contributes to the solution of the problem.
- ____ Paper is free from strings of choppy sentences.
- ____ Paper is free from excessively awkward constructions.
- ____ Paper is direct in style and approach.
- ____ Paper contains no unnecessary sections or subsections.
- ____ ______________________________
- ____ ______________________________

□ Criteria for Excellence

So OK, you can stay out of the bottom 20 percent. How do you make the top 20 percent? What are the characteristics of excellent report writing that make it stand out over 80 percent of the competition? Here the ground is not so firm. To talk about excellence in writing, we must speak of style. And style is not so easy to pin down as accuracy. Or correctness. Or completeness. But writing style, ultimately, is what makes the difference between adequacy and excellence. Evaluations of style, of excellence, are to a great extent subjective. There are multiple routes to achieving excellence in written communication; trade-offs and competing values abound. But from this morass of subjectivity and choice, six firm points that we can recognize as criteria of excellence emerge. For a report to be judged excellent, it must be:

- □ **Adequate in every respect** (see "Levels of Adequacy," above).
- □ **More than adequate in some respects** (see "Levels of Adequacy," above).
- □ **Brief, smooth, and direct** (see "Levels of Adequacy," above).
- □ **Easy to read.** The information in the report must flow into the minds of the readers with such smoothness, grace, and facility that the act of reading the report is seemingly effortless.
- □ **Interesting to read.** The report must capture and hold the interest of its readers in order to enhance the transfer of information. A boring report cannot possibly be excellent.
- □ **Easy to remember.** We read not only to gain information but to retain it. The content of an excellent report stays with readers long after they put the document away.

Notice that these criteria for excellence are subjective, with considerable room for opinion and choice. In fact, style is choice. And excellence comes from making the most of your alternatives. The degree to which your technical writing will be judged excellent is largely a matter of the range of alternatives you are aware of and the wisdom with which you choose among them.

□ The Demands for Excellence

Not every on-the-job communication situation you will encounter demands excellent communication on your part. This is one of the toughest lessons I had to learn when I got out of college, and I am willing to share this discovery with you even at the risk of offending some of my colleagues. There are situations where good enough is good enough. There are times when the boss doesn't need your report perfect, he needs it Thursday. There are occasions when the extra time and effort to achieve excellence (and believe me, it takes extra time and effort!) are counterproductive—both for you and for your employer. This is not to make a virtue of minimal adequacy (remember that the range of adequacy is a wide one, spanning 60 points or so on the scale) but rather to observe that the Law of Diminishing Returns is just as applicable to technical communications as it is to economics.

Nevertheless, there are situations that demand excellence. Here are some of them:

- □ **Large audiences.** If you have an audience of sixty, you can afford to spend an extra hour to save each reader a minute. If you have 500 readers, you can afford an extra day.
- □ **Important audiences.** If your primary audience consists of people whose time is considerably more valuable than your own,

your employer will profit if you spend some of your time in order to save their time.
- **Safety precautions.** When people's lives and well-being are at stake, nothing less than excellence can be tolerated.
- **$$$$$$.** When considerable sums of money are involved (you decide what's considerable), the extra time and effort to achieve excellence are time and effort well spent.
- **Reputations.** When your professional reputation or that of the firm or organization you represent is on the line, excellence in communication is definitely worthwhile.

Once again, the importance of a piece of writing comes into play—particularly in terms of the size and nature of your readers and the nature of the information you are communicating. This is a simple lesson, yet one often not well learned: the more important a piece of writing, the better it ought to be!

Achieving Adequacy and Achieving Excellence

In putting this book together, I have assumed that each of you as a student writer will make progress during the semester or term that you use it; that your writing will, in fact, improve. Second, I have assumed that each of you starts at a different place. Some (although few, I hope) will struggle to achieve minimal adequacy. Most will start above the threshold of adequacy and will quickly achieve those two- and three- and fourfold improvements needed in order to produce writing that is excellent. A few of you may be in a position to move directly into considerations of writing style that lead to excellence. But wherever you start and however far you progress, keep these three observations in mind:

- Anyone can produce adequate written communication.
- If you start with adequate written communication and keep improving it, sooner or later it will be excellent.
- There are algorithms for adequacy, but excellence is a matter of heuristics; that is, adequacy can be achieved by following the rules, but to achieve excellence you must be willing to experiment, to try new things, to accept new ideas, to suspend disbelief, and to risk traveling where the charts are poor.

Although there is some overlap, Parts One, Two, and Three of this text deal with basic adequacy in technical communication, while Part Four deals with aspects of excellence. The *usefulness* of your technical reports is largely a matter of how thoroughly you analyze your communication context and respond to your analysis. The remainder of Part One, "Preliminaries," discusses this aspect of your writing. The *accuracy* and *completeness* of your reports is dependent on your data-gathering, analysis, and presentation techniques. Part Two, "Basic Writing Techniques," addresses these. Most of what you need to know about sound *organization* for routine technical writing is presented in the earlier chapters of Part Three, "Types of Written Communication." And finally, Chapter 7 offers suggestions for spotting and correcting *common errors* in mechanics, usage, and grammar.

As you move up through the range of adequacy toward excellence, you will want to study the sections on "Revising for Orderliness," "Revising for Completeness," and "Revising for Brevity" in Chapter 5, "Report-Writing Procedures." The later chapters in Part Three, "Types of Written Communication," discuss varieties of technical writing that are common in contexts that demand excellence, and the five chapters of Part Four, "Revising for Excellence," get into the meat of fine-tuning your writing to achieve excellence. It is these chapters on style that will help you the most in moving your writing from "good enough" to "superior." As your writing style improves—as you throw off bad habits and develop good ones—you will find yourself more often and more easily producing reports that are judged to be excellent. The road to excellence is a long one and there are no shortcuts, but the route is well marked. Here are some salient signposts:

1. Start with an adequate report; don't strive for excellence with something that is inherently unacceptable.
2. Analyze your own personal writing style: find out where your strengths and weaknesses lie.
3. Be able to recognize the common stylistic weaknesses that plague the writing of professional people and eliminate these weaknesses from your own writing.
4. Concentrate first on your paragraphs. Construct them soundly and develop them fully.
5. Next, focus on your clauses and sentences. Develop the variety and sophistication necessary to communicate effectively about complex ideas.
6. Incorporate cohesive devices into your writing—make it hang together.
7. Finally, improve your diction. Choose words that are familiar, direct, concrete, and specific.

A final word of caution: Don't strive for excellence on the first jump out of the chute, and don't be disappointed if you don't achieve excellence on the first papers you submit. Build a sound foundation of adequate writing and then (and only then) improve the best of your writing through successive revisions until you have achieved the excellence you seek.

Writing Exercises

1. Select a recent school paper you have written—an essay, a term paper, or a lab report. Evaluate it according to the "Checklist for Adequacy." How does it measure up?

2. If your own writing is less than adequate, list the areas of your shortcomings and specific items that you need to control or improve upon.
3. Begin to build up your own collection of excellent writing and unacceptable or minimally adequate writing. As you read and study during the course of the present term or semester, keep an eye out for writing that you, personally, judge to be particularly good or particularly bad. Run off copies and save the copies. (Be sure to identify the source.)

Writing about Writing

1. Find a sample of writing that you judge to be excellent and one that you judge to be minimally adequate. Describe the differences as you perceive them. Attach copies of the samples to your description.
2. Write an essay on the concept of "adequacy and excellence." Direct the essay to fellow students who have not taken a course in technical writing.

3

Analyzing the Communication Context

You can't design and build a tool without knowing who is going to use it and what it will be used for. Similarly, you can't design and write a report without knowing who your audience will be and to what use your information will be put. Fortunately, you always have some notion of audience and purpose when you write; you will run into trouble only when your notions are too limited or when they fail to line up with those of the people who review and approve your writing. It behooves you, then, as a communicator of technical information to take some specific steps to determine audience, purpose, and other design criteria for each of your communication contexts and to document your analysis.

It is axiomatic that nothing you write on the job will have an audience of just a single person (even copies of business letters are kept on file), and little that you write will be directed at a homogeneous audience. In fact, virtually all of your written communication will be read—for one reason or another—by at least a dozen different people with different backgrounds, different motives, different levels of interest, and different purposes for reading. This chapter suggests a number of techniques with which you can analyze your communication contexts and identify their component parts, so that you can be better equipped to design and write effective reports—reports that do indeed serve as instruments of change.

To begin your analysis, you must formulate answers to some of the classic questions of journalistic writing: Who?, Why?, and What?: you must specify audience, purpose, and type of information. Let's start with audience.

□ Varieties of Audience for Technical Communication

In one sense, each person who reads what you write is unique. Each reader brings his or her own interests, background, needs to the communication situation. But in another sense, each person who reads what you write represents a group of people who share common interests, backgrounds, and needs. For technical information, as we have defined it, we can identify (following Thomas E. Pearsall's lead) four groups who share enough in common that we can call them types of audiences:

- □ Executives and administrators
- □ Technical experts
- □ Technicians
- □ Laypeople

These groups are, of course, very much like the groups of writers we looked at in Chapter 1. This shouldn't be surprising; there are only so many kinds of people who work in business, industry, and government. The executives and administrators here correspond to the top end of the managers and supervisors of Chapter 1, with lower-level administrators (foremen, shift supervisors, and so on) serving more often as immediate audiences rather than primary audiences (see "The Report Design Worksheet," below). We make no distinction between researchers and other technical experts as audience members, but treat them as a single group. The technicians here are pretty much the same as the technicians of Chapter 1. Laypeople, who rarely write technical reports, bulletins, or articles but often read them, are a new group here. Let us look at each of these audience groups in turn, considering both the group's characteristics and its purposes for reading.

Executives and Administrators

The executives and administrators in an organization are those responsible for policy formulation and the decision-making associated with implementing policy. In general, their areas of concern are personnel management, fiscal management, and program management. As a group, they include people with such titles as department head, department or division manager, vice-president, president, bureau chief, and the like. For our purposes here, we will also include members of boards, councils, commissions, vestries, and the like as executives and administrators.

What characteristics can we identify in such readers? First of all, without exception they are in a hurry; they are pressed for time. They cannot afford the luxury of a close and contemplative reading of a proposal or technical report. They are consequently interested in the gist of a report—the conclusions and recommendations. In general, you can assume that executives and administrators understand the basic principles of science and technology and are familiar to a degree with the special language and terminology of a particular field. But you cannot assume that every executive or administrator is a technical expert. Just because Mr. Johnson is Manager of Engineering Design doesn't mean that he is necessarily familiar with the latest developments and current thinking with respect to solid-state electronics and printed circuitry. His education and background as an engineer might be in mechanical rather than electrical engineering. Furthermore, even if his background is in electrical engineering, he has probably been away from it for a number of years and can no longer be considered a practicing electrical engineer. Do not assume, then, that executives and administrators are technical experts; quite the contrary, assume that executives and administrators are *not* experts in your particular field of expertise.

The main reason executives and administrators read reports and other technical documents is to aid them in their decision-making (although occasionally they seek technical information to justify a decision that has already been made). Accordingly, your best strategy with this type of audience is to present your technical information in such a way as best to serve the decision-making function. (To a much more limited extent, executives and administrators assess the performance of those working under them in an organization through their reports.)

Technical Experts

The group of technical experts (including researchers) is characterized by people with advanced degrees in a field or discipline, people with bachelor's degrees and several years of experience in a field or discipline, or people with professional certification or licensing in a field or discipline. With this particular audience for technical information, you would be wise to consider two subgroups: experts in *your* field or discipline and experts in *another* field or discipline.

As writer you share a broad base of mutual understanding with technical experts—particularly experts in your own field. You can assume that this type of audience has a detailed and current knowledge of the theory, developments, and special language of your field. You don't have to explain the background and go into the details of what you present. You can assume that technical experts will be quick to grasp the significance of what you write. All in all, experts in your own field are probably the easiest type of audience to communicate with. They will, however, be highly critical readers—quick to find errors in your judgment, inconsistencies in your data, and holes in your arguments. Technical experts will be interested not only in your conclusions but also in the methods by which you reached those conclusions.

Technical experts will have one of two purposes in reading what you write: either to seek application of your information to their own work or to assess the value and credibility of your technical information. Under the first purpose, technical experts may read a report of your work so that they may avoid having to redo that work, or they may seek in your technical data foundation or support for their own work. They may wish to learn of new methods, being more interested in your reported techniques than in your results and conclusions. Finally, technical experts—just because they are experts—often read technical literature to seek general information for possible future application in their work. When experts read to evaluate, it is generally because they have been called in by management to assess and judge your presentation of technical information. They will look to the completeness, accuracy, and clarity of your technical data as it relates to your conclusions and recommendations.

Technicians

The third audience you may have for your technical information consists of technicians in your field. Technicians include people with such functions as installers and operators of equipment, repairmen, lab technicians and assistants, draftsmen, clerical support people, foremen and shift supervisors. Their backgrounds include apprentice programs, vocational and technical schools, baccalaureate degrees with limited experience, or years of real-world experience as a substitute for formal training.

Technicians will have a somewhat narrower base of mutual understanding than technical experts, but they will have the same excellent knowledge of technical vocabulary and special techniques peculiar to your field. There is no need to talk down to technicians. A word of caution, though: technicians will look to you as the technical expert, and they will tend to believe what you tell them; you must be scrupulously honest. Also, it may be necessary at times to point out the relevance of your information so that technicians can see how what you present relates to their needs.

A major purpose technicians will have in reading your technical information will be to receive instruction for the performance of their duties. They will be seeking procedural guidelines; installation, operation, maintenance, and servicing instructions; or design specifications. Technicians may also be seeking background information that will be useful for the performance of their duties or technical information for advancement in position.

Laypeople

It is tempting to define a layperson as anyone who lacks expertise in your field. It is dangerous, however (as well as counterproductive), to treat the administrator, the technician, and the expert in another field the same way that you treat the uneducated general reader. A much better operational definition is "any person who doesn't fit in one of the first three categories of technical audience." In addition, it is a good idea to distinguish between the knowledgeable layperson and the unsophisticated layperson.

The knowledgeable layperson is much like the technical expert in another field; in fact, sometimes it is hard to tell the difference. As a model for this group, you might consider the person who has a college education in some field other than science or technology. Members of this audience will have a broad range of interests. They will have more time to read than administrators do; they will tend to be contemplative readers. You can assume that such readers have a command of basic interdisciplinary technical vocabulary and an understanding of theories and principles common to all fields of science and technology. For the unsophisticated layperson, however, you can assume no technical knowledge or expertise at all. (In point of fact laypeople often bring a broad range of expertise to the communication situation, but you cannot *assume* such a background on the part of all lay readers.)

In general, laypeople—both knowledgeable and unsophisticated—read technical information to explore its relationship to their ordinary day-to-day activities. They are interested not so much in the technical data per se, as in what impact the technical information has on their lives. They may be seeking application of your field to theirs, or they may just be seeking general knowledge for self-fulfillment. Whatever their reason for reading what you write, however, you cannot assume a high level of motivation. You may therefore have to use various techniques to build and maintain a level of interest. Writing for the lay audience or the general public is the most difficult and the most demanding of all technical writing. For this reason, most communication of technical information to the general lay public is handled by professional technical writers and editors rather than by the technical experts directly.

The Combined Audience

Often (more often than not, actually) the audience for your report, proposal, or instructions will consist of representatives of two or more of these groups. You will find yourself having to communicate your information jointly to executives and experts, or experts and technicians, or decision-makers and the general public. Moreover, individual readers may fall simultaneously into two categories. The rancher on the school board, for instance, may function as an executive decision-maker but may be more of a knowledgeable layperson by education and experience. A junior engineer in your firm may be a technical expert by educational criteria but a technician by job category. As you analyze your audience, keep in mind that real-world communication situations are complex and that rarely will you be communicating technical information to a homogeneous group for a single, simple purpose.

□ The Many Faces of Purpose

"Why am I sitting down to write?" you must ask yourself. You are not writing to express yourself. You are not writing a report because you would rather write a report than do some more research or design. You are not writing to impress your readers with your knowledge of the subject at hand (at least you ought not to be!). You are writing to communicate technical information that someone else needs. Your purpose in writing, then, is your reader's purpose. Your goal is to meet your reader's needs. But since you have multiple and disparate readers, you will have multiple and disparate purposes. Furthermore, you may well have primary, secondary, and tertiary purposes—some of them audience-centered and some of them, perhaps, more self-serving. Consider the several writing purposes that might be associated with a recommendation report you submit to top management in your firm:

- □ You want to convince the decision-makers to accept your recommendation.
- □ You want to convince the technical experts that you indeed know what you're talking about.
- □ You want to convince your immediate superior that your work is sound and that you possess skills that merit recognition and promotion.
- □ You want to save time for the administrators and executives (because time is money).
- □ You want to save time for yourself (because there are things you would rather do than write reports).

Or consider the several writing purposes that might be associated with a letter to a customer or client requesting payment for an account past due:

- □ You want to get the customer to pay the bill.
- □ You want to retain the future business of the individual, though; to keep him or her as a customer.
- □ You want to convince your boss (who will read your letter) that you are fair and honest, that you have the customer always in mind, but that you are neither creampuff nor pushover.

You must analyze your purpose, just as you analyze your audience, to identify its components and begin to arrange them in some order of priority. Each communication context will be different, of course, but in general you should be able to identify and specify the following purposes:

- □ The primary purpose your major audience segment has for reading.
- □ Any secondary purposes for reading that this group might have.
- □ Primary and secondary purposes minor audience segments might have.
- □ The ostensible purpose of your communication—and the "real" purpose, if that is different from the ostensible purpose.
- □ The secondary purposes in communication that benefit your firm or organization.
- □ Any personal purposes or goals you might have that do not conflict with the more direct purposes of your writing.

□ Types of Technical Information

The third consideration in analyzing a communication context is the type of information you are going to be communicating. In general, avoid putting different kinds of information or information about different subjects in a single report; instead, write separate reports. Keep in mind that a report isn't a sink for you to dump all the information you happen to have into; it is a working tool that must serve the needs of your readers. Don't tell readers things they already know. Don't tell readers things they don't need to know. Don't tell readers things they don't care about knowing. Stick to the point. Don't try to make a report or a letter or a memo serve double duty. Don't ask someone for a favor and then thank him or her for it in the same letter. Don't put details about several different projects in the same memo. Avoid writing the kind of a report that would wind up being filed in two or three different places under two or three different headings. Avoid combining temporary, transient information and permanent or long-lived data in the same document. Give heed, in other words, to how your technical information is going to be used.

□ The Report Design Worksheet

Your tasks in analyzing the communication context are: (1) to identify the various segments of your audience and record certain information about these readers, (2) to identify your various communicative purposes and arrange them in some order or priority, and (3) to document your analysis so that you have it available for reference. As David E. Fear and others have suggested, you can accomplish all of these tasks easily by using a report design worksheet similar to the one illustrated in Figure 3–1.

In the first section, "Preliminary Information," you record aspects of

REPORT DESIGN WORKSHEET

PRELIMINARY INFORMATION

What is to be done? ______________________

Who is it to be presented to and when? ______________________

Restrictions (format, length, etc.): ______________________

READER ANALYSIS

Audience:	*Immediate*	*Secondary*	*Primary*
Position or title:	________	________	________
Relationship to author:	________	________	________
Technical expertise:	________	________	________
Attitude toward subject:	________	________	________
Effect of report on readers and organization:	________	________	________
Personal Characteristics:	________	________	________
READER'S PURPOSE			
Why will he read the report?	________	________	________
What should he know and be able to do after reading the report	________	________	________

WRITER'S PURPOSE

Desired effect(s) of the report: ______________________

____

____

Figure 3–1 Report Design Worksheet

the report-writing assignment as it is given to you. (If you're not clear about the *what* and the *who* and any restrictions that may exist, you had better find out before you proceed.)

In the next two sections, "Reader Analysis" and "Readers' Purposes," you record certain data about audience and purpose according to the three audience segments that will exist for everything you write on the job:

- □ Your primary audience
- □ Your secondary audience
- □ Your immediate audience

Your primary audience consists of the people directly affected by the information you convey: those who will make a decision based on your information, those who will follow the instructions or procedures you outline, or those who will use your information in some phase of their jobs. Your secondary audience consists of the people indirectly affected by your information: those who must implement a decision that is made; those who benefit when proper procedures are followed; or those who review for technical, legal, fiscal, or policy implications what you have written. Your immediate audience is the person to whom you deliver your written document and through whom you transmit it to your primary and secondary audiences.

Consider the simple case of a letter you write in response to an inquiry for pricing information on one of your company's products (see Figure 3–2). Your primary audience is the management decision-makers in the firm that requested the information—the people who will decide whether or not to purchase your company's product. It includes but is not limited to the person to whom you address the letter. (It is possible—in fact, fairly common—that your addressee won't be part of your primary audience at all. The person to whom you address the letter may be serving the clerical function of mere transmitter of the correspondence.) Your secondary audience could be rather broad: cost accountants and technical experts in the firm requesting the information, the ultimate users of your product in the firm requesting the information, pricing people within your own company, and possibly policy-makers within your own company. The immediate audience is your boss, who will check your correspondence over before it is mailed. So even a simple letter might be read by a dozen or more people with a more than casual interest in the information it contains.

A more complex situation might involve a proposal for architectural design services that you submit to a local school district. Here, your primary audience would be the members of the school board: the actual decision-makers. Included in the secondary audience might be the school district business manager and the school superintendent, an engineer retained by the school district, interested patrons of the district, teachers and students, and, of course, top people in your own architectural firm. Your immediate audience is once again your immediate supervisor.

We can see, then, that the total audience for a technical document is sizable and complex, with different segments of the audience having different interests and different levels of expertise. The most common failing of reports and other technical documents is that they are too restrictive in scope and are aimed at too narrow an audience. To avoid this pitfall, pay particular attention to what each reader or group of readers should know and be able to do after reading your report. If you ever hope for your reports to have an effect on your organization, you must be able to specify what those desired effects are. The more detailed you are at the analysis stage, the easier it will be for you to write an effective report.

In the final section of the Report Design Worksheet, "Writer's Purposes," you record any additional purposes you as writer might have that are not specified in section three. Be honest with yourself; if you have some secondary but very real purposes—even ulterior motives—specify them. The Report Design Worksheet is not part of the report; nobody will ever know. But if you're honest about purpose here, your ultimate task of writing the report will be easier.

SUPERIOR PERFORMANCE SHOP
1422 Front Street
Ashland, WI 83456

October 29, 1983

Mr. John Harrison, Manager
Custom Manifolds, Ltd.
2525 Burley Drive
San Leandro, CA 94577

Dear Mr. Harrison:

We are interested in stocking your line of high-performance manifolds and headers for 1976 through 1984 Toyota Land Crusiers.

Can you quote us a dealer net price for three sets each for each of the Crusier models we are interested in? Also, can you tell us what delivery dates you can guarantee and what terms of payment you will need?

We look forward to opening an account with you, as there is a growing interest in high-performance modifications for off-road vehicles here in northern Wisconsin.

Very truly yours,

Chris Davenport

Chris Davenport,
Parts & Service Manager

Figure 3–2 Inquiry for Pricing Information

□ Success Strategies for Student Writers □

When you analyze an on-the-job communication situation, the process is largely one of discovery—you *find out* answers to the various questions of Who?, Why?, and What? The classroom writing situation, however, is different; here the process is largely one of invention—you *think up* answers to the various questions of Who?, Why?, and What? If you question your purpose honestly, you come up with answers like "To pass the course," "To get a good grade," "To impress my instructor," or maybe "Darned if I know." However, if you fail to ask questions of purpose, you wind up dumping enough junk in a report to make five pages worth, putting it in some semblance of format, turning it in, and hoping for the best.

The answer to this dilemma is to play a role as you consider a communication context. Simulate. Pretend. Exercise your imagination a bit. First of all, perceive yourself not as a student but as a graduate, certified technical expert. Project yourself into the kind of job you hope to have five or ten years from now. Assume that you have been asked to write a particular report because you are the resident expert on the subject; there is no one available who knows more about the subject than you do. Second, give yourself a realistic audience with realistic purposes for wanting to read what you write, for wanting to receive and use the information you communicate. This will involve projecting yourself into various audience roles. Put your executive decision-maker hat on. What kind of report would *you* want to receive? Put your technical expert hat on. How would *you* evaluate someone else's report? Put your technician hat on. What kind of instruction manual would *you* prefer to read?

Remember also that as student writer you are in charge of the situation; you can mold it and shape it to suit your immediate needs. On the job, you will have to take each communication situation as it comes; with the classroom writing assignment, you can create a situation that maximizes your chances of writing an effective yet manageable report. Select a subject you have more than a little knowledge of. This may be something from your major field of study, something from past employment experience, or something from your hobbies or avocational interests. But whatever you do, don't fake it. Specify your fictitious readers in some detail, but select the detail so that you have a real difference among primary, secondary, and immediate audience segments. Keep the purposes of these fictitious readers fairly clear and uncomplicated. In your classroom writing assignments, learn how to deal with straightforward communication situations; you will get into complex and cloudy communication situations soon enough.

Finally, consult with your instructor as to what he or she feels is an appropriate hypothetical situation. Look at detailed writing assignments. If possible, review earlier student papers that have dealt with reasonable and realistic writing situations. And if you have questions on a particular classroom writing assignment, write out a fairly detailed description of your hypothetical situation and your role in it and ask your instructor to review it. Make things easy for yourself. Above all, don't go out of your way to make things difficult for yourself; there are enough problems involved in learning to write effectively as it is.

□ The Flow of Information in Organizations

Information is the lifeblood of business, industry, and government. Indeed, the vitality of an organization can be gauged by sensing the pulse and flow of information within it and into and out of it. As a professional in some field of science or technology, you will be immersed in this flow of information: you will receive information, you will transmit or relay information, and with increasing frequency as you mature in your job you will generate new information. From your position in an organization, the information you generate will flow in a number of directions (see Figure 3–3): upward, downward, or laterally. A good final step in analyzing a communication situation is to consider the vectors of information flow—the directions and distances.

Upward communication—communication directed to individuals who rank higher in an organizational hierarchy than you do—most commonly serves as input for decision-making. Less often, upward communication serves to document organizational accountability. The farther upward your information goes, the less it serves a documenting function and the more it serves a decision-making one. In fact, once your

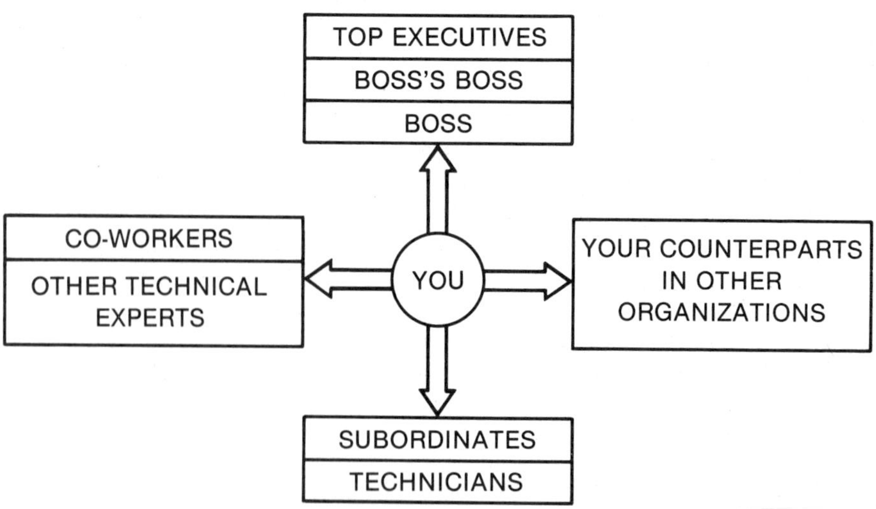

Figure 3–3 Directions of Information Flow

information goes beyond your immediate supervisor, it serves almost exclusively a decision-making function. Your primary audience for upward communication will consist mainly of executives and administrators.

Downward communication has two major functions: (1) to transmit instructions and outline procedures, and (2) to provide background information—to let people know what's going on. Of the two, the former is more common, but the latter is perhaps more important. Your primary audience here will be mainly technicians.

The functions of lateral communication are more difficult to specify. A good bit of lateral communication serves a courtesy function—to keep others posted as to what is happening in the organization so that they may keep abreast of new developments. Some lateral communication functions specifically to provide others information they need in order to perform their jobs. Often, though, information is just broadcast laterally on an FYI (For Your Information) basis on the off chance that someone might derive some benefit from it. Your primary audience for most lateral communication will be other technical experts, either experts in your field or experts in some other field of science or technology.

Information going outside the organization is even more difficult to pin down. It may serve in decision-making (a proposal or an environmental impact statement), it may provide instruction (an operation and maintenance manual), it may provide specific information to help people in their jobs (a technical bulletin or a response to a request for information), or it may serve a public relations function (an annual report or a newsletter). The corresponding primary audiences for each of these functions would, of course, be executives and administrators, technicians, technical experts, and laypeople respectively.

□ Developing Design Criteria

Throughout this chapter I have stressed the process of analysis. We mustn't lose sight of the fact, however, that analysis is a useful exercise only inasmuch as it leads to the process of design. Once you have analyzed a communication context, you must then design a communication vehicle to meet the needs of that context. The key parameters of analysis are: (1) audience, (2) purpose, (3) type of information, and (4) vectors of information flow. These parameters combine to give you design specifications for such variables as:

- □ The basic type of report you will write.
- □ The amount and type of information that will go in the main body of your report and in appendixes.
- □ The level of language you will use in various parts of your report.

- □ The style of writing you will use in various parts of your report.
- □ The illustrations and other graphic material you will use in your report.
- □ The degree of revision and final polishing you will give to your report.
- □ The number of copies of your report you will generate and the distribution list for them.

Here are some things you can do to help yourself in developing design criteria:

- □ Specify the type of document or publication (see Chapter 1, "Writing in Business, Industry, and Government") and the appropriate form and format (see Chapter 4, "Report Forms and Formats"). Write out your specifications in a sentence or two.
- □ Specify the organizational pattern you will follow—either a suggested general outline from an appropriate chapter in Part Three, "Types of Written Communication," a specific outline from some outside source, or a specific outline you generate yourself. If you plan to depart from a suggested general outline, justify the departure with several sentences of explanation.
- □ From the list in Figure 3–4, select ten adjectives that best describe the tone and approach appropriate for your given commu-

Adjectives for Tone and Approach		
abstract	detached	obscure
ambiguous	devious	orderly
authoritative	dogmatic	personal
balanced	direct	positive
blunt	dull	precise
boring	emotional	pompous
breathless	general	remote
brisk	humorous	rude
bureaucratic	interesting	scholarly
chaotic	"it"-centered	serious
chummy	"I"-centered	simplistic
clear	impressive	slow
cocky	informative	specific
concrete	messy	straightforward
condescending	muddled	stuffy
confident	negative	subjective
confusing	neutral	superficial
convincing	natural	"you"-centered
courteous	objective	wishy-washy
dense	obsequious	

Figure 3–4 Adjectives for Describing Tone and Approach

nication context (see Chapter 21, "Style at the Word and Phrase Level," and Chapter 17, "Technical Articles," for more on tone and approach).

- □ Specify the appropriate level of technicality of your words and phrases (see Chapter 21, "Style at the Word and Phrase Level"). Are Level-One technical terms appropriate? Level-Two technical terms?
- □ Specify the appropriate degree of revising and polishing for the communication context. Is basic revision for adequacy sufficient (see Chapter 5, "Report-Writing Procedures")? Is first-pass revision for excellence in order? Or does the communication context demand your fullest efforts toward excellence (see Chapters 18 through 22 in Part Four, "Revising for Excellence")?
- □ Estimate the extent of your total audience by specifying the number of copies of your document that will be duplicated and the distribution list for each. Also, consider the group of people who will review or approve your manuscript (see Chapter 5, "Report-Writing Procedures").
- □ List any other aspects of the communication context—audience, purpose, or type of information—that might be pertinent to the level of adequacy or excellence of what you are to write (see Chapter 2, "Adequacy and Excellence").

Writing Exercises

1. Suggest some differences in the way you would handle a report directed to knowledgeable laypeople on one hand and technical experts in another field on the other.
2. Suggest some differences in the way you would handle a report directed to researchers in your field on one hand and other technical experts in your field on the other.
3. List and describe some differences (in addition to those mentioned in the text) between what a technical expert as *technical expert* might look for in a report and what a technical expert as *manager* might look for.

Writing about Writing

1. Write detailed descriptions of three separate communication contexts you might encounter in your field or discipline. Distinguish clearly between primary, secondary, and immediate audiences. Discuss purpose and type of information as well. Identify clearly what members of each audience segment will do with the information you present.
2. For one of the communication contexts you described in Assignment 1 (or another that you describe in detail) write a detailed analysis of the communication context, as outlined above.

3. Select some piece of technical writing from your field (other than a journal or magazine article, a technical bulletin, or a textbook). Reconstruct as best you can the communication context. Analyze the piece of writing in terms of how well it was tailored to meet the situation. Suggest any changes that would make the writing more effective. Attach a copy of the piece.
4. Write an essay on the topic "Audience analysis as a means of making writing more effective."

□□ TWO □□

Basic Writing Techniques

A dilemma that faces every teacher of technical writing is whether to present writing forms first and then techniques or to present techniques and then forms. It is difficult for students to study techniques without some forms to practice on, but it is equally difficult to practice forms of writing without some knowledge of techniques. In the classroom, a teacher can ease the burden of this dilemma a bit by presenting both simultaneously, but in a linearly ordered textbook one must choose in some way between techniques and forms. I have chosen to begin with basic techniques in Part Two, to present the various forms of technical writing in Part Three, and then to return to techniques—advanced techniques—in Part Four. There is some merit, however, in your first skimming the material in the six chapters of Part Two, then skimming the chapters of Parts Three and Four, before returning to Part Two for more careful reading and study.

Another word of caution. Not all of the chapters in Part Two contain techniques as such. Some present basic information and ideas that you as a writer of technical documents should know. Chapter 4, "Report Forms and Formats," is one such chapter—mostly information. Chapter 5, "Report-Writing Procedures," and Chapter 6, "Handling Manuscripts and Drafts," deal mostly with techniques. The remaining three chapters of Part Two are mixed bags of techniques and information. Collectively, though, these six chapters offer you the basic information and techniques you need to write consistently adequate, acceptable reports and other technical documents. The emphasis in this part is on adequacy and efficiency, although the later sections of Chapter 5, "Report-Writing Procedures," and much of the material in Chapter 8, "Graphics and Illustration," and Chapter 9, "Documentation," go a way beyond *minimal* adequacy. And some may say that Chapter 7, "Correctness in Usage and Mechanics," goes beyond minimal adequacy as well. If you find this the case, read the chapters for what they are worth and to know *what is in them*. Then you can return to them and use them as reference when you reach that point in the course at which you are ready to go beyond minimal adequacy yourself.

□ 4 □

Report Forms and Formats

This chapter offers an overview of some of the different things that get called reports. Here (and throughout) the term *report* broadly refers to a variety of written documents—including proposals, literature reviews and state-of-the-art papers, instructions, procedures manuals, environmental impact statements, and even certain types of magazine and journal articles. Many people use the term much more narrowly, but there is little consistency as one moves from one organization or discipline to another. In fact, even within the narrow use of *report* there is little consistency. My best advice is to become familiar with the characteristics of basic report types and not to put too much stock in what different people call them.

Any system of classifying types of reports has to consider both *what a report looks like* and *what a report does*. When we use labels like "short report," "memo report," or "formal report" we're concerned with the appearance, or *form*, of the report. When we use labels like "feasibility report," "research report," or "progress report" we're concerned with the purpose, or *function*, of the report. There is no one-to-one correspondence between report form and report function; we have variations in both categories. In Chapter 1 we looked at the major functional varieties of technical writing—correspondence, external publications, and internal documents. Here we consider some of the variations in report form and format.

☐ The Fill-in-the-Blanks Report

The fill-in-the-blanks report starts with a blank form (see Figure 4–1). Your job as report writer, basically, is to fill in the blanks. A good many laboratory reports are of this type, particularly in working laboratories (as opposed to school laboratories). Fill-in-the-blanks reports are used widely in business and industry for monthly summaries, inspection and travel reports, and other oft-repeated kinds of communication. They are generally used for primarily numerical (rather than primarily verbal) information. Figure 4–2 shows a completed fill-in-the-blanks report.

When writing (or maybe I should say completing) a fill-in-the blanks report, remember that the blank form was devised as a convenience for you and not as a device to intimidate you. Use the blank form so long as it is convenient to do so, but don't be a slave to it. If a blank calls for information that isn't pertinent, enter *NA*, for "not applicable." If you have pertinent information to report for which there isn't a blank, use a separate sheet of paper. Similarly, if a blank isn't big enough (and it seems that most of them aren't), continue your report on a separate sheet of paper. The form should work for you. You shouldn't be working for the form.

☐ The Letter or Memo Report

Many reports are published in the form of a letter or memo. (Remember, a memo is just a letter sent internally within an organization.) The letter or memo report (see Figure 4–3, page 49) is typically short—one to five pages, although it may be longer. It may have headings and subheads but often doesn't. And it may have accompanying tables or figures but usually doesn't. Since letter reports and memo reports often become permanent records, their tone is often a bit more formal than the routine business letter (see Chapter 10, "Technical Correspondence"), but here as elsewhere, tone is established by audience and purpose, not by report format. If you have a short report that is basically free from bulky appendixes or illustrative material, consider a letter or memo report. If you're not sure a letter or memo form would be appropriate for your report, you probably should consider a formal or semiformal report with an accompanying cover letter.

☐ The Formal Report

The formal report, like formal men's wear, has evolved over the years to become a fairly rigid convention. There are certain elements, or components, that *must* be present in a report for it be called formal. If any of these elements are missing, the report is something other than a formal

WEEK OF ______

OPERATIONS AND MAINTENANCE NAVY

BUDGET	AUTHORIZED YTD	OBLIGATED/EXPENDED YTD	BALANCE
STAFF TRAVEL			
PHONE			
PURCHASE SERVICES			
SUPPLIES			
CIVILIAN PERSONNEL			
REPRO & PRINTING			
VEHICLE			
OTHER			
TOTAL			

RESERVE PERSONNEL NAVY (CLOTHING)

BUDGET	AUTHORIZED YTD	OBLIGATED/EXPENDED YTD	BALANCE
INITIAL ISSUE			
ORGANIZATIONAL			
CLEANING/ ALTERATIONS			
REPLACEMENT			
TOTAL			

UNIVERSITY BUDGET

BUDGET	AUTHORIZED YTD	OBLIGATED/EXPENDED YTD	BALANCE
SALARY			
OTHER EXPENSE			
TOTAL			

Figure 4–1 Representative Form for a Fill-in-the-Blanks Report

WEEK OF 31 AUG 81

OPERATIONS AND MAINTENANCE NAVY

BUDGET	AUTHORIZED YTD	OBLIGATED/EXPENDED YTD	BALANCE
STAFF TRAVEL	3,100.00	2,303.63	796.37
PHONE	7,000.00	5,827.72	1,172.28
PURCHASE SERVICES	2,400.00	2,351.58	48.42
SUPPLIES	8,900.00	8,869.16	30.84
CIVILIAN PERSONNEL	13,500.00	10,707.49	2,792.51
REPRO & PRINTING	400.00	132.80	267.20
VEHICLE	4,500.00	3,502.55	997.45
OTHER			
TOTAL	$39,800.00	33,694.93	6,105.07

RESERVE PERSONNEL NAVY (CLOTHING)

BUDGET	AUTHORIZED YTD	OBLIGATED/EXPENDED YTD	BALANCE
INITIAL ISSUE	14,600.00	12,379.20	2,220.80
ORGANIZATIONAL	50.00	-0-	50.00
CLEANING/ ALTERATIONS	2,100.00	1,005.75	1,094.25
REPLACEMENT	50.00	-0-	50.00
TOTAL	16,800.00	13,384.95	3,415.05

UNIVERSITY BUDGET

BUDGET	AUTHORIZED YTD	OBLIGATED/EXPENDED YTD	BALANCE
SALARY	4,786.00	0	4,786.00
OTHER EXPENSE	7,028.00	2,530.66	4,497.34
TOTAL	11,814.00	2,530.66	9,283.34

Figure 4–2 Completed Fill-in-the-Blanks Report

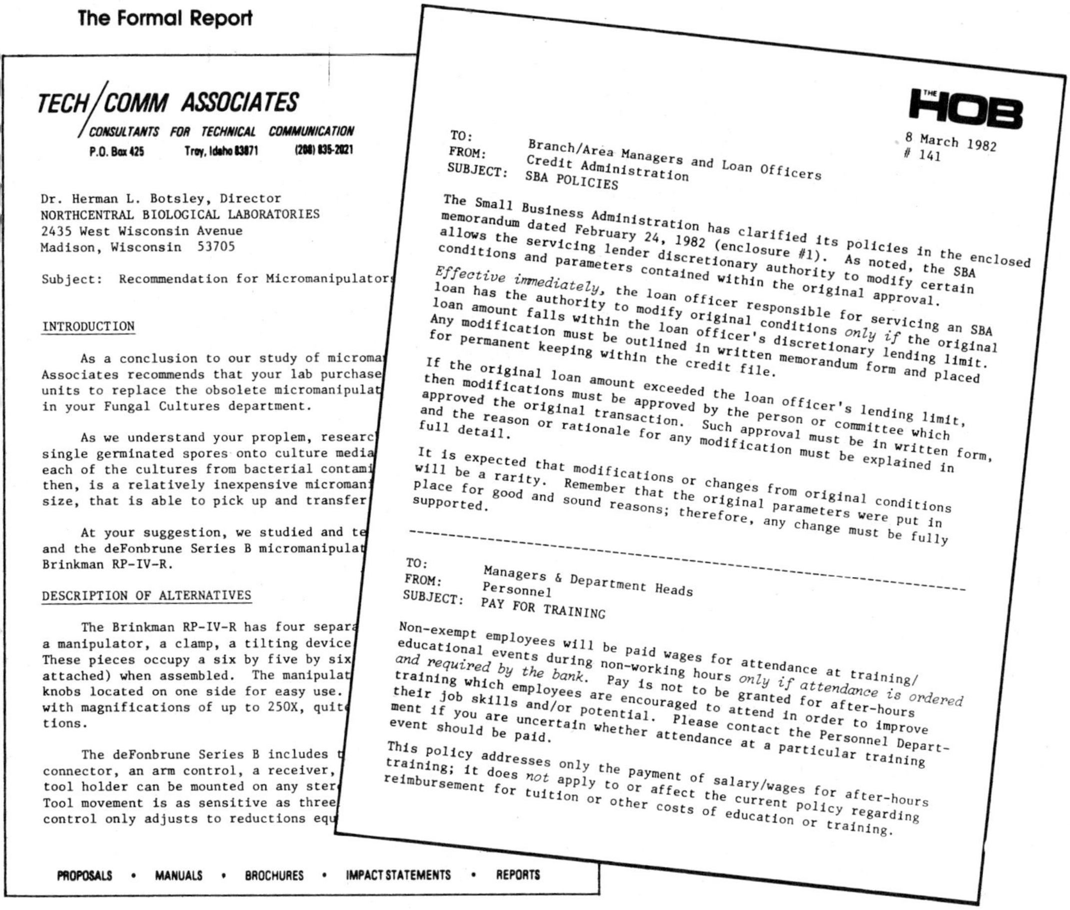

TECH/COMM ASSOCIATES
CONSULTANTS FOR TECHNICAL COMMUNICATION
P.O. Box 425 Troy, Idaho 83871 (208) 835-2021

Dr. Herman L. Botsley, Director
NORTHCENTRAL BIOLOGICAL LABORATORIES
2435 West Wisconsin Avenue
Madison, Wisconsin 53705

Subject: Recommendation for Micromanipulator

INTRODUCTION

As a conclusion to our study of microma
Associates recommends that your lab purchase
units to replace the obsolete micromanipulat
in your Fungal Cultures department.

As we understand your proplem, researc
single germinated spores onto culture medi
each of the cultures from bacterial contam
then, is a relatively inexpensive microman
size, that is able to pick up and transfer

At your suggestion, we studied and te
and the deFonbrune Series B micromanipulat
Brinkman RP-IV-R.

DESCRIPTION OF ALTERNATIVES

The Brinkman RP-IV-R has four separ
a manipulator, a clamp, a tilting device
These pieces occupy a six by five by six
attached) when assembled. The manipulat
knobs located on one side for easy use.
with magnifications of up to 250X, quit
tions.

The deFonbrune Series B includes
connector, an arm control, a receiver,
tool holder can be mounted on any ster
Tool movement is as sensitive as three
control only adjusts to reductions eq

PROPOSALS • MANUALS • BROCHURES • IMPACT STATEMENTS • REPORTS

THE HOB

8 March 1982
141

TO: Branch/Area Managers and Loan Officers
FROM: Credit Administration
SUBJECT: SBA POLICIES

The Small Business Administration has clarified its policies in the enclosed memorandum dated February 24, 1982 (enclosure #1). As noted, the SBA allows the servicing lender discretionary authority to modify certain conditions and parameters contained within the original approval.

Effective immediately, the loan officer responsible for servicing an SBA loan has the authority to modify original conditions *only if* the original loan amount falls within the loan officer's discretionary lending limit. Any modification must be outlined in written memorandum form and placed for permanent keeping within the credit file.

If the original loan amount exceeded the loan officer's lending limit, then modifications must be approved by the person or committee which approved the original transaction. Such approval must be in written form, and the reason or rationale for any modification must be explained in full detail.

It is expected that modifications or changes from original conditions will be a rarity. Remember that the original parameters were put in place for good and sound reasons; therefore, any change must be fully supported.

TO: Managers & Department Heads
FROM: Personnel
SUBJECT: PAY FOR TRAINING

Non-exempt employees will be paid wages for attendance at training/educational events during non-working hours *only if attendance is ordered and required by the bank*. Pay is not to be granted for after-hours training which employees are encouraged to attend in order to improve their job skills and/or potential. Please contact the Personnel Department if you are uncertain whether attendance at a particular training event should be paid.

This policy addresses only the payment of salary/wages for after-hours training; it does *not* apply to or affect the current policy regarding reimbursement for tuition or other costs of education or training.

Figure 4–3 Representative Letter and Memo Reports

report (see below). In other words, a formal report is a report that meets certain conventions of form. The label *formal* has little to do with style of writing or tone of language; a formal report is not necessarily stuffy or pompous or "formal" in tone and style. It just has all the necessary parts.

We can divide these parts into two groups—front matter and back matter:

- Front matter—Cover letter
  - Cover (optional)
  - Title page
  - Table of contents
  - List of illustrations (optional)
  - Executive summary
  - List of definitions and symbols (optional)
- Back matter—Bibliography
  - Appendixes
  - Index (optional)

Cover Letter

The cover letter (or cover memo), also known as the letter of transmittal, is the first thing a reader should see on picking up your formal report. Its primary purpose is to provide you with a vehicle for saying things to *some* members of your audience that you may not want to say to *all* members of your audience. (You send *different* cover letters to different members of your total audience.) Suppose you've written a proposal for professional services to be submitted to a prospective client—say, the city of Pottsville, Pennsylvania. You'll be sending copies of the proposal to your immediate supervisor, the vice-president of your company, the mayor of Pottsville, the city engineer, and each of the four councilmen. Here are some possible cover letters:

To: Chuck Bronson, Supervisor
From: Jan Stevenson
Subject: Pottsville Proposal

Here's the Pottsville proposal. The section on "Project Management" doesn't reflect our new corporate structure, but I didn't feel that would make any difference to the Pottsville people. Note the graphs on pages 6 and 7—Harold Crouch did an excellent job on them. . . .

To: Michael C. Rhodes, Vice-President
From: Janene W. Stevenson, Municipal Services
Subject: Proposal for Engineering Services for Pottsville, PA

Attached is a copy of the proposal for engineering services for the City of Pottsville. The city council has scheduled your formal presentation for next Tuesday evening, at 7:30. I have prepared a schedule of paving costs and a list of available survey crews; they are enclosed as appendixes D and E to your copy of the proposal. Please let me know if there is any additional information you will need for the presentation. . . .

The Honorable James S. Whittom, Mayor
City of Pottsville
121 City Hall
Pottsville, PA 15792

Dear Mayor Whittom:

H & J Engineering is pleased to submit a proposal in connection with the City of Pottsville's North Lake Subdivision development. Enclosed are five copies of the proposal for you and the other city council members. (We are sending John Yost's copy separately, as it has some additional technical data appended to it.) Michael Rhodes, Vice-President of H & J Engineering, will make a formal presentation at your meeting Tuesday night and will answer any questions you might have at that time. . . .

Mr. John C. Yost, P.E.
City Engineer
224 City Hall
Pottsville, PA 15792

Dear John:

Here is your copy of H & J's proposal for work needed in the North Lake Subdivision. I have included the soil sample results and hydrologic profiles, as you requested. Mike Rhodes will make the presentation Tuesday night, as John Bremmerton is out of town. If you need any additional data before the formal presentation, please call me. . . .

Cover

Longish reports—a dozen or more pages—are usually distributed with a cover of some kind (see Figure 4–4, page 52); shorter reports are usually distributed without covers (or, as we say in the trade, are self-covered). The cover serves to identify and to protect. For brochures, proposals, catalogs, and other documents that have a certain promotional aspect to them, the cover serves to attract as well. To protect, the cover should be sturdy and firmly attached to the report. To identify, the cover should present at least the following information:

- □ Title of the report
- □ Name (and address) of the organization publishing the report

If a report has no title page (see below), the cover should also provide information about the author(s) or compiler(s) of the report, the date of publication, and any identification numbers. All covers should be attractive. Covers for promotional documents often use two or more colors of ink, photographs or graphic designs, and/or a rendition of the corporate or organizational logo to enhance their visual appeal. Since reports spend a fair amount of time face down, many organizations use graphic elements on the back cover as well.

Title Page

The title page of a formal report (see Figure 4–5, page 53) serves to identify the report and provides all necessary bibliographic data so that someone could prepare a complete bibliographic citation for the report (see Chapter 9, "Documentation"). For a full-dress formal report you should always include a title page, even if your report also has a cover. The items to include on your title page are:

- □ Title of the report
- □ Author(s) (or editor or compiler) of the report
- □ Organization (and division or section) publishing or distributing the report (with mailing address)

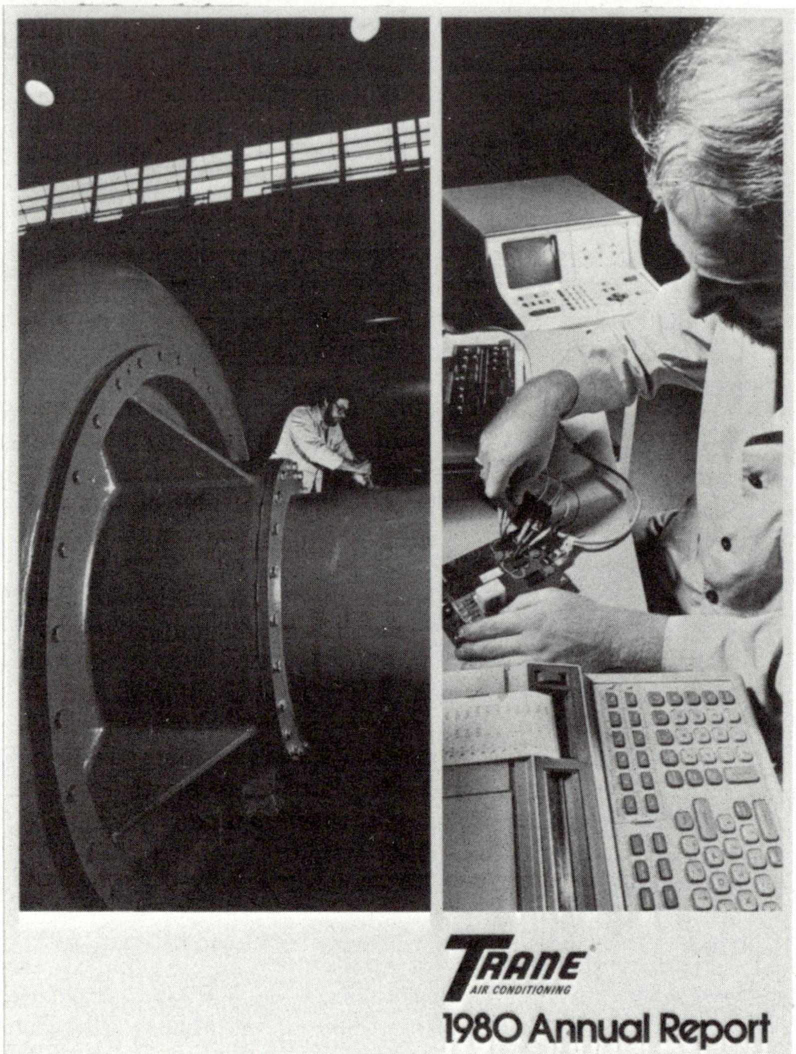

Figure 4–4 Representative Cover for a Formal Report (The Trane Co.)

PERCOLATION TESTS IN THE

PROPOSED REDWOOD SUBDIVISION

Prepared for

EVERGREEN BUILDERS, INC.

Wayland, New York

By

EMPIRE SOILS TESTING SERVICE

422 Springwater Street
Elmira, New York 14901

July 22, 1983

ESTS Report No. 83-101

Figure 4–5 Title Page of a Formal Report

- □ Date of publication or issuing of the report
- □ Form number, order number, etc. of the report (if any)
- □ Person (or organization) for whom the report was prepared (if appropriate)
- □ Copyright, distribution, or classification data (if appropriate)

Table of Contents

A table of contents is appropriate for reports of fifteen to twenty pages or longer. With shorter reports, it's just as easy for your readers to thumb through the pages as to look something up in the table of contents. If you use a table of contents, list all chapter and section headings as a minimum and list subsection headings if you feel they would help your readers find crucial information. Figure 4–6 shows a convenient format for a typewritten table of contents.

List of Illustrations

A list of illustrations (and particularly a list of tables) is useful in a report if the illustrative or tabular data are such that readers may want access to them independent of the report's main text. Again, though, such a list is of limited value in reports shorter than fifteen to twenty pages. The list of illustrations (see Figure 4–7, page 56) can appear on the same page as the table of contents if there is room; otherwise put it on a separate page following the table of contents.

Executive Summary

An executive summary or introductory summary (see Chapter 13, "Abstracts, Summaries, and the Like") is a component of a formal or semiformal report that is designed to give busy administrators, executives, or decision-makers the gist of a report so that they don't have to read the whole report in order to decide whether or not they want to read the whole report. Typically, an administrator will read the executive summary and then turn the body of the report over to technical experts or administrative assistants for detailed study. Alternatively, the administrator will read the executive summary to determine which portions of the report he or she wants to read in detail.

Your executive summary should be the last part of the report you write—not the first. It should represent your clearest and most effective writing; remember, it is the first thing most people will read and the only thing that some people will read. Concentrate on the what, who, and how much of your report. Leave the why and the how for the main body. Limit yourself to a single typewritten page for reports of a dozen pages or fewer, and limit yourself to 5 or 10 percent of total report length (exclusive of

TABLE OF CONTENTS

Figure 4–6 Suggested Format for Typewritten Table of Contents

TABLE OF CONTENTS

LIST OF ILLUSTRATIONS

Figure 4–7 List of Illustrations with a Table of Contents

appendixes) for longer reports. Avoid choppy paragraphs in your executive summary, but at the same time avoid loose-jointed and rambling paragraphs. Strive for cohesiveness and continuity (see Chapter 19, "Style at the Paragraph and Section Level"). Figure 4–8 is a representative executive summary.

EXECUTIVE SUMMARY

Brown's Interiors of Kansas City will design, provide, and install six interior window coverings needed in Myklebust's Men's Apparel for a cost of $1260.00. Myklebust's is in need of aesthetically pleasing window coverings that will efficiently block sunlight entering the store late in the afternoon. In this proposal, Brown's describes the most effective window coverings, outlines its design and installation techniques, presents a timetable for the services to be provided, and presents the firm's past experience and credentials. Brown's Interiors is pleased to present this proposal and is looking forward to providing Myklebust's with the finest of professional interior design services available in the area.

Figure 4–8 Representative Executive Summary

List of Definitions and Symbols

It is best, of course, to avoid obscure terminology and unusual symbols in your reports. If you must use a few, it is best to define or explain them at the point you first use them, but there is little need for a list of definitions and symbols in most reports. However, if you find yourself writing a report that does contain a good many strange terms or symbols, the place to define them is at the front of your report—immediately ahead of your introduction. Your list of definitions and symbols, if you have one, should be the final item in the front matter of your formal report.

Bibliography

Your bibliography, or list of references, can appear either before or after your report's appendixes or even as one of your appendixes. It is, of course, a list of complete bibliographic citations for all documents you refer to in your report—plus, if appropriate, additional documents that your readers might find useful. See Chapter 9, "Documentation," for a detailed discussion of bibliographic conventions and formats.

Appendixes

Information that is vital for your primary audience (see Chapter 3) must go in the main body of your report. Almost anything else is a candidate for inclusion as an appendix. This might include information that is of secondary interest to your main audience, information that is of interest to your secondary audience, and background and supplementary information of any kind. The trend nowadays is definitely toward greater use of appendixes. Yesterday's fifty-page report is tomorrow's ten-page report with forty pages of appendixes.

For a management report with executives and administrators as primary audience and technical experts as secondary audience, some types of data that should appear in appendixes rather than in the main body of your report are:

- ☐ Calculations and derivations
- ☐ Detailed design criteria or specifications
- ☐ Copies of research, laboratory, or data reports on which you are basing a proposal or recommendation
- ☐ Resumes of personnel
- ☐ Lists of clients, jobs, and projects
- ☐ Copies of correspondence pertinent to your report
- ☐ Detailed tables, charts, and graphs
- ☐ Highly technical descriptions, analyses, or discussions

Whenever you include such data in appendixes, be sure to refer your readers to the appendixes at appropriate places in the main body of your report ("See Appendix A for a more detailed discussion" or "Refer to complete resumes in Appendix B") and, if appropriate, summarize the appended information in the main body of your report.

For a research or data report with technical experts as primary audience and either executives, technicians, and/or laypeople as secondary audience, consider presenting some of the following types of information as appendixes:

- ☐ A general, plain-language discussion of background and theory
- ☐ A glossary of terms
- ☐ Detailed descriptions of alternatives (for example in a feasibility report)
- ☐ Bench notes, calculations, and detailed findings that even the technical experts wouldn't be vitally interested in.

If you have a single appendix, label it "Appendix." If you have more than one, label them "Appendix A," "Appendix B," and so on; better yet, include a short descriptive title: "Appendix A—Calculations," "Appendix B—Resumes," "Apendix C—Background of the Sedimentation Problem in Howell Lake."

Index

Indexes are used primarily in manuals or data reports that are likely to be used as reference documents. Furthermore, their use is limited to longish documents—twenty pages or more. If a document needs to be indexed, it's probably better for someone other than the author to do the job. As author you tend to be rather narrow in your view of what topics readers might look under to find certain categories of information. A good index has many cross-reference entries in it; ideally, an index should have an entry for whatever anyone might conceivably look under. If you have a document that needs indexing, get some professional help. Preparing a good index is not a trivial chore, and a casual index is little better than no index at all.

Page Numbering in a Formal Report

The first page in the main body of your report should be page 1, and you should number subsequent pages of the main body with Arabic numerals. Use lowercase Roman numerals for pages of front matter (i, iv, vi, etc.). Don't put a page number on your title page, but reserve a number for it; that is, number the following page ii. Similarly, you may leave the page number off the first page of the main body and number subsequent pages starting with 2.

Pagination for back matter gets a bit more complicated. One option is to continue Arabic numbering throughout your bibliography, appendixes, and index. More often, though, appendix pages are given a letter-plus-number designation: A-1, A-2, A-3. If you follow this convention, put your bibliography and index ahead of your appendixes and number the bibliography and index with the Arabic sequence of the main body of your report. Or alternatively, include your bibliography and index as your last two appendixes. If any appendix is a reprint of a document with its own page numbers, retain the original pagination and put in your own page numbers in square brackets either above or below the original ones.

□ The Semiformal Report

A semiformal report (as Theodore A. Sherman astutely suggested) is best regarded as a formal report with something missing. You don't always need a bibliography; you don't always have material for appendixes; not every report needs a cover letter. On the other hand, you may want a title page, or a table of contents, or an appendix without all the other paraphenalia of a full-scale formal report. Feel free. In fact, probably the majority of the reports you write on the job will fall into the semiformal category. Just treat each element of front matter or back matter as you would for a formal report and collate each element in the order you would for a formal report.

□ Variations in Report Function

In Chapter 1 and 3 we discussed different functions of reporting largely in terms of audience and purpose. It is useful as well to look at report function in terms of the kind of information being reported. Basically, there are only three things you can report on:

- □ The status of a *project* you are accountable for
- □ Certain *activities* you've been engaged in
- □ *Data* you've been asked to gather or generate

These, then, suggest three major functional varieties of reports ("reports" here in the narrow sense): project reports, activities reports, and data reports. Many (perhaps most) of the reports you write will serve two or even all three of these functions, but let us look at the functions separately before we look at them in combination.

Project Reports

In today's technological society, virtually all human productivity in businesses, industries, and government agencies is generated by groups of people working on projects. And project reports are the fuel (or perhaps the exhaust) of the project machine. Project reports (see Chapter 12) are read mainly by administrators to (1) keep current on the status of work in their departments, (2) estimate future work loads and personnel needs, and (3) prepare higher-level management reports. They may also be read by other technical experts to seek valuable interactions between what you're doing and what they are doing, and to prevent duplication of effort. Some common report labels that fall in the project report category are:

- □ Progress reports
- □ Status reports
- □ Weekly, monthly, quarterly reports
- □ Methodology sections in research reports

Such reports generally focus on things that have been done, things that remain to be done, problems that have been encountered, and noteworthy contributions (good or bad) of various individuals and groups.

Activities Reports

Unless you're self-employed, you will be asked on a regular basis—usually monthly—to report how you spent your time during the month (week, year) immediately past. It goes with the job—just like payroll deductions. In addition, if you engage in any special activities (travel, inspections, training programs, tours, and the like) you will probably be asked to report on them separately. Your immediate supervisor, and perhaps other administrators as well, will read your activities reports primarily to estimate future work loads and personnel needs and secondarily to evaluate you as an employee and to assess your performance. Chapter 12 has suggestions on writing activities reports. Here are some common report labels that would fall in the category of activities reports:

- □ Monthly activities report
- □ Travel (or trip) report
- □ (Traffic) Accident report
- □ Conference report

Reference Data Reports

As its name implies, a reference data report is a report that presents data you have gathered or generated and that is used by others for some reference purpose such as general enlightenment or education, specific instruction or training, or input data for some project they are pursuing. The reference data report is at once the type of report you have the most familiarity with from your college writing and the type of report you will have the least occasion to write when you get on the job. In form, the undergraduate term paper is a good model for reporting data you have *gathered* (although for other reasons the undergraduate term paper is not a good model for *any* type of report; see Chapter 1). On the other hand, the journal article (see Chapter 17, "Technical Articles") is a good model for reporting data you have *generated*.

One interesting audience (of sorts) you should keep in mind for reference data reports you might write is the scientific or technical editor. Actually, you as technical expert will have relatively little occasion to communicate directly with unsophisticated laypeople or the general public. (I should say *write* rather than *communicate*; you will have many occasions to speak to the general public.) You may have occasion to be the original author or generator of technical information that has the general public as its ultimate audience, but in the majority of cases such writing will be mediated through a professional writer or editor. Or to look at the situation differently, among those who read your data reports and technical papers may be editors who read what you write so that they might disseminate your technical information to a wider audience.

Management Data Reports

Most of the reports and other documents you will be asked to write in on-the-job situations will be management reports of one kind or another. Project reports and activities reports are, in a very real sense, management reports. So are all the proposals, feasibility reports, state-of-the-art papers, recommendations, resumes, and impact statements that you write. All of these present data primarily for management decision-making purposes. One of the key facts that you as technical expert must keep in mind is that reference reports and management reports are *totally different* with respect to organization, tone, content, and writing style. Topics, or subject matter, may be similar, but audience and purpose—and your analysis of and reactions to audience and purpose—are as different as gnats and gnomes.

In the course of writing this book I visited or corresponded with several hundred managers of technical divisions in several dozen scientific and technical fields. The overwhelmingly most frequent and serious communications shortcoming they reported on the part of recent college graduates was *inability to write a management report.*

Reports with Combined Functions

As I suggested earlier, many if not most of the reports that you write will be functionally complex. A progress report, for instance, may be primarily a management project report but may contain sections on new data or results you have turned up. A research report may be primarily a reference data report but may serve also as a report of your research activities. One or more of your laboratory reports (reference data) may become appendixes in a feasibility report (management data). Or the "programmer's manual" that you write on the heels of a six-month project developing a new computerized cost-accounting system may become a veritable stew of history and state-of-the-art, user's instructions, management recommendations, hardware feasibility analyses, activities reports, and a lengthy narrative of how you came to put the whole package together.

Resist the temptation to homogenize all your information in a single document because it all deals in some way or other with the same technical topic. Consider separate documents for separate audiences and separate purposes. Or at least organize your information in separate sections for separate audiences and separate purposes. When you should be logical, don't be merely chronological. Don't justify when you should be explaining; don't explain when you should be instructing; and don't instruct when you should be analyzing. Don't tell your readers things they already know, and above all don't tell your readers (especially administrators and executives) things they don't really care to know. Package your information according to audiences and *their* purposes, and label your packages. Work at it a little bit. Remember Richard Brinsley Sheridan's admonition: "Easy writing's curst hard reading."

Writing Exercises

1. Collect several examples each of the four major varieties of report formats. Can you find any reports that do not fit the classification

scheme presented here? If so, develop a revised classification scheme.

2. Collect several examples each of the five major functional varieties of reports. Can you find any reports that do not fit the classification scheme presented here? If so, develop a revised classification scheme.
3. Compare the reports you have collected with those collected by several classmates. Can you find any correlation between report form and report function? If so, write a brief report describing the correlation.

□ 5 □

Report-Writing Procedures

This chapter consists of a set of hints, tips, and suggestions to get you started in putting reports together and get you on the way toward writing acceptable reports and moving your reports up the scale of quality toward excellence. (The five chapters of Part Four, "Revising for Excellence," take up where this chapter leaves off.) Basically, there are five phases involved in putting a report together: two that are important before the actual writing begins—

1. Data gathering
2. Preliminary organization

and three that involve the actual writing itself—

3. Initial draft preparation
4. Revision
5. Final inspection

□ Data-Gathering

In some senses, data-gathering is a college-writing concept that has relatively little application in on-the-job professional writing. As a college student writing a term paper, you start off with minimal knowledge of a subject, garner certain additional information on the subject from published sources in the library, and then write a summary of what you have learned. As a technical expert on the job, however, you will already possess most of the information you need for a report, and the bulk of the remaining specific data you need is readily available from your daily activities, research you have conducted, or feasibility studies you have performed. In a very real sense your problem will not be to find enough data for a five-page report but rather to limit and select from the data you have at hand so that it will fit in a five-page report. There will be occasions, however, when it will be necessary to gather data for a report, so it is good to know something of data-gathering techniques.

Stephen E. Pauley suggests four methods of obtaining technical data and information:

- □ Direct observation
- □ Testing and experimentation
- □ Consultation and interviews
- □ Library research

Direct observation is probably the most common method of gathering technical data. If you are writing instructions for the disassembly of some piece of machinery, you take it apart yourself to determine the best method. If you are writing a feasibility analysis of two manufacturing processes, you go out where the processes are being performed and observe them actually being done. The rough notes you take during this direct observation then become your data.

Testing and experimentation is nothing more than observation under controlled conditions, often in a laboratory. By eliminating or controlling extraneous variables you can increase the reliability of your observations; that is, you can be more confident that the measurements you take indeed reflect aspects of the phenomenon you are investigating. Often, research of this sort will be your main activity and writing up the results will be an ancillary activity. In such cases, though, remember that a written presentation of your data will have to be made, and gather and record your research data in such a way that your writing activities will be made easier and your communication of research results more effective.

Consultation and interviews with other technical experts is a third method of gathering technical information. Here, you are going directly to the source of information rather than relying on secondary reports and publications. Some tips on interviewing techniques: First, do some preliminary reading so you have some idea of what your expert is going to be talking about. Second, tape-record the interview if possible, or have someone else with you to take notes. It is difficult to conduct a good interview and take notes at the same time. Third, go to the consultation session or interview with some specific questions to ask. Remember, you are gathering data; you are not going on a fishing expedition. Finally, let the consultant or interviewee do most of the talking; you can listen to yourself any time.

Library research is the fourth method of gathering technical data. Since you have been checked out on library research in earlier composition courses, we will not labor the point here but will consider a few supplementary suggestions. First, become familiar with the various computerized indexing and abstracting techniques available on your college campus. When you get on the job, find out what computerized resources your firm or organization has and practice a couple of dry runs with them. Computer-based data storage and retrieval systems are growing exponentially—particularly in science and technology. Within the decade there will be little else.

A second suggestion is to utilize the services of a librarian, particularly a company librarian, or a technician within your department to do the legwork. A literature-search specialist is better at what he or she does than you are, and such a person is probably paid less than you, so your firm or organization saves money two ways. And you get more and better information.

Third, when you get down to the journal articles themselves, avoid getting carried away with note cards and bibliography cards. Use a copy machine instead. Read an article over briefly, and if it looks like it has useful information in it, run a copy. Then go back into it later on, underlining and abstracting the important data. This saves you time, eliminates the possibility of error in note-taking, and gives you a file of literature on a particular topic that may well prove valuable in the future. One suggestion when copying articles: many journals put a bibliographic citation like the one shown in Figure 5–1 on the first page of each article. Check to see that each article you copy has one or write one in—when you make the copy. Otherwise you wind up with a stack of useful articles but no clear idea where they came from.

□ Preliminary Organization

It is a peculiar fact of life that people working in those most highly organized areas of intellectual pursuit—science and technology—tend to produce writing that is among the most unorganized there is. The subject matter of technical writing is organized; only the writing itself suffers from lack of organization. This entire subsection can realistically be summarized in a single sentence: *Let the organization inherent in your work carry over into your writing*. Your task as a technical writer is not to *impose* organization on your data but rather to *discover* the organization that is already there.

There are three essential characteristics of good organization—or of things that are well organized:

- □ All related items are together in one place.
- □ All unrelated items are separated.
- □ Packages of related items are clearly labeled.

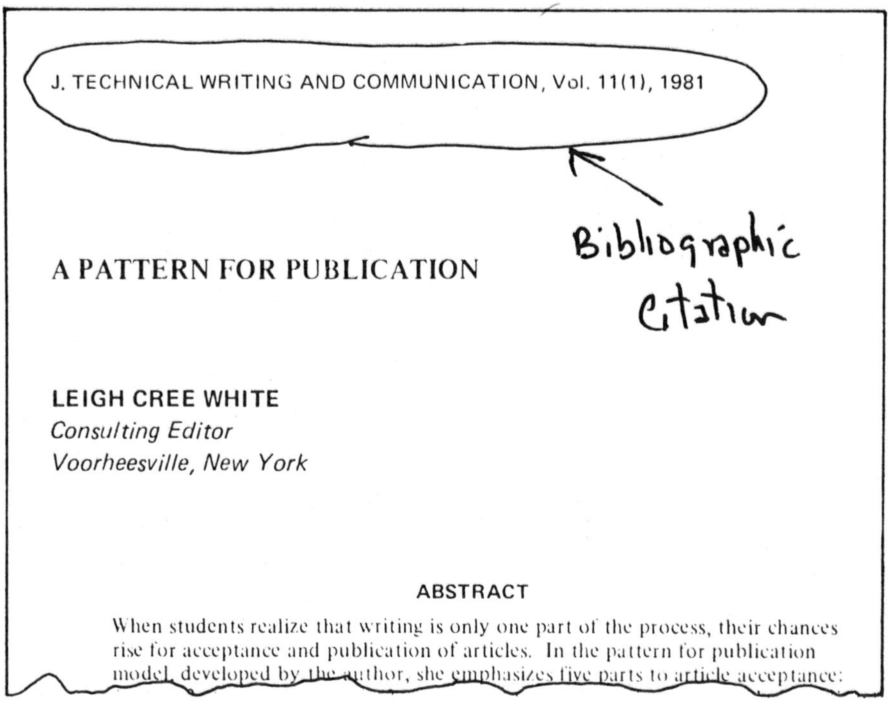
J. TECHNICAL WRITING AND COMMUNICATION, Vol. 11(1), 1981

A PATTERN FOR PUBLICATION

LEIGH CREE WHITE
Consulting Editor
Voorheesville, New York

ABSTRACT

When students realize that writing is only one part of the process, their chances rise for acceptance and publication of articles. In the pattern for publication model, developed by the author, she emphasizes five parts to article acceptance:

Figure 5–1 Bibliographic Citation on the First Page of a Journal Article

These characteristics of good organization apply to anything—not just to information. They apply to a well-organized file cabinet, a well-organized workbench, a well-organized kitchen cupboard, or what have you. Let's see how they apply to a well-organized report.

First, a well-organized report has all related items of information gathered together in a single section of the report. Second, it does not have extraneous bits of information buried in an otherwise coherent and homogeneous section. And third, a well-organized report has each section and subsection clearly labeled by heading and subheading as to its contents. A good set of instructions on, say, planting a lawn will list all the materials and equipment needed in a single section—usually at the beginning—so the readers do not have to study the entire document to find out what they will need to get the job done. The section on preparing the seedbed will tell everything about the preparation and will not include information on repairing Rototillers or getting rid of dandelions. Finally, good lawn-planting instructions will have each phase of the operation clearly set out by heading and subhead: Leveling the Subgrade, Preparing the Topsoil, Preparing the Seedbed, Fertilizing, Planting, Watering, and so on. The goal of the instruction writer is not only to put all necessary information in the report (see Figure 5–2), but also to put that information in the report in such a way that it is easy for the readers to get

Figure 5–2 The Ins and Outs of Information

the information out of the report. Never write a report in such a way that your audience has to read all the way through it just to find out if it contains the information they need!

Procedures for Achieving Sound Organization

How does a writer of technical information go about achieving the sound organization good writing demands? Since most of the writing you will do on the job will be fairly standard types of reports, the key first step is to follow standard organizational patterns or general outlines. (This textbook, for instance, gives general outlines for standard types of reports, proposals, and other documents. There are many other sources as well.) This is not to say that you must blindly follow rigid outlines for every writing situation; what you should do is follow the general outline and touch all the bases unless you can specify a clear and compelling reason for departing from the standard pattern. In other words, start out by making things easy for yourself rather than inviting unnecessary difficulty.

There will be occasions, however, when the standard outlines just don't fit your writing situation. Perhaps you are asked to write a special type of report that isn't covered in the handbooks and textbooks. In such cases, your strategy should be to go through the files of your firm or organization to find earlier documents that come as close as possible to fitting your immediate situation. Or ask your boss to suggest something that can serve as a pattern to follow. Then follow it. Again, make things easy for yourself.

Finally, if you find yourself completely at sea with a communication context for which there are no prototypes at all, you can use the following procedure to come up with a workable organizational structure:

1. Perform a detailed communication context analysis (see Chapter 3) and *write out* your analysis.
2. On several sheets of paper, jot down in random order as they occur to you the specific items of information that you might include in the

report. (Don't limit yourself to *categories* of information like "Background," "Costs," "Benefits," and "Conclusions," but also list specifics like "grade-staking," "three-man crew," "need 4-wheel drive truck," and "sandy-loam soil with pockets of clay.")

3. Gather related items of information together.
   a. Think of having each of these items of information written out on an index card, and think of having a set of empty shoeboxes in front of you. (You don't have to do this physically, but keep this card-and-shoebox image in mind.)
   b. Select an item of information (index card) and place it in an unlabeled category (empty shoebox).
   c. Select a second item of information and ask yourself whether or not it is related to the first. If it is, place it in the same category (the same shoebox); if it isn't, place it in another category (another empty shoebox).
   d. Proceed in similar fashion through your entire list of items of information. (Keep in mind that as you do this additional items of information will occur to you. Just add them to your list and keep going.)
4. Separate out unrelated items of information. After you have completed step 3, go through each category of information (each shoebox) and recheck the items. If you find something that doesn't seem to fit, pull it out and place it in a separate pile on the side.
5. Think of some appropriate labels to put on each of the shoeboxes. Select a label that is descriptive of the contents; don't start with labels and make the contents fit the labels but let the nature of your categories of information generate the labels.
6. Consider all the items of information in each shoebox: Does each item jibe with the label? If so, fine. If not, either remove the item of information from the category or change the label. If you wind up with a pile of miscellaneous items of information that don't quite fit anywhere, don't just sprinkle them around here and there—put them all in a single box and label the box appropriately: "Miscellaneous."
7. If necessary, go through the contents of each box and follow the same procedure to generate some subcategories of information. Then use these subcategories to generate sublabels. Your set of labels and sublabels, then, becomes your information-based outline.

Finally, you must give some thought to the arrangement of boxes and the balance between and among categories of information. Is there any inherent ordering to the boxes? Is some body of information dependent on some other body of information? Do some things have to be understood before others can be? Is there a well-recognized pattern that might be appropriate? Should information be presented chronologically, or spatially, or in order of importance? (Here, consider what you learned in high school or freshman composition about methods of development. Your method of development, essentially, is the order in which you present information.) As a rule of thumb to achieve proper balance, your longest section should be no more than twice (or at most three times) as long as your shortest section, and your miscellaneous section, if you have one, should be your smallest section, or at least one of your smallest. It certainly shouldn't be your largest, for your miscellaneous section contains unorganized information, and if it is one of your largest sections most of your report is unorganized. If you run into problems with balance, consider joining small sections together to make big ones or breaking large sections apart into smaller ones. (Be sure to change your labels appropriately.)

This completes what is often called the "prewriting phase" of preparing a written document. We have devoted quite a bit of space to prewriting, but this is as it should be, because if you are diligent in your prewriting, the actual writing and revision phases will move right along. On the other hand, whenever student writers complain that they are inefficient at writing, investigation almost invariably shows that they failed to do their homework with data-gathering and organization.

□ Initial Draft Preparation

The third phase of report-writing is preparing the initial draft. And let me stress *initial* draft. Regard everything you write at this stage as something you are going to have to write over again. Don't strive for perfection the first time around; just get it down on paper. The biggest timekiller is staring at a blank sheet of paper or worrying for endless minutes over the phrasing of a single sentence or paragraph. Your primary job in this phase is just getting your ideas into words—worry about form and style later. If you happen to hit on something in your initial draft that is acceptable so far as form and style go, great. Use it the way it is. But don't count on this happening very often. Plan on having to rewrite everything.

Pick your own most productive time of day to do your initial draft-writing, and block out a sufficient amount of time for the task. A good first approximation is about a half hour per typewritten page of draft. If you have trouble getting started, go back and do some more data-gathering or organization. To be effective in draft writing, you should have all your information at hand and be eager to write it down. Follow the suggestions in Chapter 6, "Handling Manuscripts and Drafts," for initial drafts—write on only one side of a sheet of paper, double- (or triple-) space, and leave wide margins. Make it easy to revise and rewrite. Consider using a tape recorder or dictating machine for getting your initial ideas down, and then have the recording transcribed for your initial draft. Many writers recommend this technique highly.

Another tip for initial-draft preparation: don't start at the beginning, with your introduction. Your intro is often the hardest thing to write, and it is also the most important because it is what your readers see first. Delaying buys you two things: it lets you start with the easiest section first—the one you know the most about or have the best handle on—and it lets you turn to the introduction after you have a chance to warm up. If you are like most of us, you tend to write a little better after you have cranked off a page or two. Similarly, don't be afraid to leave gaps in your initial draft. List a topic and then write in "to come later" or "more to come." If you get stuck on something, don't fret and worry over it—go on to something else. Get whatever you can down on paper and then come back to the hard parts. One of the greatest fallacies perpetrated on student writers is the notion that you must write something in precisely the order you expect your reader to read it.

The key to success with initial-draft preparation is easy, rapid writing. The words should pour onto the paper. You accomplish this by having all your information at hand and having a sound organizational structure for it, by having a clear picture of the people who will read what you write and what they will do with the information you convey, by picking a productive time and setting for your efforts, by starting on the easy parts of your writing first, and by refusing to allow yourself to stop and polish your writing for form and style. That is a proper function of the next two phases of report writing.

□ Revision

Effective revision is the single most important factor that separates good writers from poor writers. Many student writers wish to learn the secret of producing excellent writing without revising—the secret of turning out prize-winning prose on a single pass. There is no secret. There is no way. In no other walk of life do we expect our first efforts to be our best efforts, and writing isn't any different. The secret of excellent writing is learning how to generate a sound product on your first pass and then learning how to apply a set of directed efforts to that product in order to make improvements in it. And improvement leads inexorably to excellence. Most of the directed efforts involved are collectively called *revision*.

The primary suggestion for the revision phase of report-writing is to give yourself time for a breather—at least a full day, better yet, two. Do something else. Get away from your report for a while. Presumably you have done all you could with the initial draft; it is unreasonable to expect yourself to turn right from drafting to revision and be particularly productive. In your initial draft you worked with the passion of a creative artist; with your revision you must work with the cold detachment of an editor, a quality control inspector. And you need time to switch roles.

When you do get down to actual revisions, do your major revisions first. It is counterproductive to change a few sentences in a section for minor reasons and then come back and do a wholesale revision of the entire section. Also, if you revise for things like word choice and punctuation and then come back and revise for sentence structure, what you often wind up doing is introducing word-choice and punctuation errors into your revised sentences.

One of the first major revisions you might want to consider is reorganization—either of the entire paper or of one or two sections in it. Often what seems like a good organizational pattern before you start writing turns out to be less than effective once you actually get into the drafting. If your paragraphs are tight and well-structured, this reorganization may well be just a matter of shifting entire paragraphs around. This is the time to get the scissors and tape out; there is no need to recopy a paragraph simply to move it—just cut your initial draft apart and rearrange the pieces.

A related major revision to consider is that of adding or deleting large blocks of information; that is, revising for completeness and revision for brevity. Adding information is usually no problem, but deleting significant portions of what you have worked so hard to write is painful indeed. But if the information doesn't fit, or isn't pertinent, or isn't worth the space you have devoted to it, you must get rid of it. Just remember that a good writer fills his or her own wastebasket rather than the readers' wastebaskets.

A third major item to consider in the revision phase is your graphics and illustrations. If you haven't included any graphic aids during your initial drafting, look over each section of your report to see if an illustration or two would help. Also, if you find sections that present quite a bit of numerical information—dimensions, capacities, costs—consider using a table or graph to communicate this information (see Chapter 8, "Graphics and Illustration"). Once you get your graphic aids keyed to particular sections, check your figure and table references to see that you have enough of them and that they are in the most useful places so far as your readers are concerned.

A fourth major revision is revising for writing style. Stylistic revision, of course, is the stuff that excellence is made of. At the same time, however, a particular communication context might call for a limited amount of stylistic revision to move a report from minimally adequate to middling adequate, or maybe even to above-average-but-still-not-quite-excellent. Some first-pass revision techniques for stylistic revision are discussed below. Full-scale revision for excellence is the subject of Chapters 18 through 22 in Part Four, "Revising for Excellence."

Minor revisions consist of checking for things like punctuation and spelling, grammar and usage, and overall smoothness and appropriateness of word choice. In other words, your minor revisions are the things you were taught to look for in earlier composition courses. A useful device to help you with minor revisions is a checklist like the one in Figure 5–3. Don't underestimate the value of checklists; the most experienced commercial pilot uses one on every takeoff. Develop your own checklist, one tailored to suit your particular needs. As you get writing assignments back with comments on them, note particular writing problems that you have and add these items to your own personal checklist. For all subsequent writing assignments, be sure that you are not making the same mistakes over and over again.

Don't try to go through your entire draft in a single pass, attempting to correct everything at once. Make multiple passes through your draft,

REVISION CHECKLIST

- ___ Each page is free from spelling and typographical errors.
- ___ Each page is free from dangling or misplaced modifiers.
- ___ All subjects agree with their corresponding verbs; all pronouns agree with their corresponding antecedents.
- ___ Each page is free from errors in grammatical parallelism.
- ___ Each page is free from sentence fragments.
- ___ Each sentence has proper end punctuation.
- ___ All introductory phrases or clauses are set off with commas.
- ___ All independent clauses are joined with semicolons or comma-plus-conjunctions.
- ___ All interruptive elements are set off with a *pair* of commas, dashes, or parentheses.
- ___ All possessives have apostrophes; no plurals have apostrophes.
- ___ All lists are introduced with colons rather than semicolons.
- ___ All periods and commas that appear with quotation marks are inside the quotation marks.
- ___ All abbreviations and symbols follow conventions.
- ___ All numbers and numerals follow conventions.
- ___ All titles of major publications and all unassimilated foreign words and phrases are underlined.
- ___ Each page is free from unnecessary capitalization.
- ___ All figures and tables have proper captions.
- ___ All headings and subheads are in proper format.
- ___ Each page has a page number properly located and sequenced.

Figure 5–3 Sample Revision Checklist

checking for a single item on a single pass. Go through your paper once for spelling, looking only at spelling. Go through it again for punctuation, looking only at your punctuation. Go through it a third time for dangling participles, looking only at phrases containing participles (-*ing* words). And so forth. The particular order you use—like the particular items on your checklist—is a highly personal thing that will vary from person to person. The important thing to do, though, is to develop and use a systematic procedure for making multiple-pass revisions. By doing this you will both save yourself time and improve the quality of your writing. Which is what learning to write is all about.

Revising for Orderliness

If you have followed a standard outline for your report, you don't have to worry too much about orderliness, or organization. You can be pretty well assured that you have all the right parts and they're in the right order. This is the chief value of standard outlines. On those occasions you have to depart from the standard outline, however, or for those situations that have no standard outline, the key guideline is order of importance *to your readers*.

If there is some inherent ordering to the information you are presenting—if certain things must be understood before others can be comprehended—make sure you have presented the basic data first or have provided adequate cross-references so that your readers can gain access to the basic information first. In management decision-making documents, present your best evidence first. If administrators aren't going to read your whole report (and many of them won't), make sure what they do read is the best and most convincing of what you have to say. In particular for feasibility reports, see that you have led off each criterion subsection with a discussion of the alternative you wish to recommend.

When you write reference material—instruction manuals, archival data reports, state-of-the-art papers, and the like—consider not only what information you put into the document but also how people will get the information out. One thing you can be sure of: no one will read your document in sequence, from front to back. (Have you ever read a dictionary that way? Or an encyclopedia or any other reference work?) People will use your document only when they have a need for information—most often only a small portion of the total information that you present. First, they will expect to find any general topic they might be interested in listed in your table of contents or your index—or at least flagged with a heading or subhead. Second, they will expect to find all you have to say about that topic pretty much in one place in your document—they won't like being cross-referenced all over your document, and they certainly won't like having to skim through your entire document just to see if you buried some interesting or useful tidbit of information somewhere else. And third, they will expect to find *only* information on the topic they are interested in in any given section or subsection; they won't be all that keen on wading through a bunch of interesting but unrelated data. Think of the computer analogy. Assume that your readers have at their disposal a computerized information retrieval system that will identify specific topics in your document on some sort of key-word basis and then will print out all the sections in which these key words appear, but *only* those sections in which the key words appear. In other words, all the readers get is what you have flagged. Is each of your sections sufficiently orderly and self-contained that the readers with such a computerized information retrieval system would get all the data they need but only the data they need?

In your management decision-making documents, have you kept your introduction, body, and conclusion all separate? Check to see that you have limited your introduction to background information and scope of the document; save your data presentation and analysis for the main body. In the body, see that in every case you have presented data before you begin to analyze it, and see that you have compartmentalized and separated the various aspects of the problem you are addressing and the solution you recommend for it. Keep in mind that not all readers—not even all of the decision-makers themselves—will be interested in reading everything you have to say. Your readers will want to be able to retrieve information from your report or proposal according to *their* backgrounds and interests and ignore other topics at *their* discretion. Your job is to provide a sufficiently well-organized (and well-labeled) document for each reader to be able to do just that.

Revising for Completeness

It is easy to leave things out. It is easy to assume that since you know something, your readers will know it too. It is easy to let fear of insulting your readers' intelligence lead you to starting way out in the area of new or unfamiliar information (see Figure 5–4, page 76) rather than overlapping with your readers' knowledge and familiarity as you should, as in Figure 5–5. It's easy to forget how little others might know—even the boss. As you judge a report for completeness, you *cannot* apply your own personal standards of familiarity with and knowledge of your subject. You're the expert. By definition, then, *all* your readers know less about your subject than you do. (If there's someone else in your organization that knows more about what you're reporting on, he or she will be assigned to write the report—not you!) So don't be afraid of starting well back in your readers' area of familiarity with your topic. If you tell them things they know and agree with, they will nod their heads in agreement, and you will begin to build up a base of confidence. This base of confidence will serve you well when you lead your readers out into the unfamiliar. The only way you can get into trouble starting well back in the known is never to get out of it, as in Figure 5–6 (page 77).

Figure 5–4 Beginning Way out in the Unfamiliar

For routine activities reports and archival data reports, completeness is pretty much a matter of making sure you haven't forgotten anything. Rarely do we *choose* to leave some crucial item of data out of a routine report, of course; we just forget to include it. So go down your checklists and standard outlines item by item. For fill-in-the-blanks reports, make sure you've filled in all the blanks. Compare your routine reports with other similar reports. Have you included all the items of information other people did? For archival data reports, better to err in the direction of too much than of too little. Think of the computer programmer's concept of "capturing data." Seize it and record it, before it gets away.

Completeness in correspondence is a considerably greater problem. In fact, follow-up letters requesting (and supplying) additional information account for well over half of the working world's total correspondence. Letter writers (like a lot of other writers) never heard that it is quicker to do it right than it is to do it over. If you are writing a letter of request, specify (preferably by numbered list) everything you wish to have. If you're replying to a letter of request, read the original and your response together, and literally check off the items of information requested as you read through your response. Periodically, go through your correspondence files and look at all the follow-up letters (either yours or somebody else's) that request additional information. Since most of your correspondence involves recurring situations, you can use past lapses as a guide to completeness in your current correspondence.

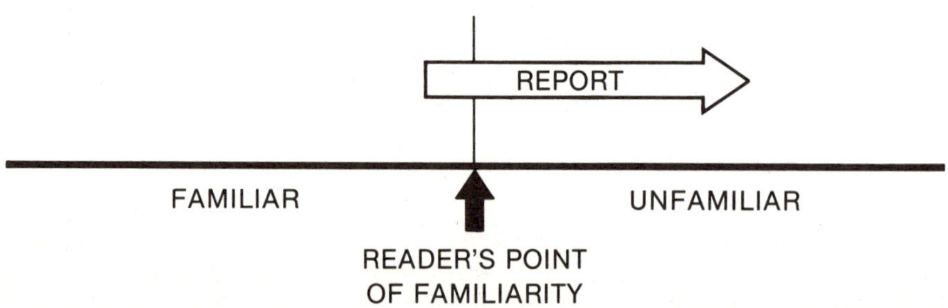

Figure 5–5 Beginning in the Familiar and Moving to the Unfamiliar

Figure 5–6 Staying in the Familiar

And then there are instructions! Incomplete instructions probably account for more confusion than any other single category of lapse in information completeness. It is easy to see why. Typically, instruction writers are so intimately familiar with the process or procedure they're writing about that they just can't imagine what it's like *not* to be familiar with it. So they leave little things out. Or maybe big things: "You mean you've got to jack the car *back down* before you drive off?" After you've written the initial draft of a set of instructions, have someone unfamiliar with the procedure you're discussing attempt to perform it by following your directions. Don't get another expert to review your instructions; find a novice. You'll find out where your holes are that way. As you write instructions, picture your readers as intelligent but very literal-minded robots—of the type we see all too frequently on TV these days. Assume that your readers will do what you tell them to, but only and exactly what you tell them—no more and no less. That way, your instructions will be complete. (In more than twenty years in the business, I've never heard anyone complain that a set of instructions was too complete!)

With management decision-making documents, most problems with lack of completeness arise from either failure to provide sufficient background information on the problem or failure to discuss the implications of what you propose or recommend. You cannot assume that all—or even most—of your readers will have your depth of knowledge and familiarity with details concerning the problem that you're attempting to solve, and similarly you cannot assume that your readers will grasp immediately—without any help or direction from you—the full implications of the data and information that you present. You don't have to lecture to people in a condescending fashion, but for goodness' sake give them a hint or two as to what is most significant in what you are about to say or what you have just said.

Revising for Brevity

In point of fact, most novice report writers err in the direction of too much information rather than too little. It's always tempting to tell all you know about a particular subject—maybe even a little more than you know. And it's always tempting to assume that if you're interested in certain aspects of a subject, everyone is interested. Besides, it's easier for you to dump all the information you have into your report than it is to sift and winnow. It takes time to be brief. The seventeenth-century French mathematician-philosopher Blaise Pascal ended a letter to a friend this way: "I have made this letter longer than usual because I lack the time to make it short." Keep in mind, however, that the goal of being brief is to save time, not paper. A report that is incomplete may save paper, but it is a waste of time.

Now, by brevity I don't mean boiling the unnecessary words out of phrases and sentences (although that is important and is discussed in Chapter 21, "Style at the Word and Phrase Level"); I mean eliminating whole paragraphs and sections that are repetitious or tangential or otherwise fail to contribute directly to your readers' needs for information. Of course it's tempting to include everything you know about a subject. Somebody might find it interesting. But interesting isn't enough; it's also got to be pertinent to the purpose of your report. I asked George F. Bailey, Manager of the Fuel and Technical Studies Department of Washington Public Power Supply Systems, about the kinds of revisions he made most to the reports his technical staff prepared. His reply: "Taking out redundancies and eliminating data that is scientifically interesting but not germane to the problem at hand."

In revising for brevity, you must approach your report from the position of the readers in your primary audience. You must develop a little empathy. You can't just put yourself in their position. You know what you know, and they don't know what you know. The whole question of pertinence and relevance of information must be answered from a point of view different from your own. And the question boils down simply to "Who cares?" and "Why?"

Lack of brevity can be a problem with any kind of report, but it is a particular problem with reports of research information—including, of course, articles for scientific and technical journals. A literature survey and state-of-the-art report is an essential component of any article reporting research (see Chapter 17, "Technical Articles"), but don't let your literature survey become a compendium of minutiae. In particular, avoid a general statement followed by a string of references:

> Many ducks and geese are found in Minnesota./4, 6, 7, 8, 12/
> Alloying steel makes it strong. (Smith, 1934; Jones, 1945; Johnson, 1952; Brown; 1967)

Stick to the important citations, the germane references. More people will be impressed with the *quality* of your bibliography than with its length. And don't get carried away describing your methodology in excruciating detail, either, unless your methodology is unique or controversial. If you're following standard procedures with standard equipment, say so and let it go at that. Don't get hung up on the model number of your radio tracking collar or the nameplate serial number of your circulating pump. Who cares?

With progress and status reports, stick to aspects of the project itself. Don't get into a presentation of data your project turned up (put that in a separate report), and don't let your "Special Problems or Difficulties" section become a proposal for solving the problems or difficulties. Similarly, don't let your management decision-making reports—proposals, justifications, and feasibility reports—get sidetracked into interesting but essentially nonproductive cul-de-sacs. Cut that stuff out and save it for another report.

Don't try to revise for brevity immediately after you have completed your initial draft; you're still too much in love with what you've just written. Get away from it for a day or two and come back to it with the icy eye of a publisher who has to pay $250 a page. Start by looking for the particular trouble spots mentioned above. Then look at each section and at each paragraph within each section. Ask yourself these questions:

- □ Is anyone besides me interested in this?
- □ Is it likely that anyone in my primary audience doesn't know this already?
- □ Is this information crucial to the understanding of the rest of my report?

If you answer no to any of these questions, get rid of the section or paragraph in question. If you answer maybe, move the information to an appendix. Don't fudge. Unless you can answer yes to all three questions, get it out of the main body of your report.

□ Final Inspection

As an industrial engineer will tell you, quality control inspection is vitally necessary to any production facility that hopes to turn out consistently acceptable products. And final inspection is likewise necessary to any writer who hopes to turn out consistently acceptable reports. Final inspection occurs after you have your final revised manuscript redrafted or retyped and just before you send your final revised manuscript out for review and approval (see Chapter 6, "Handling Manuscripts and Drafts").

You check for two things during your final inspection: (1) errors and mistakes that might have been introduced during the final retyping, and (2) errors and mistakes that have been there all along but that you have overlooked. What allows you to spot such errors that you have missed earlier is the fact that you have been away from your manuscript for the several days that it takes to be typed. This gives you the detachment necessary to find mistakes in your own writing; it is a very human trait to be blind to your own errors when you read over something you have just written, so the longer you stay away from it the more effective you can be as an inspector. Here are some things to check for during your final inspection:

- □ Typographical errors
- □ Misspellings of peoples' names or the names of firms
- □ Mistakes in numerical data; figures that don't add up
- □ Words or whole phrases missing, or words or phrases repeated
- □ Inconsistencies in abbreviations or the use of symbols
- □ Missing page numbers or pages out of order
- □ Inconsistencies in terminology—particularly inconsistencies between body copy and figure captions or callouts
- □ Inconsistencies between section headings and table of contents
- □ Common errors, like *it's* for *its*, *respectively* for *respectfully*, or *forward* for *foreword*

Also during your final inspection you may find words that don't quite fit, phrases that aren't quite smooth, or sentences that don't quite sound right. Resist the temptation to ignore such little things because you begrudge the time for retyping. If you take the time to do it right, you won't have to take the time to do it over. Mark the things that need changing and then change them or have them changed. Often, neatly done hand corrections are quite acceptable on a review manuscript.

One last point: Your final inspection must be the last thing you do. If your inspection leads to changes that require retyping of all or portions of your report, inspect it one more time when you get it back. Keep inspecting it until it passes.

□ Success Strategy for Student Writers □

Things get confused when you have to do your own typing. Roles get mixed up. Revising runs into retyping, and proofreading runs into final inspection. Try this: Complete your editing and revising on one day, wait until the following day to do your typing, and then—most important—wait another full day, twenty-four hours, to do your final inspection. Never turn in a classroom writing assignment the same day you finish typing it or the morning after you finish it. That extra day and a thorough final inspection will probably be worth a letter-grade improvement on every paper you write.

Writing about Writing

1. Team up with a classmate (preferably one whose major is the same as yours) and interview a faculty member in your discipline about some aspect of his or her current research:
   a. Prepare a set of interview questions.
   b. Conduct the interview, taking written interview notes.
   c. Prepare a written summary of the interview.
2. Determine what computerized indexing and abstracting services are

available at your college or university library. Write a report on your findings to be submitted to the class.

3. Survey printed material distributed by your campus library or interview a subject area librarian to determine what special references and resources are available to students in your discipline. Write a report on your findings to be submitted to other students majoring in your field.
4. Consult a dozen or so style guides or technical writing textbooks on the subject of the organization of proposals. Note similarities and differences, evaluate the various approaches, and identify the approach you feel is best. Write a report on your findings to be submitted to the class.
5. Do the same for the organization of feasibility reports (or recommendation reports or analytic reports, as they are sometimes called).
6. Do the same for the organization of instruction manuals.
7. Describe the situations and specific procedures for initial draft preparation that work best for you. Cover such things as location, time of day, lead time before deadline, length of time spent, equipment needed, and the like.
8. Write an essay on the topic "The advantages of adequate lead time for writing and revising technical documents."
9. Review your thinking on the concept of adequacy and excellence. If it has changed significantly since you studied Chapter 2, write a new essay on the topic.
10. Write an essay on "The advantages of multipass revision over single-pass revision."

6

Handling Manuscripts and Drafts

As you gain an appreciation of the importance of revision and multiple drafts in your writing, you gain as well an appreciation of the extra work involved in achieving excellence in written communication. This chapter is designed to help you cut down a little on that work; or, more precisely, to help you find ways to channel your energies away from the drudgery of writing and toward the more creative aspects. In the world of business and industry, the ability to make *efficient* use of your time is prized almost as highly as the ability to be *effective* in your work and in your communication about your work.

Preparing the Initial Draft

As Chapter 5 indicated, you should write the initial draft as if you were going to write it over, since most of the time you will. So make things easy on yourself by making the draft physically easy to revise. Write on one side only of a sheet of paper. This will enable you physically to cut up the draft and rearrange portions of it without having to recopy them. If you type, double-space. If you write longhand (see Figure 6–1), write on every other line. This leaves room to strike out words and phrases and write new ones in above. Similarly, leave a wide right margin—at least two inches—for additions and comments you might want to incorporate

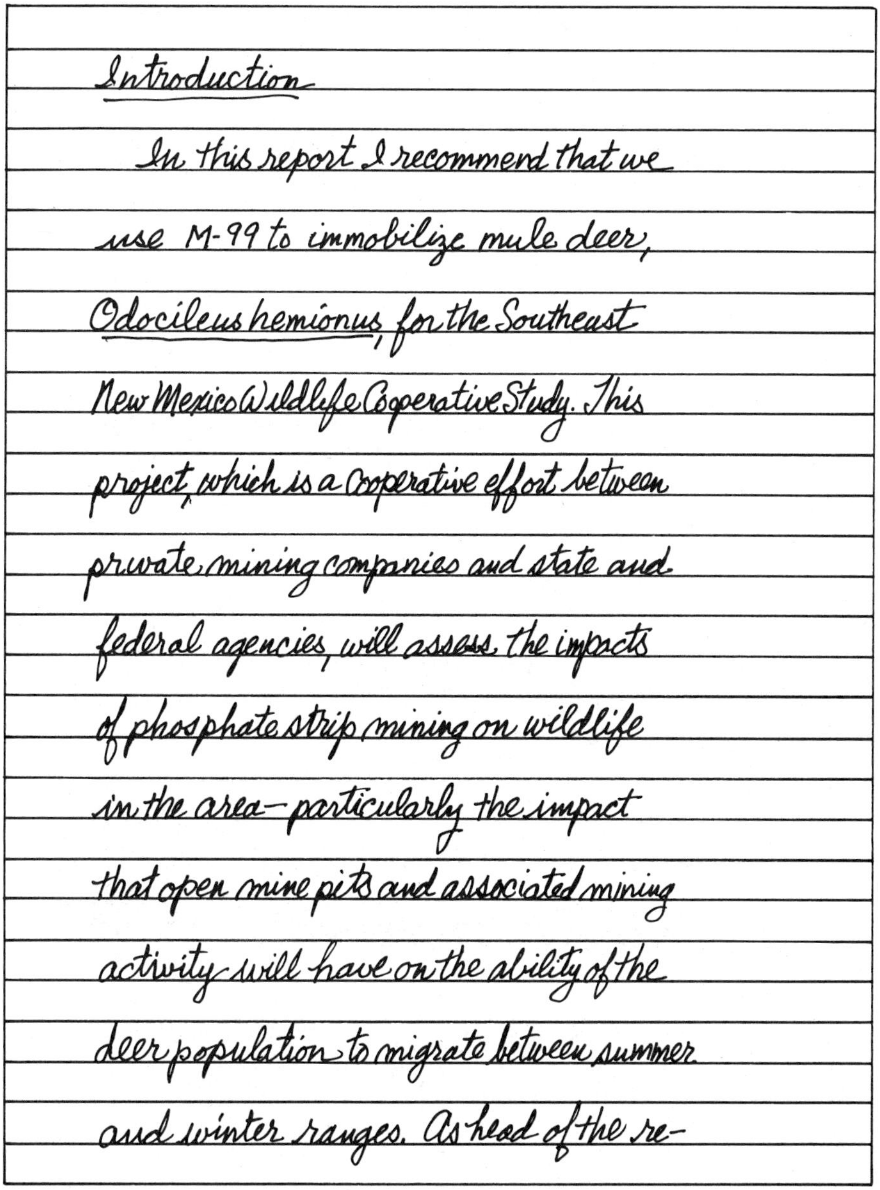

Introduction

In this report I recommend that we
use M-99 to immobilize mule deer,
Odocileus hemionus, for the Southeast
New Mexico Wildlife Cooperative Study. This
project, which is a cooperative effort between
private mining companies and state and
federal agencies, will assess the impacts
of phosphate strip mining on wildlife
in the area—particularly the impact
that open mine pits and associated mining
activity will have on the ability of the
deer population to migrate between summer
and winter ranges. As head of the re-

Figure 6–1 Suggested Form for a Longhand Manuscript

during revision. (As a reminder, draw a light pencil line down the sheet two inches from the right edge of the paper.) Don't erase. Scratch out or strike over. Don't interrupt your train of thought to try to make your initial draft look pretty. No one's going to see it but you and your typist, anyway. Remember that the initial draft is just a tool that you use to build your final product.

□ A Short Course in Typing

If you type at all—even hunt-and-peck at ten or fifteen words a minute—it will probably pay you to type at least initial drafts. Once you're doing that, you might as well type revisions as well. This section offers a few hints to save time and make your end product a little neater.

Paper

Use a moderate-weight, bright-white opaque paper for typing. A good choice is 20-pound white offset paper. Check with an office supply store or a commercial printer or your college's duplicating center. Avoid rag bond (it's too expensive) or light-weight semitransparent paper (it's too awkward to handle) or erasable bond (it smudges like mad). Mimeograph paper and spirit duplicator paper work pretty well, although they are somewhat more expensive than offset paper.

Use a backup sheet in your typewriter. It not only protects the platen (roller) from excessive wear but also cushions the keys and makes for a better page image. Use colored paper or heavy mylar for your backup sheets. That way you won't get them mixed up with your regular typing paper and wind up with indentations all over page six of your final draft.

Margins and Centering

For elite type set your left margin at 12 and your right margin at 85. For pica type, use 10 and 68. To tell which you have, roll a piece of paper into the typewriter and check the page gauge. If the page is 100 spaces wide you have elite type; if it is 85 spaces wide you have pica type. (In preparing a manuscript for publication by others, follow the publisher's suggestions for side margins. Publishers usually want wider margins so they can mark questions for the author or instructions to the typesetter.) For your top margin, soft-roll your paper until it lines up with the typed-line guide. Then roll it up six clicks to get a one-inch margin. If six clicks give you only half an inch, you have one of the newer typewriters with half-space capability; in that case roll it twelve clicks.

To keep from going off the page at the bottom, draw a light pencil mark in the lower left margin of your sheet about an inch from the bottom. When you come to the pencil mark, stop typing. When you remove the sheet from the typewriter, erase the pencil mark and no one will ever know.

To center things horizontally, move the carriage until the character indicator is at the center of the page (50 for elite, 42 for pica). Then depress the backspace once for *every other* character in the title you want to center. For instance to center "Chapter 8—Instructions," you would depress the backspace key for *C, a, t, r, 8, -, I, s, r, c, i,* and *n*. Try it; it works!

Numbers

If your typewriter doesn't have a numeral 1, use a lowercase ell. For zero use the numeral 0 rather than a lowercase oh. Avoid things like 2 × 5 = Io when 2 × 5 = 10 is just as easy.

Making Corrections

The best advice for making corrections in your initial draft copy is don't. Strike over or cross out. When striking over use a row of slashes (///////) rather than a row of x's (xxxxxxx). With the slashes you will be able to read what you struck out (in case you change your mind); with the x's you won't. For final typing, there are a number of things available that beat erasing. One of the handiest is a small bottle of opaque white paint (see Figure 6–2). You just paint out mistakes and type over them. Another useful correction aid is a roll of white tape that is just the width of a line of type. This is quite useful for covering up a major portion of a line or even several lines. Finally, there are several varieties of powder-backed paper that can be inserted under the paper guide on your typewriter. As you strike the incorrect letter a second time, powder is transferred from the

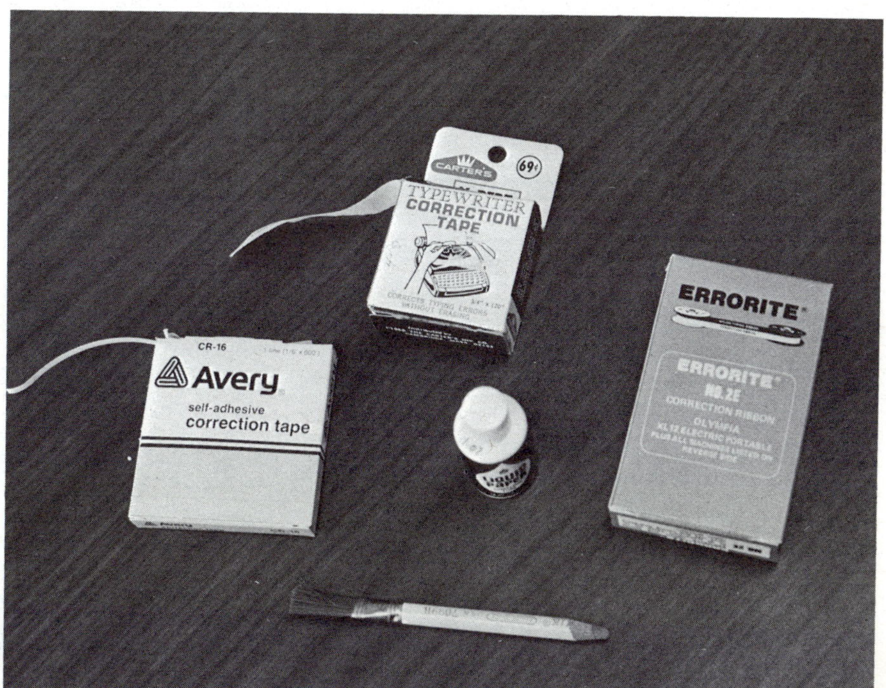

Figure 6–2 Aids for Making Corrections to a Typed Manuscript

back of the correction paper to your typed original to cover up the mistake. This yields neat corrections, but it is a bit cumbersome to use. The neatest and most convenient means are the self-correcting typewriters, but these tend to be a little expensive.

□ Marking a Draft for Typing

On the job, of course, you won't be typing your own final drafts but giving them to a secretary or typing pool. If your firm or organization has a style guide for different types of documents, become familiar with it and use it. It will save a lot of time for both you and the typist. If not, make use of standard proofreaders' marks (see Figure 6–3), particularly those for delete, close up, insert, new paragraph, spell out, italics, transpose, lowercase, capitals, and restore words crossed out. To indicate the insertion of new material, place a caret *under* the typed line at the point of insertion (see Figure 6–4, page 88) and write the material to be inserted either above the typed line or in the margin. Or if the material to be inserted is extensive, write "Insert Copy A" in the margin and type (or write) the new material on a separate sheet of paper headed "Copy A." Paperclip the separate sheet to the sheet on which the insertion is to be made so the typist can find it easily. To indicate deletion, draw a circle around the word or words to be deleted and attach a pigtail (deletion symbol) to the circle (Figure 6–4). For general information, such as line spacing, number of copies, format for headings and subheads, location of illustrations, and so on, attach a memo with instructions.

□ Revising Without Retyping

Revising is a matter of making additions to, deletions from, or changes in what you have already written. Retyping (or redrafting) is a matter of running what you have written and revised through the typewriter one more time to improve its appearance. Typically, the writer revises and a typist retypes; as a student writer, however, you will be doing both—perhaps simultaneously. It is good, though, to keep the two functions conceptually separate and not confuse them. In particular, there are a number of ways you can revise your technical writing papers without retyping them.

For minor changes like punctuation and spelling, just make the corrections neatly by hand with a black ballpoint or felt-tipped pen. For more extensive changes, type the new material on a separate piece of paper, cut it to fit over the old material, and tape it in place. (Use the frosted mending tape rather than shiny cellophane tape: it's neater.) If your revision involves major rearrangement of information, prepare several blank sheets with page numbers typed in, cut the initial draft

Proofreaders' Marks

Operational Signs

- Delete
- Close up; delete space
- Delete and close up
- # Insert space
- eq # Make space between words equal; make leading between lines equal
- hr # Insert hair space
- ls Letterspace
- ¶ Begin new paragraph
- no ¶ Run paragraphs together
- □ Move type one em from left or right
-] Move right
- [Move left
-] [Center
- Move up
- Move down
- = Straighten type; align horizontally
- || Align vertically
- tr Transpose
- (sp) Spell out
- stet Let it stand
- Push down type

Typographical Signs

- lc Lowercase capital letter
- cap Capitalize lowercase letter
- sc Set in small capitals
- ital Set in italic type
- rom Set in roman type
- bf Set in boldface type
- wf Wrong font; set in correct type
- X Reset broken letter
- Reverse (type upside down)

Punctuation Marks

- Insert comma
- Insert apostrophe (or single quotation mark)
- Insert quotation marks
- Insert period
- (set) ? Insert question mark
- ;| Insert semicolon
- :| Insert colon
- |=| Insert hyphen
- M Insert em dash
- N Insert en dash

Figure 6–3 Standard Proofreaders' Marks

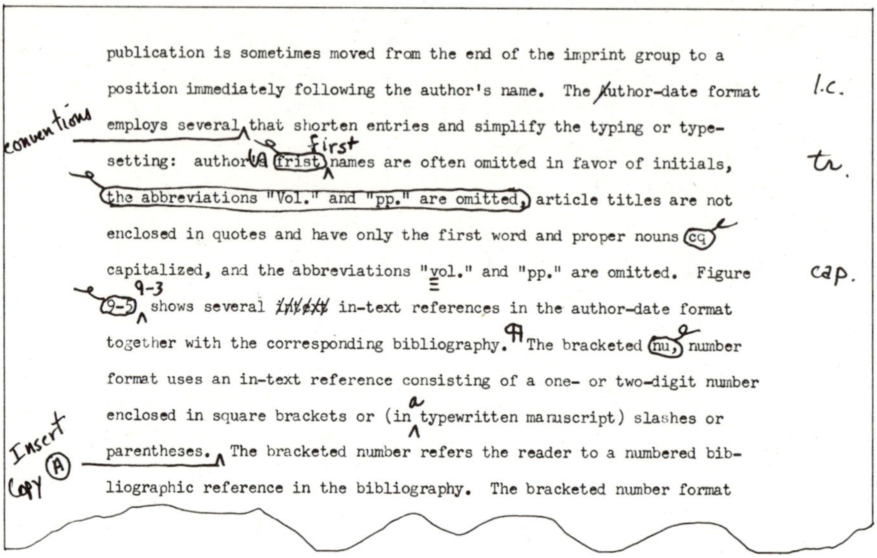
publication is sometimes moved from the end of the imprint group to a position immediately following the author's name. The Author-date format employs several that shorten entries and simplify the typing or type-setting: authors frist names are often omitted in favor of initials, the abbreviations "Vol." and "pp." are omitted, article titles are not enclosed in quotes and have only the first word and proper nouns cq capitalized, and the abbreviations "vol." and "pp." are omitted. Figure 9-5 shows several in-text references in the author-date format together with the corresponding bibliography. The bracketed nu, number format uses an in-text reference consisting of a one- or two-digit number enclosed in square brackets or (in typewritten manuscript) slashes or parentheses. The bracketed number refers the reader to a numbered bib-liographic reference in the bibliography. The bracketed number format

Figure 6–4 Manuscript with Proofreaders' Marks

apart, rearrange the paragraphs or sections, and tape them in place. If, early in a paper, you have major revisions that change the length of the copy significantly, you can save retyping all of the later pages either by stopping in the middle of the last sheet on which you have revisions and leaving the bottom blank (if you have deleted material), or by inserting an extra sheet or sheets between two numbered pages (if you have added material). If you insert, say, two sheets between pages 3 and 4, number the sheets "3A" and "3B." The key to all such revisions is neatness. If your end product is still fairly attractive to the eye, chances are it will be acceptable; if it's a mess, you had probably better retype it.

□ Electronic Word-Processing Systems

Virtually every firm or organization of any size has an electronic word-processing or text-editing system (see Figure 6–5), and increasingly these systems (or terminals connected to these systems) are being made available to professional staff for initial-draft preparation and revision. Such systems make typing obsolete. The keyboard for a word-processing system is virtually identical to that of a typewriter; if you can use one you can use the other. Your copy comes up on a cathode ray tube screen, however, rather than on a sheet of paper. Making corrections is simple—you just type over what you have written. Also, you can insert or delete letters, words, or whole lines with just a couple of key strokes. What's

Figure 6–5 Representative Word-Processing System

more, most word-processing systems have additional time-saving features that will enable you to just about double your typing speed.

The real payoff comes, though, in revisions. Your initial draft is stored in a disk or tape file as well as printed out. When revision time comes, you can call up your initial draft on the screen, change only what needs changing, and then have a clean copy printed out. In this way it takes less than a half-hour to revise *and retype* a five-page paper! Not only that, many word-processing and text-editing systems have syllabication and spelling routines, and some even can help you with your grammar and punctuation. Word processors and text editors are designed to take the drudgery out of writing and revising and to leave you as writer free to concentrate on the more creative aspects of effective written communication.

□ Integrating Artwork

Chapter 8, "Graphics and Illustration," discusses the importance of visual communication in technical writing and reviews some techniques for handling artwork. Here we will look at some techniques you can use to assure that illustrations are coordinated with the initial and review drafts and integrated properly into the final reproduction and distribution copy of your report or article.

As you write the initial draft you will have some idea of the line drawings, photographs, tables, or graphs you wish to include in the final product. At appropriate places in the draft, put in references to these, but leave the figure or table number blank—"(see Figure)"; "as shown in Table ,"—so you can add illustrations later without having to change numbers. When you begin the revision, one of your early steps will be to review what you have written to see if additional graphics will enhance the effectiveness of communication. As you do this, add figure references as appropriate, put in the figure numbers, and generate a list of figure numbers and captions. As one of the last steps in your revisions, insert figure locators into the copy as shown in Figure 6–6. These should appear in your intermediate and review drafts to give both you and your reviewers an indication of where illustrations will be located with respect to the copy.

In the larger firms and organizations or with the more substantial pieces of technical writing that are to be printed outside, your responsibility as writer for the integration of artwork ends with the review draft. For other situations, however, you as writer will have to see that the final reproduction copy of your report or article is typed (or typeset) in such a way that sufficient room is left in appropriate places for the illustrations (see "Handling Artwork" in Chapter 8). For anything that is to be printed commercially or duplicated on a small offset press, the printer will insert artwork as one of the last steps before printing. Even for the report that is to be reproduced by photocopier or electrostatic copier, illustrations can be generated on separate sheets of paper and then taped or cemented into place. About the only graphic element that should be generated on the same sheet as typed copy is a typed table.

□ Proofreading and Copyreading

Proofreading and copyreading are functions that are performed at several stages in the handling of a manuscript to guard against the inadvertent introduction of typographical and other minor errors into the manuscript and to eliminate such minor errors as might have crept in at some earlier stage. *Proofreading*, as the term is used in the publications industry, is a clerical function in which two people (one of them often being the typist) read through a newly typed draft of a manuscript and the most recent earlier draft of the manuscript. One person reads the original aloud—word-for-word and number-for-number—often verbalizing punctuation marks and spelling out names and difficult words. The second person silently follows along with the newly typed (or typeset) draft, making sure that the second is identical in every detail to the first or marking

7-2

(dates on which you started and finished individual tasks) and the hours each day you devote to individual tasks. That way, you've got something to work with at the end of the month.

// Figure 7-3 //

Reports are always after the fact. You prepare last week's activities report on Monday or Tuesday of this week; you fill out July's activities report on the 4th or 5th of August. This situation makes it tempting to fudge a little bit. You really completed that Johnson report on the 1st, and those last two sections of guard rail went in on the 2nd, but why not claim them for last month? What the heck, they're finished aren't they? I mean, it's not like you're claiming something is completed that really isn't. Resist the temptation. Sooner or later, you're going to come up short. In fact, if you do anything at all, fudge a little bit in the other direction--save those tasks completed on the 29th or 30th for next month. That way you'll have a little cushion to fall back on if you have a lean month.

Another sort of activities report is the report of special activity--something out of the ordinary that isn't part of your regular, routine duties. The travel report (see Figures 7-4 and 7-5) or trip report, as they are sometimes called, is perhaps the most common sort of special activities report. Another one that falls in this category is the accident report (Figures 7-6 and 7-7). The secret of success with the travel report is, once again, keep notes. If your trip lasts several days, write up a set of notes on your day's

// Figure 7-4; Figure 7-5 //

activities each evening. As soon as you get back to your desk, write up your travel report--even if it isn't due until the end of the month. Memories have a way of failing. With accident reports, be complete, be accurate, and be

// Figure 7-6; Figure 7-7 //

truthful--or at least consistent. And keep a copy of the report for yourself! That way, if questions ever arise you'll know what you said. In fact, it isn't a bad idea to keep copies of *all* of your activities reports; you never know when they might come in handy.

ARCHIVAL DATA REPORTS

Some of the data reports that technicians and technical experts generate find immediate use for planning and decision-making or serve as input for

Figure 6–6 Manuscript Page with Figure Locators Keyed in

corrections where necessary. *Copyreading,* on the other hand, is an editorial function in which a single person (often the author of the manuscript) reads silently through a newly typed draft, checking for clarity of wording, sequencing of elements, consistency of terminology, and the like. The terms are frequently confused, with proofreading being used to refer to the function more properly called copyreading—particularly in organizations and situations where proofreading as such is not done.

As a student writer (and often your own typist) you will be responsible for both functions and you will be tempted to perform them both simultaneously. Try to resist the temptation. If you can, get a second person to proofread with you; perhaps you can exchange proofreading services with another student in the class. If you must read alone, work with both drafts in front of you and try going through the manuscript sentence-by-sentence from *back to front*. Once you have done your more-or-less mechanical proofreading and made necessary corrections, read through the newly typed draft from *front to back* as if you had just picked it up for the first time. Or better yet, arrange your schedule so that a day intervenes between proofreading and copyreading (see Chapter 5).

□ Review and Approval Procedures

On the job, virtually everything you write will have to be reviewed and approved before it is released for distribution. *Review* is a process in which you as writer solicit comments and suggestions from others on something you have written. *Approval*, on the other hand, is a process whereby your superiors give their OK for something you have written to be released. Like revising and retyping or proofreading and copyreading, these are conceptually distinct functions that are sometimes done simultaneously and are often confused with one another. Anyone can *review* your writing—a superior, a co-worker, a subordinate, or even someone not in your organization at all. But only a superior in your organization (or a client) can *approve* what you have written. To *review* what you have written a person must read it. But your boss can (and sometimes will) *approve* your report or proposal without even looking at it. By doing so, he is saying that he trusts your work and is willing to accept accountability for it sight unseen. Finally, as writer you are free to accept or ignore *review* comments as you see fit; you are obliged to respond to *approval* comments, however, for if your boss suggests certain changes and then signs off on your report, he is giving his approval *as corrected*.

Informal Review

All excellent writers solicit informal review comments on every major document they write. (That's one of the things that makes them excellent writers.) This is usually a matter of offering a draft copy of your manuscript to the person at the next desk, or to a friend, or to someone you know who has a reputation of being a good writer, and asking for reactions to your manuscript and suggestions for improvement. As a student writer, you can solicit informal reviews from your spouse or roommate, from your fellow students in the technical writing class, and, most importantly, from your instructor. Cultivate the habit of heeding what people say. It will make you a better writer.

Formal Review

Many firms and organizations have established procedures for formal review. The final draft of a report or other document will be circulated among several people in the organization who have special knowledge of the contents of a document or special interest in the organizational change the document will effect. You as writer may identify these people, or they may be identified for you. Typically, the review draft of the document will be prefaced with a cover sheet (see Figure 6–7) giving reviewers a place to put their signatures or initials and the date of their review.

RUPERT O & M MANUAL

PRELIMINARY DRAFT REVIEW COVER SHEET

The attached material is being routed for your review and comment. It is a preliminary draft only -- the information may be incomplete, and there may be mistakes and inaccuracies. Please feel free to mark up the copy and write comments in the margins. Or if you prefer, attach written comments on a separate sheet. Your initials on this cover sheet do not imply approval of the copy -- this is a review only. Approval copies of the entire manual will be routed at a later date.

MANUAL SECTION:______________________________

Organization	Individual	Initials	Date	Comments
TECH/COMM	Skip Stratton			
H & V	Jack Hammond			
	Ted Sorensen			
	Art Smith			
RUPERT	Bill Peterson			
	Lyle Kaymeyer			
H & V	Jack Hammond			
TECH/COMM	Skip Stratton			

Figure 6–7 Cover Sheet for Formal Review

Two procedures for formal review are commonly used: sequential review and simultaneous review. With sequential review, a single copy of the review draft is circulated to one reviewer after another, with all reviewers making their comments on that copy. With simultaneous review, a separate copy of the review draft is sent to each reviewer for his or her comments, and then the writer and editor consolidate the various comments onto a single copy. If relatively few changes are generated during the review process, simultaneous review is the faster of the two and is preferred. On the other hand, if the suggested changes are extensive—and particularly if they are contradictory—the time required to consolidate comments and resolve conflicts among the reviewers will quickly eat up the time saved in the review process itself. Furthermore, sequential review has the added benefit of allowing each successive reviewer to read not only the copy itself but also the comments that earlier reviewers have made. If the sequential review is set up such that higher-ranking people review later, potential conflicts can often be resolved in the review process itself. All in all, simultaneous review is the more common of the two, but sequential review is probably more effective in the long run.

Approval

Approval is a formal management act in which a superior accepts accountability for work one of his or her subordinates has performed. With reports and other documents, approval is often combined with formal review, and an approval routing sheet (see Figure 6–8) is almost invariably used. If you as a writer have your work circulated for formal approval, it is a good idea to keep on file a copy of the signed approval sheet and the draft that was circulated with it. That way, if questions arise in the future about your report or other document, you can produce the necessary records to show who is accountable for the questionable information.

Responding to Review and Approval Comments

In the best of all possible worlds, review and approval comments will each be significant, creative, and constructive, and your immediate reaction will be to shower profuse thanks on the reviewers. In the real world, review and approval comments will often strike you as petty, obstructive, and perhaps even malicious. You labor for weeks over a piece of writing and some clown in another department who doesn't really understand the problem anyway spends fifteen minutes to tear it apart. And you have to put up with that sort of nonsense! It is only human to react with anger and frustration; in fact, to fail to do so would be to confess a lack of pride in your work. You will be tempted to fire off a scathing memo to the offending reviewer, with copies to his boss, your

APPROVAL ROUTING FOR MULTIPLIED EXPRESSIONS

SUBJECT ________ DATE ________

ORIGINATOR ________ PHONE ________ DEPT. ________ LOC. ________

FINAL APPROVAL DEADLINE ________ RELEASE DATE FOR PUBLICATION ________

ROUTING			APPROVAL		
PLANT BLDG. FLOOR	DEPARTMENT	ATTENTION	INIT	DATE	REMARKS

PROPOSED DISTRIBUTION	
CHECK	GENERAL CLASSES
	GENERAL PUBLIC
	DEALERS — DOMESTIC
	DEALERS — EXPORT
	MANUFACTURER — CUSTOMERS
	AUXILIARY EQUIPMENT MANUFACTURERS
	ALL EMPLOYEES
	ALL SUPERVISORS
	SEE ATTACHED LIST
	OTHER CLASSES (SEE "NOTES")

NOTES:

01-84842 (44 01)

Figure 6–8 Approval Routing Sheet (Caterpillar Tractor Co.)

boss, and the president of the company. This, like so many temptations, is a good one to resist.

Here is a better program of response. First, read through all the review comments without reacting or responding to any. Next, sort all the comments into three categories: those constructive changes with which you agree, those changes that seem merely alternate (not necessarily better) ways of saying things, and those changes that you can't live with—changes that alter or destroy your intended meaning, that introduce hopelessly awkward or unclear wording, or that introduce inconsistencies with or contradictions of things you have said elsewhere. Third (if you must), write out responses to the "can't live with" comments—but do so on scratch paper and file them away privately (and securely) for later disposition. Never—I emphasize NEVER—send off a written response to negative comments the same day you receive the comments. Give yourself a grace period of a day or two to cool off.

Accept the first two categories of change. If the suggested changes are in fact improvements, be grateful for them. If you don't feel strongly one way or the other about suggested changes, accept them with equanimity. Remember, you have only so many fights coming to you; save your fights for the issues that really matter. For the changes you can't live with, decide whether it is better to argue for the original wording or to suggest an alternate, compromise wording. Then marshal your supporting arguments, go privately to the reviewer who suggested the unacceptable change, and attempt to work out a mutually acceptable solution. By following this program you will generate good will (which is always in short supply) and you will find that end products of your writing efforts will increasingly be recognized as examples of excellent technical communication.

Writing Exercises

1. Try to arrange with the chairman of your academic department to have one of your reports for this class typed by one of the departmental secretaries—just as if you were a faculty member. If you are successful, write a brief report on your experiences vis-à-vis the handling of manuscripts and drafts.
2. Try to arrange to produce one of your reports for this class on a word-processing or text-editing system. If you use such a system, do the initial keyboarding yourself so that your preliminary draft is in machine-readable form. Do all of your subsequent revising by machine.

Writing about Writing

1. Conduct a survey to determine the various kinds of word-processing and text-editing systems that are used on your campus. As best you can, determine the advantages and disadvantages of each. Write a report to the chairman of your academic department recommending a system to be used.
2. Interview the publications groups or editorial staff of several nearby firms or organizations to determine the review and approval systems they use for internal documents and external publications. Write a report for the class.
3. Interview several senior secretaries or administrative assistants or heads of publications production groups—both on campus and off—to determine recommended practices for handling manuscripts and drafts. Summarize your findings in a report to the class.

7

Correctness in Usage and Mechanics

One of the characteristics of an acceptable report (or other piece of technical writing) is that it is essentially correct in grammar, mechanics, and usage. That is to say, an acceptable report conforms to the conventions of edited Standard American English and to the format conventions of the organization the writer is representing. Or, put another way, an acceptable report is essentially free from the common errors and mistakes we learn about in high school English and freshman composition. Some writers can churn out pages of correct prose with hardly the blink of an eye. Others of us have to work at it a little bit. This chapter is for writers who have to work at correctness.

Perhaps an ultimate goal in gaining writing skills is to develop a sufficient number of good habits so that you can write correctly in your initial drafts or correct them through revision so that your final drafts are essentially error-free. The purpose of this chapter is to help you gain that more modest goal. I have not attempted a thorough and systematic review of English usage and mechanics; instead I have focused on the common problems that seem to crop up over and over again in student writing. I have also taken a problem-oriented approach; that is, I have tried as much as possible to move from problem to solution rather than just inventorying correct forms and letting you match the solutions to your problems on your own. You may well profit from a quick scanning of this chapter—if for no other reason than to see what's in it—but its main value will probably be as a reference, a source of help when you begin to encounter problems with correctness in your writing.

□ Editing for Correctness

You have to find mistakes before you can correct them. In my experience, most students who have problems with correctness have more trouble spotting errors in their writing than correcting the errors once they are aware of them. Here are some things you can do:

- □ Acknowledge the fact that you have difficulty with correctness in your writing. Accept the fact that you're going to have to spend extra time and effort overcoming your difficulties. Don't just ignore the problem and hope it will go away.
- □ Systematically identify and inventory the particular problems and classes of problems that you, personally, experience. Make a list. Every time you get a paper back with correctness problems noted on it, add the new items to your list and add a check after items that are already on your list.
- □ With each paper that you write, use your personal list of problems as an editing guide. When you get to the stage of revision at which it is appropriate to edit for correctness (see Chapter 5), go down your list item by item and make a separate pass through your entire paper for each item.
- □ Work through your paper from back to front. When you're editing at the word level, read each sentence word-for-word backward. That way you'll be able to focus on each word in relative isolation rather than losing it in the syntax of your sentence. Similarly, when you're editing at the sentence level, read the last sentence in a paragraph and then the next-to-last and then the third-to-last and so forth. One of the biggest problems with editing for correctness is seeing what your mind knows *should* be on the paper rather than what actually *is* on the paper. Working from back to front helps overcome this problem.
- □ Take advantage of informal reviews. Ask someone whose writing skills you respect to look over your paper for mistakes. (In the classroom, trade off with someone else who has problems with correctness. Often it's easier to spot errors in someone else's writing than in your own.) Get your help wherever you can find it—as long as you stop short of having someone else do your work for you.
- □ Whenever you make changes in a paper, recheck for correctness. Make sure you haven't introduced new errors in your added or altered material, and make sure what you have added or changed fits in with what was already there. Read each change

in the context of several sentences before it and several sentences after it.

- Don't overlook the value of final checking and final polishing (see Chapter 5). After you have done all the editing you feel necessary, put your paper aside for a day or two and then come back to it fresh. You'll be surprised what you'll find.

Common Problems with Grammar

We have to start somewhere, so let's start with grammar. In point of fact, college students have relatively few problems with grammar, but the same old problems keep coming back over and over again. (If you have difficulty with any of the terms or concepts in this section, see Chapter 18, "Grammar for Engineers.")

Subject-Verb Agreement

We all know that subjects and verbs are supposed to agree in number—singular subjects go with singular verbs, and plural subjects go with plural verbs—and most of the time we don't have any trouble. But there are some problem cases.

One problem case is the clause in which you have a good many words between your subject and its verb:

Wrong Jack and Jill, Jill being the smarter of the twins, is coming to the party later on.

Wrong The general, who not only has a command of his own but also oversees the activities of the commanders of six other battalions, are going to be in for a tough fight.

What happens, in effect, is that by the time you get to your verb you have forgotten what your subject was. To spot this kind of subject-verb agreement problem, read directly from subject to verb:

Right Jack and Jill . . . *are* coming to the party later on.

Right The general . . . *is* going to be in for a tough fight.

Better yet, recast the clause to reduce the distance between subject and verb, generally a more effective writing style anyway.

A similar problem occurs when a singular noun is followed by a phrase that begins with such words as *together with, in addition to, besides,* or *along with:*

Wrong Jack, together with Jill, are coming to the party later on.

Wrong The general, in addition to the six other commanders, are going to be in for a tough fight.

Such phrases do not influence the number of the subject; the verb must still be singular:

Right Jack, together with Jill, *is* coming to the party later on.
Right The general, in addition to the six other commanders, *is* going to be in for a tough fight.

Finally, watch out for the following problem pronouns, each of which takes a singular verb when used as the subject of a clause: *each, each one, every, everyone, everybody, anybody, nobody, no one, none, either,* and *neither.*

Right Either Jack or Jill *wants* to go to the party.
Right None of us *is* going to the party.
Right Each of the secretaries *is* going to get a raise.
Right Neither of the secretaries *has* earned a raise.

Pronoun-Antecedent Agreement

Like subjects and verbs, pronouns and their antecedents are supposed to agree in number. Most problems occur with possessive pronouns that refer back to one of the problem pronouns (or phrases containing them) as antecedents:

Wrong Each of the soldiers must clean their own rifle.
Wrong Every student must check with their advisers.
Wrong Everybody bring their books tomorrow.
Wrong None of the birds fell out of their nest.

As you recall, each of these problem pronouns is singular and each must take the singular form of a possessive pronoun referring to it.

Right Each of the soldiers must clean his own rifle.
Right Every student must check with his adviser.
Right Every student must check with his or her adviser.
Right All students must check with their advisers.
Right Everybody bring his or her book tomorrow.
Right None of the birds fell out of its nest.

This problem has been compounded in recent years by the aversion of many writers toward allowing the masculine pronouns *he, him,* and *his* to stand generically in reference to people of both sexes. But if you prefer to avoid gender-specific language in your writing, don't fall into the sloppy habit of using an incorrect plural pronoun just to avoid a *his* or a *him.*

Grammatical Parallelism

The general rule of parallelism is that semantically parallel ideas ought to be expressed in grammatically parallel forms. An important corollary is that ideas *not* semantically parallel shouldn't be expressed in grammatically parallel forms. The four major problem areas are lists, successive headings and subheads, conjoined nonfinite clauses, and series of three or more items.

Actually, we can treat items in lists and successive headings and subheads together. Since such entries in a report are by definition parallel semantically, your job is to check to see that they're all in the same form grammatically:

Wrong
1. Sorting the cards.
2. Check to see that the machine is on.
3. Loading the deck.
4. Card jams from mutilated cards.
5. Put each deck together with its printout.

Right
1. Sorting the cards.
2. Checking to see that the machine is on.
3. Loading the deck.
4. Clearing jams from mutilated cards.
5. Matching decks with printouts.

Right
1. Sort the cards.
2. Check to see that the machine is on.
3. Load the deck.
4. Clear jams from mutilated cards.
5. Put each deck together with its printout.

As you can see, it doesn't make too much difference which grammatical pattern you follow, as long as you follow some pattern. If you find a mixed bag of full sentences, phrases, imperative statements, participles, or what have you, pick a pattern (usually the predominant pattern) and change nonparallel items into the standard you have selected. The same goes, of course, for headings and subheads; they should be in grammatically parallel form as well—usually noun phrases or participial clauses.

Similarly, you should always make sure that conjoined nonfinite clauses are both (or all) in the same form—either the infinitive form or the participial form:

Wrong John likes dancing and to go to parties.

Right John likes dancing and going to parties.
Right John likes to dance and to go to parties.

Wrong Harold is responsible for setting the traps, collecting the specimens, and to go to town every Friday for provisions.

Right Harold is responsible for setting the traps, collecting the specimens, and going to town every Friday for provisions.

Right Harold is expected to set the traps, to collect the specimens, and to go to town every Friday for provisions.

A special problem arises with conjoined infinitive clauses. It is possible to let the domain of the first infinitive marker, *to*, extend over all the clauses:

Right Harold is expected to set the traps, collect the specimens, and go to town every Friday for provisions.

Careless writers, however, will occasionally try to put infinitive markers on the first and third (or some other combination) of the conjoined infinitives. This is an error in parallelism:

Wrong Harold is expected to set the traps, collect the specimens, and to go to town every Friday for provisions.

To be correct, you should put the infinitive marker on *all* the items in the list or *only* the first one. A similar situation exists with the definite article and a series of noun phrases.

Wrong The stove consists of the base, fuel tank, burner, and the wind screen.

Right The stove consists of the base, the fuel tank, the burner, and the wind screen.

Right The stove consists of the base, fuel tank, burner, and wind screen.

Again, use the definite article with all of the items or only the first.

The final common problem with parallel construction is different from all the others. This is the situation in which you have three ideas that are *not* semantically parallel but you try to cram them into a grammatically parallel form anyway. This occurs most often when you have two items, say B and C, that are parallel to one another and a third item, say A, that is parallel to the *pair* of items B and C, as in Figure 7–1. The proper way to express such a set of items is: A and B and C. But since this construction has the superficial form of a series of three items each parallel with each of the others, as in Figure 7–2, careless writers are tempted to express the construction in series form: A, B, and C. This is wrong. Let's look at some examples:

Wrong John cleared the table, washed the dishes, and then he straightened up the living room.

Wrong Stick-up lamps are inexpensive to purchase, easy to install, and they can be used almost anywhere.

Wrong Aspirin substitutes are effective, safe, and they don't upset your stomach.

Right John cleared the table and washed the dishes, and then he straightened up the living room.

Right Stick-up lamps are inexpensive to purchase and easy to install, and they can be used almost anywhere.

Right Aspirin substitutes are effective and safe and they don't upset your stomach.

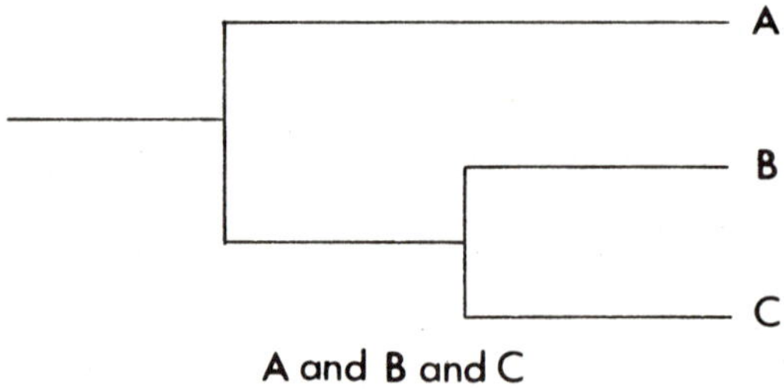

Figure 7–1 One Item Parallel to Two Others

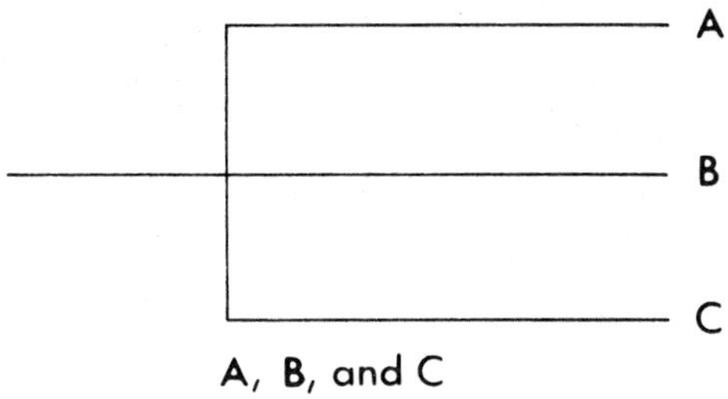

Figure 7–2 Three Items Parallel to One Another

Dangling Modifiers

Writers of mature prose use nonfinite clauses extensively, and they use a fair number of them in introductory subordinate positions:

Right After considering all the alternatives, I recommend a pipe culvert for the stream crossing.
Right Following his tracks for two days, the searchers finally caught up with the lost hunter.
Right By going to double shifts, the crew completed the project on schedule.

Since these introductory nonfinite clauses have no overt grammatical subjects of their own, the subjects of the independent clauses serve conceptually as the subjects of the nonfinite clauses: "I" is the subject of "considering," "the searchers" is the subject of "following," and "the crew" is the subject of "going." Everything is fine in these examples, because the semantics of the situations allow the grammatical subjects of the independent clauses to serve as the conceptual, or logical, subjects of the nonfinite clauses. But careless writers often use introductory nonfinite clauses in constructions where the subject of the independent clause cannot possibly serve as the logical subject of the introductory clause:

Wrong Running along the beach, my ice cream cone fell in the sand.
Wrong Working swiftly, the experiments were completed by midnight.
Wrong By going to double shifts, the project was completed on schedule.

The "ice cream cone" cannot be "running"; nor can "the experiments" be "working swiftly"; nor can it be "the project" that was "going to double shifts." In constructions such as these where the grammatical subject of the independent clause cannot logically serve as the conceptual subject of the introductory nonfinite clause, the nonfinite clause is left dangling with no possible conceptual subject. The reader of such a sentence has to stop and figure out what the writer had in mind. Usually readers *can* figure out what the conceptual subject of the dangling modifier is supposed to be, but they shouldn't have to. Your semantics and your grammar should line up in one-to-one fashion so your readers don't have to stop and figure out what it is you were trying to say.

To spot dangling modifiers in your writing, scan for *-ing* verbs (since most of the problems occur with participial clauses) and check to see if the grammatical subject of the independent clause to which the *-ing* verb is attached can logically serve as subject of the *-ing* verb. If not, either change the syntactic structure of the independent clause:

Right Running along the beach, I dropped my ice cream cone in the sand.
Right Working swiftly, we completed the experiments by midnight.
Right By going to double shifts, the crew completed the project on schedule.

or add an overt subject to the *-ing* verb clause and make a full finite clause of it:

Right As I was running along the beach, my ice cream cone fell in the sand.
Right We worked swiftly to complete the experiments by midnight.
Right Since the crew went to double shifts, the project was completed on schedule.

One final observation about dangling modifiers. Most of the problems with dangling modifiers arise when the independent clause is in the passive voice. As you strive to eliminate unnecessary and unwarranted passive clauses from your writing (see Chapter 20, "Style at the Clause and Sentence Level"), you will also eliminate a good many dangling modifiers—or at any rate situations where dangling modifiers commonly occur.

Sentence Fragments

Sentence fragments arise most frequently when writers punctuate phrases or nonfinite clauses or subordinate finite clauses as full, stand-alone sentences:

Wrong Harold went to the hardware store. And to the drug store.
Wrong Having been manufactured overseas, brought into this country as modular components, and assembled at a small plant located on the East Coast before shipment to the marketing areas in the Midwest. The units were inexpensive.
Wrong The control rod jammed and allowed the cable to snarl. Although the entire setup had been checked thoroughly.

The cure is simple. Since almost always such sentence fragments are related semantically to the sentence preceding them or the following sentence, the error can be corrected merely by changing the punctuation in such a way as to attach the sentence fragment to an adjacent sentence:

Right Harold went to the hardware store and to the drug store.
Right Having been manufactured overseas, brought into this country as modular components, and assembled at a small plant located on

the East Coast before shipment to the marketing areas in the Midwest, the units were inexpensive.

Right The control rod jammed and allowed the cable to snarl, although the entire setup had been checked thoroughly.

Sentence fragments are sometimes hard to spot when you read over your paragraphs in normal order, but they are easy to spot when you read the sentences of a paragraph in reverse order because you can focus on each sentence (or what you have punctuated as a sentence) in isolation.

□ Problems with Spelling

Although poor spelling is perhaps the least significant of the correctness problems, it is the most visible; hence it tends to get a great deal of emphasis. Spelling habits are established early in life, and if you're still a poor speller by the time you're in college it's probably too late to change. What I will stress in this section, then, is not how to avoid spelling mistakes in the first place but rather how to spot and correct spelling mistakes through revision and final polishing. I will also discuss the closely related problem of typographical errors.

Editing for Spelling and Typos

The key to finding spelling and typographical errors in your writing is looking at each word in isolation. And one of the best ways to do this is to read your manuscript word-for-word backward. Start at the end of your paper (or at the end of a section) and look at each word, pronouncing it to yourself—and for longer words or problem words spelling it to yourself as well. If you find a word that looks funny (or sounds funny) or if you find a word you're not sure of, consult a dictionary for proper spelling. (Or better yet, buy yourself a secretary's spelling guide to 20,000 common words—available at most office-supply stores. These little books aren't cluttered with definitions, so they're easier and quicker to use.) Such a procedure is time-consuming, but if you know you have a spelling problem and want to do something about it, you're going to have to invest the extra time.

Keeping a Spelling List

If you're like most poor spellers, you keep misspelling the same words over and over again. You can combat this tendency by making a list of all the words you misspell and keeping the list taped to the wall in front of your typewriter. Then, as you write, you can use the list as a reference. Each time you get a paper back with spelling errors noted on it, add the misspelled word (spelled correctly, of course) to your list. Periodically, as your list becomes cumbersome, rearrange it into alphabetical order, and as you learn the correct spellings of various words, drop them from your list. Once again, this is a time-consuming procedure but, like reading backward, it works. Try it.

□ Common Problems with Punctuation

Punctuation marks, for the most part, are signals of syntactic structure. Problems with punctuation come about from (1) failing to signal certain structures and (2) signaling the wrong structures. Of these two types of problem, the second is far more serious than the first. Let's look at some common syntactic structures and the punctuation problems that often arise with them.

End Punctuation with Sentences

The most common end punctuation problem is the syntactic question that is followed by a period rather than a question mark. This problem occurs most often when the question in question is a thinly veiled imperative, anyway:

> **Wrong** Can you have the report on my desk by noon tomorrow.
> **Wrong** Why don't you stop by the store on your way home.
>
> **Right** Can you have the report on my desk by noon tomorrow?
> **Right** Why don't you stop by the store on your way home?

Even though such sentences have the force of an imperative or command, they require a question mark at the end because of their syntactic structure.

Another end punctuation problem is the sentence that ends with an abbreviation. Many writers are tempted to use two periods—one for the abbreviation and one for the end of the sentence. One is sufficient.

> **Wrong** The bill was from ARCO Engineering, Inc..
> **Right** The bill was from ARCO Engineering, Inc.

Clause-Level Punctuation

Most punctuation problems occur with clause-level construction. Let's look at some of the common troublemakers.

Independent Clauses. Two independent clauses in a sentence are properly joined either with a comma followed by a coordinating conjunction, or with a semicolon. Don't try to stick two independent clauses together with just a comma:

> **Wrong** Ellen was careful with the experiments, the results were satisfactory.

Right Ellen was careful with the experiments, and the results were satisfactory.
Right Ellen was careful with the experiments; the results were satisfactory.

A sentence adverbial does not carry the weight of a coordinating conjunction; it cannot appear with just a comma before it. Sentence adverbials should appear with a semicolon before and a comma after (or occasionally a comma before and a semicolon after, if the sentence adverbial is more closely related to the first clause than to the second):

Wrong Ellen was careful with the experiments, therefore the results were satisfactory.
Wrong Ellen was careful with the experiments, therefore, the results were satisfactory.
Right Ellen was careful with the experiments; therefore, the results were satisfactory.
Right You can join independent clauses with sentence adverbials; however, you should use a semicolon before the sentence adverbial and a comma after it.

The sequence of comma followed by conjunction (", and" or ", but" or ", or") signals the coordination of two full independent clauses each of which has a surface structure subject. When the second of two clauses has an implied subject identical to the surface structure subject of the first clause, the comma is inappropriate:

Wrong John ran up the stairs, and opened the door.
Wrong The volume of goods entering or leaving a country is an indicator of productivity, and is used by many economists as a measure of strength.

If you find yourself writing such sentences, correct them either by deleting the comma:

Right John ran up the stairs and opened the door.
Right The volume of goods entering or leaving a country is an indicator of productivity and is used by many economists as a measure of strength.

or by adding a surface structure subject (usually a pronoun) to the second clause:

Right John ran up the stairs, and he opened the door.
Right The volume of goods entering or leaving a country is an indicator of productivity, and it is used by many economists as a measure of strength.

Remember: The comma and the pronoun have to appear together. If you have a comma, you need a pronoun; if you have no pronoun, the comma before the conjunction is a false signal.

Occasionally one punctuation signal can be replaced by another if such replacement is needed for clarity. For instance, if you are joining independent clauses that have internal commas of their own, you may wish to promote the comma of your comma-conjunction sequence to a semicolon to signal to your readers where the main clause juncture is:

Confusing Johnson, who owns the computer facility, is vice-president, but Smith, who knows how to run the computer, is president.

Clearer Johnson, who owns the computer facility, is vice-president; but Smith, who knows how to run the computer, is president.

Similarly, if your independent clauses are short and there is no possibility of confusion, you may downgrade a semicolon joining them to just a comma:

Right Harry sets them up; Jack knocks them over.
Acceptable Harry sets them up, Jack knocks them over.

Or if you have two short clauses joined by a comma-conjunction sequence, you may delete the comma (in effect, downgrade the comma to nothing):

Right Harry sets them up, and Jack knocks them over.
Acceptable Harry sets them up and Jack knocks them over.

Introductory Subordinate Clauses. By convention, subordinate clauses that precede an independent clause are separated from the independent clause by a comma. This is true regardless of whether the subordinate clause is finite or nonfinite:

Right Because she was happy, Helen smiled.
Right Being happy, Helen smiled.
Right Since he had gotten an A on his thermodynamics exam, Jack treated himself to a night off.
Right Having gotten an A on his thermodynamics exam, Jack treated himself to a night off.

Similarly, introductory words and phrases are, by convention, separated from main clauses by commas:

Right In addition, he can play the drums.
Right Next, let us consider the entropy of the system.

Trailing Nonfinite Clauses. A finite subordinate clause that follows an independent clause need not be set off with a comma, but a nonfinite clause that follows an independent clause should be:

Right Jack treated himself to a night off since he had gotten an A on his thermodynamics exam.
Right Jack treated himself to a night off, having gotten an A on his thermodynamics exam.

Wrong Jack treated himself to a night off having gotten an A on his thermodynamics exam.

Interruptive Elements. Interruptive elements are set off from the clauses in which they appear by a *pair* of punctuation marks: a pair of parentheses, a pair of dashes, or a pair of commas:

Right Jack (the man in the front row) gave me the free tickets.
Right Jack—the man in the front row—gave me the free tickets.
Right Jack, the man in the front row, gave me the free tickets.
Right The fat man (who owns the sports car) played the tuba in the band.
Right The fat man—who owns the sports car—played the tuba in the band.
Right The fat man, who owns the sports car, played the tuba in the band.

Nobody has trouble with parentheses, and few people have trouble with dashes; but there are all kinds of careless writers who want to put in one comma or the other but not both:

Wrong The fat man, who owns the sports car played the tuba in the band.
Wrong The fat man who owns the sports car, played the tuba in the band.

Conceptually, the interruptive element needs two punctuation marks—one in front of it and one behind it—and you should never try to get away with just one. A point of confusion arises, however, when the interruptive element appears at the end of a sentence. In such cases, the parentheses behave properly:

Right I got the free tickets from Jack (the man in the front row).

but with dashes and commas, the second of the pair of punctuation marks is subsumed by the sentence-final period:

Right I got the free tickets from Jack—the man in the front row.
Right I got the free tickets from Jack, the man in the front row.
Right The tuba was played by the fat man, who owns the sports car.

If you have trouble punctuating interruptive elements (or parenthetical elements, as they are often called), use parentheses for all your interruptive elements in your initial draft and then substitute commas or dashes, following the rules outlined above.

Minor Punctuation Problems

Most punctuation problems—certainly the most serious ones—occur with clause-level structures, but there are a few other problems that crop up with an annoying frequency. Let's look at them.

Possessive Apostrophes. The possessive of most English nouns is formed by adding an apostrophe and an *s* to the noun:

Right	John	John's hat
Right	cat	the cat's tail
Right	experiment	the experiment's results
Right	bridge	the bridge's rail

Often, careless writers get the *s* right, but they forget the apostrophe:

Wrong Johns hat
Wrong the cats tail
Wrong the experiments results
Wrong the bridges rail

Further confusion arises with plural possessives. Here, the convention is to add an *s* (to form the plural) and then an apostrophe:

Right	cats	the cats' tails
Right	experiments	the experiments' results
Right	bridges	the bridges' rails

Exceptions to this convention are the irregular plurals:

Right	children	the children's books
Right	women	the women's center

But most of the confusion arises with the third person pronoun *it*. *It* forms its possessive with just an *s*—no apostrophe at all:

Right The horse broke its leg.

Wrong The horse broke it's leg.
Wrong The horse broke its' leg.

The form *it's* occurs *only* as a contraction of *it is*—never as a possessive. The form *its'* never occurs (except incorrectly). (Or for nitpickers, as a citation form, as in the preceding sentence.) If you have trouble with *its* and *it's*, just remember that the apostrophe is appropriate if and only if you can substitute the words *it is*. If you can't substitute *it is*, use *its*. And never use *its'*.

Introducing Lists. Introduce a list with a colon—not a semicolon:

Wrong The main parts of a computer are; the central processor, the memory, and the input/output devices.
Right The main parts of a computer are: the central processor, the memory, and the input/output devices.

On your typewriter, the colon is a capital semicolon; there is a significant difference between the two. While we're on the topic of colons, always use a colon (never a semicolon) after the salutation of a business letter:

Wrong Dear Mr. Johnson;
Right Dear Mr. Johnson:

Using Quotation Marks with Other Punctuation. By convention, commas and periods appear inside quotation marks, colons and semicolons appear outside quotation marks, and question marks and exclamation points appear inside if they are part of the quoted material and outside if they are part of the matrix sentence. Most problems occur with commas and periods:

Wrong It was the Boise River—"River of the Woods".
Right It was the Boise River—"River of the Woods."

Wrong Frank Sinatra, "Old Blue Eyes", has been singing for forty years.
Right Frank Sinatra, "Old Blue Eyes," has been singing for forty years.

Wrong I read your article, "Carbon Monoxide Poisoning", in the latest issue of *Environmental Concerns.*
Right I read your article, "Carbon Monoxide Poisoning," in the latest issue of *Environmental Concerns.*

Wrong Smith, H. R. "Rotating Crops in Arid Lands", *Applied Agronomy,* 22:4 (1982), 15–22.
Right Smith, H. R. "Rotating Crops in Arid Lands," *Applied Agronomy,* 22:4 (1982), 15–22.

□ Conventions with Mechanics

This final section covers some of the conventions governing the more mechanical aspects of writing technical reports and other documents. The key word here is *conventions*. There are no hard-and-fast rules for most of the topics here; each organization, professional society, and publishing house has its own set of conventions. These are some guidelines as to what most of the people do most of the time, but keep in mind that you will find exceptions in quite respectable publications. When you get on the job, you'll have to find out what the local ground rules are and follow them. In the meantime, if you follow the guidelines suggested here, you'll be right more often than you're wrong.

Abbreviations and Symbols

Abbreviations and symbols are used to save space, but rarely is the space saved worth the possible confusion introduced by their use. Hence, the general rule is to avoid abbreviations and symbols in running text—particularly in manuscripts that are typewritten rather than set in type and printed commercially. As you prepare your typewritten manuscript, spell everything out; you can always come back through in a later revision and introduce abbreviations and symbols to conform to local conventions.

As with most general rules there are a few exceptions:

- □ Use the dollar sign ($) and pound sign (£) in running text. Write "$14" rather than "14 dollars" and "£22" rather than "22 pounds Sterling."

- □ Use the percent symbol (%) if you are reporting a good many percentages— say, five or more per paragraph. Otherwise, write out "percent."
- □ Use the degree symbol (°) when reporting temperatures (35° F, 20° C).
- □ Use symbols for degrees (°), minutes ('), and seconds (") for bearings in legal land descriptions and other surveying documents and for latitudes and longitudes—especially if you are reporting quite a few of them.
- □ Use the ampersand (&) when it is part of a corporate name ("John Wiley & Sons, Inc."); otherwise write "and."
- □ Use the "less than" symbol (<) when reporting probabilities in a social science document ($p < 0.05$).
- □ Use other special symbols (mathematical, chemical, astronomical, meterological) judiciously in documents written by technical experts *for* technical experts. If laypeople or experts outside your field might be reading your report, spell everything out.
- □ Use abbreviations for common organizations (CIA, FBI, GOP, IRS, NATO), especially if the abbreviation is more familiar to most people than what it stands for. Use all capital letters and no periods.
- □ Use the following common abbreviations freely: AC and DC, A.D. and B.C., AM and FM, A.M. and P.M., C.O.D., DDT, ICBM, RSVP, TNT, and UHF and VHF.
- □ Use abbreviations for metric units of measure (g, kg, mg, ml, cc). Use lowercase letters with no periods. Spell out all English units of measure, however (foot, pound, inch, yard).
- □ If you abbreviate the names of countries, include periods: U.S., U.S.S.R., U.K., U.A.R.
- □ Use abbreviations for common derived units (cfm, rpm, gpm, hp, Btu, psi). Use lowercase letters (unless one of the words is a proper noun) and no periods.
- □ Use abbreviations for Latin expressions used commonly in English: etc., e.g., i.e., vs., and viz.
- □ Use other abbreviations only if what you are abbreviating is long and used often. If you use a term only once or twice, spell it out each time. If you use a long term fairly frequently, spell it out the first time and include the abbreviation in parentheses—"National Environmental Policy Act (NEPA)"—and then use the abbreviation thereafter.

Numbers and Numerals

The convention for numbers is simple: spell out numbers one through nine and use numerals for numbers 10 and above. Again, however, there are some exceptions:

- Many publishers of books and journals will use numerals for all numbers; others will spell out numbers from one through 99 in straight narrative.
- Always spell out a number at the beginning of a sentence—regardless of size. Write "Thirty-seven people came to the party" rather than "37 people came to the party."
- Use numerals for street addresses, dates, and sums of money:

  3568 Park Avenue South
  March 9, 1983
  $4.53, $5, $5.00
- Always write out ordinal numbers (first, third, fifth).
- Use numerals for mixed numbers below ten (3½, 5.375, 7¾) but spell out fractions that stand alone (three-fourths of the money, half of the membership). Use numerals for decimal fractions (0.025, 0.56, 0.5).
- With compound numbers used as modifiers, spell out the first (or the shorter) of the two numbers:

  six 4-inch boards
  three 25-pound blocks
  425 six-inch bolts
- Use a hyphen between the number and the unit for capacities and dimensions:

  55-gallon barrel
  250-ml flask
  four-foot doorway
  six-inch ruler

Italics and Underlining

In print, titles of books, journals, magazines, newspapers, and other major publications; stressed words; and foreign words and phrases that have not been assimilated into English are conventionally set in italic type. In typescript, underlining is the equivalent of italics. The trend, however, is away from italics—especially when computerized typesetters and electronic word-processing systems are being used. Here are some guidelines to help you avoid a few of the common problems:

- Always underline the Latin names of genera and species of plants and animals:

  Acer saccharum (sugar maple)
  Pteridium aquilinum (bracken fern)
  Bonasa umbellus (ruffed grouse)
- Always underline the titles of books, journals and magazines (but not the articles in them), and newspapers.
- Underline foreign words and phrases unless they are so common as to have become part of the English language (resume, laissez

faire, et cetera, versus). If you are in doubt, consult a recent dictionary.

- □ Do not underline the following abbreviations: e.g., i.e., vs., etc.; in fact, avoid using these abbreviations whenever possible.

Capitalization

Most novice writers use too many capital letters rather than too few. In general, avoid capitalizing a word unless you are absolutely certain it should be capitalized. Frequent capitalization in running text breaks it up and makes it hard to read. Here are some guidelines:

- □ Do not capitalize the word *federal* unless it is part of a proper name: "Federal Bureau of Investigation," but "federal, state, and local laws."
- □ Do not capitalize fields of study or general programs or degrees unless they are parts of proper names: "English 317—Technical Writing," but "a course in technical writing"; "College of Forestry," but "a major in forestry"; "Master of Science in Chemical Engineering," but "a master's degree in chemical engineering."
- □ Do not capitalize collective nouns unless they are parts of proper names: "the University of Michigan," but "the university programs"; "the Board of Trustees," but "board policy"; "the Troy City Council," but "the city council meeting."
- □ Do not capitalize names of offices (except "President" when referring to a specific president of the United States): "the principal will report to the superintendent monthly on all high school activities," not "the Principal will report to the Superintendent monthly on all High School activities."
- □ Do not capitalize the names of components of a mechanism: "bolt one side of the hinge to the door assembly and the other side of the hinge to the frame," not "bolt one side of the Hinge to the Door Assembly and the other side of the Hinge to the Frame."
- □ Capitalize days of the week and months of the year, but do not capitalize seasons of the year: "Tuesday classes," but "fall semester."
- □ Capitalize the names of languages: "French class," "German workshop," and (especially) "English literature." (Don't let your attitude toward English show through with a lowercase letter.)

8

Graphics and Illustration

One of the key things that characterizes what we call technical writing and separates it from creative writing and expressive writing is the extent to which its writers use tables, graphs, and pictures to complement their verbal communication. If you have any doubt about this, consult half a dozen journals in your field and compare the frequency of illustration to what you find in such journals as *Shakespeare Quarterly*, *American Literature*, or *The Philosophical Review*. Or look at a corporate annual report. Or check an instruction manual prepared by a major automotive or heavy-equipment manufacturer. In fact, many installation and maintenance manuals or service manuals will devote 70 to 90 percent of their page area to illustration, as shown in Figure 8–1 (page 116). As a writer of technical reports, you must be aware of the importance of visual communication and you must know the conventions and techniques for incorporating graphic elements into your writing.

Illustrations serve to communicate information that is difficult to express verbally. Shapes and sizes of objects, relationships of components, magnitudes of various measurements, colors and textures, and flows and processes are among the things better communicated visually than verbally. Illustrations—especially tables and graphs—serve as convenient ways to summarize large bodies of precise numerical data.

FUEL SYSTEM
TESTING AND ADJUSTING

ENGINE

NOTE: If nozzle can not be adjusted to specifications, make a replacement of the nozzle.

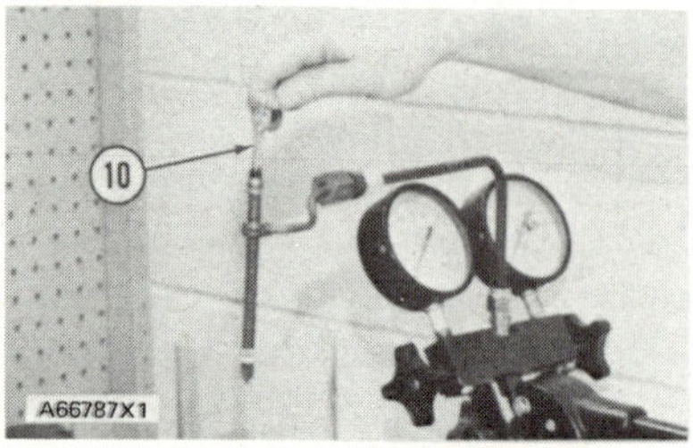

OPENING PRESSURE ADJUSTMENT
10. 5P4813 Socket.

VOP SPECIFICATIONS	16 545 to 17 690 kPa (2400 to 2600 psi)

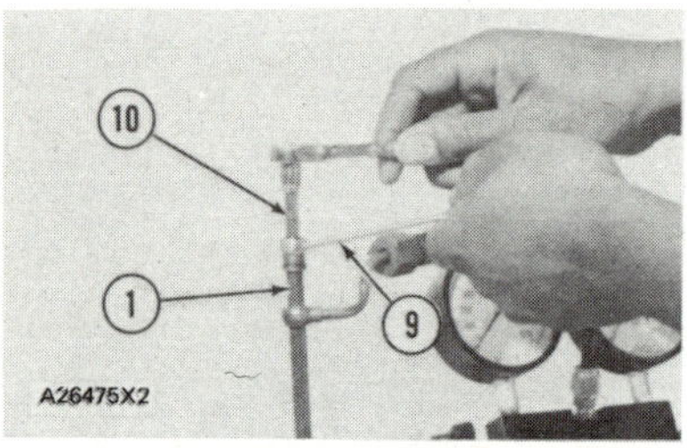

TIGHTENING LOCKNUT
1. Fuel injection nozzle. 9. 8H8505 Combination Wrench. 10. 5P4813 Socket.

7. Hold pressure adjustment screw (3) and tighten locknut (2) just enough so that pressure adjustment screw (3) will not turn.
8. After the opening pressure adjustment is made, install locknut (5) that holds lift adjustment screw (6). Make the valve lift adjustment. See VALVE LIFT ADJUSTMENT.

Valve Lift Adjustment

1. With the valve opening pressure correct, pump test oil through the fuel injection nozzle. At the same time, hold locknut (5) and slowly turn lift adjustment screw (6) clockwise until the pressure starts to increase above the opening pressure.
2. To be sure the valve is on the seat, increase the pressure 1380 to 3450 kPa (200 to 500 psi) more than the opening pressure.

CAUTION

Do not bend the valve or damage the seat by turning lift adjustment screw (6) with too much force.

NOTE: Some test oil can be at the tip of the fuel injection nozzle, but a constant flow of drops must not be seen.

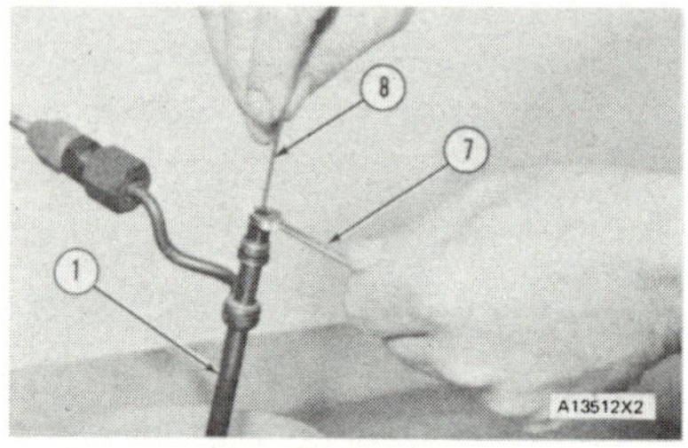

TIGHTENING LOCKNUT
1. Fuel injection nozzle. 7. 8H8502 Combination Wrench. 8. 5/64" Hex Wrench.

3. Turn lift adjustment screw (6) counterclockwise 3/4 ± 1/8 of a turn.
4. Hold lift adjustment screw (6) with 5/64" hex wrench (8) and tighten locknut (5) just enough so that lift adjustment screw (6) will not turn.

Tightening Locknuts and Cap

1. Remove fuel injection nozzle (1) from the 5P4150 Nozzle Tester and put it in 8S2250 Nozzle Holding Tool (P).

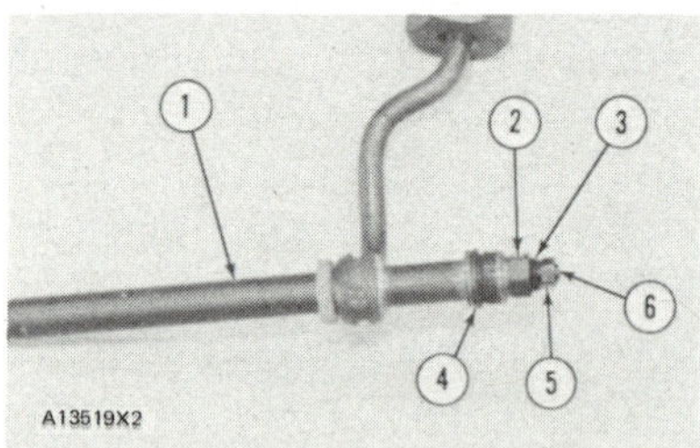

FUEL INJECTION NOZZLE
1. Fuel injection nozzle. 2. Locknut (for pressure adjustment screw). 3. Pressure adjustment screw. 4. 1H1023 O-ring Seal for cap. 5. Locknut (for lift adjustment screw). 6. Lift adjustment screw.

1-33

Figure 8–1 Profusely Illustrated Instruction Manual (Caterpillar Tractor Co.)

Finally, illustrations serve to break up large sections of print, making the entire package of information easier to comprehend. Most mediocre technical writing suffers from too little illustration; rarely do we see too much. Novice writers (and especially student writers) are not used to thinking visually as they write, nor are they willing to put forth the extra effort it takes to generate effective visuals. Let me suggest the following procedures for incorporating graphics and illustrations into your writing:

- As you write your initial draft, put in references to tables and figures wherever you think an illustration might be useful. Don't stop to prepare the illustration at that time, though. Come back later.
- As one of your passes in revision, look for additional places to incorporate visuals. Pay particular attention to sections in your draft where you had difficulty describing or explaining what you wanted to. Verbal difficulty often indicates the need for visual communication.
- Ask reviewers to comment on the usefulness of the illustrations you do have and to suggest other places where illustrations would be helpful.
- In general, err in the direction of too many visuals rather than too few. It's always easier to take them out at the last minute than to try to plug some more in at the last minute.

□ Preliminary Terms

Before we proceed with a discussion of specific types of graphics and illustration, you should have a handle on certain terms common among editors and graphic illustrators. Such familiarity will not merely help you understand the discussion; more important, it will help you on the job when you have to converse with editors or illustrators about your visuals.

Table A table consists of two or more rows and two or more columns of precise data, usually numerical. (See "Tables and Graphs," below, for examples and further discussion.)

Figure By convention, anything that isn't a table is called a figure. We use the word *figure* to refer to charts, graphs, line drawings, photographs, and various items—even tables and samples of text—that are included for illustrative purposes.

Figure caption A figure caption (see Figure 8–2) consists of the figure number plus a short title or description. Put your figure captions *below* your figures, either centered or flush with the left edge of the figure. Never have just a figure number—always include a short title or description.

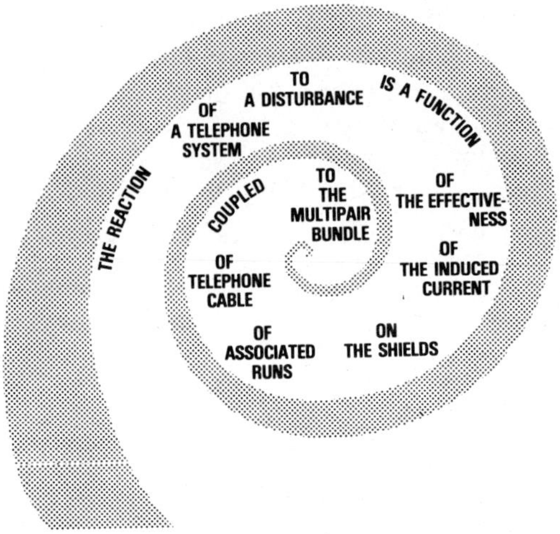

Figure 2 Maelstrom of prepositional phrases.

Figure 8–2 Figure Caption (*Journal of Technical Writing and Communication,* Vol. 9, No. 1, 1979)

You may, if you wish, abbreviate "Figure" as "Fig."; I advise against it, however. Technical writing tends to have too many abbreviations as it is, and abbreviating *Figure* rarely if ever saves any space.

Table caption Tables have captions (or titles), too (see Figure 8–3), but by convention they always appear *above* the table, again either centered or flush with the left edge of the table.

Figure (or table) reference The figure reference (see Figure 8–4) is the device you use in your running text to refer your reader to an illustration. (See "Handling Artwork," below.)

TABLE 15-2 —Evaporator Correction Factors

ΔT (F)	4	6	8	10	12	14	16	18	20	22	24	26	28	30	32	34	36
ONE PASS	-2.0	-3.0	-4.0	-5.0													
TWO PASS			0	0	-.5	-.5	-1.0	-1.0	-1.0	-1.0	-1.0						
THREE PASS					1.5	1.5	1.5	1.5	1.5	1.5	1.5	1.5	1.5	2.0	2.0	2.0	2.0

Figure 8–3 Table Caption (The Trane Company)

ELECTRICAL CONNECTIONS

All electrical wiring must comply with local and national codes. The installer must provide a fused disconnect switch which should, for safety reasons, be installed within sight of the unit.

Bring the line wiring into the unit through one of the conduit knockouts on the lower right side of the unit (see Figure 2). Splice the line wires to the unit power leads in the main junction box (see Figure 5).

Refer to Table 2 for wire and fuse sizes.

Figure 8–4 Figure and Table References

Callout A callout is a component of an illustration that identifies, or "calls out," a certain part of whatever is being illustrated. It consists of a label—either verbal (see Figure 8–5) or numerical (see Figure 8–6)—and an arrow running from the label to the part being identified.

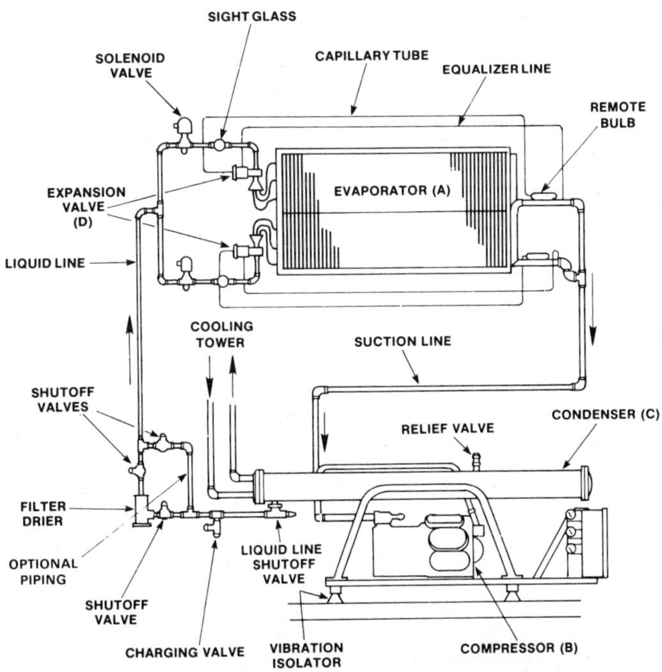

Figure 8–5 Verbal Callouts (The Trane Company)

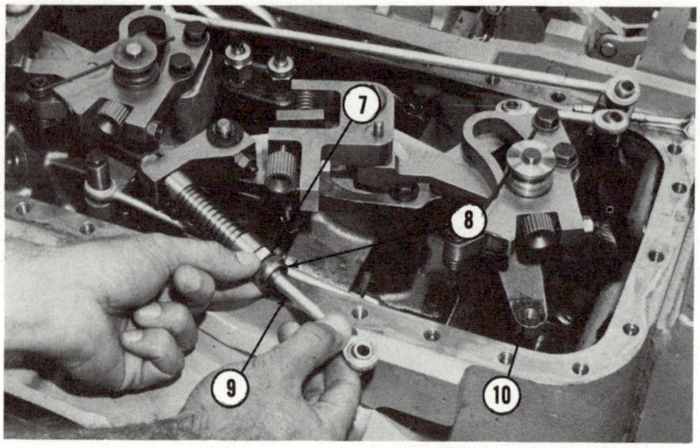

Figure 8–6 Numerical Callouts (Caterpillar Tractor Co.)

ELECTRICAL SYSTEM
OPERATION

INSTRUMENT PANEL

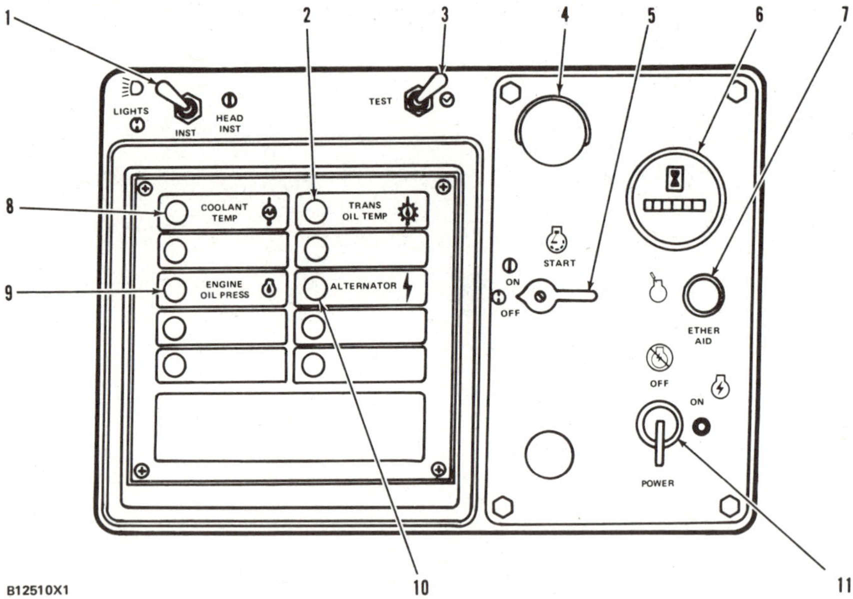

INSTRUMENT PANEL

1. Light switch.
2. Transmission oil temperature indicator.
3. Panel test switch.
4. Panel light.
5. Start switch.
6. Service meter.
7. Ether aid switch.
8. Coolant temperature indicator.
9. Engine oil pressure indicator.
10. Alternator indicator.
11. Power (disconnect) switch.

Figure 8–7 Legend (Caterpillar Tractor Co.)

Legend A legend is a list of definitions or explanations of symbols or abbreviations used in an illustration. Graphically, a legend may be either part of the illustration itself or part of the caption. A common type of legend (see Figure 8–7) is one that identifies numerical callouts.

□ Photographs

Photographs are used extensively in instruction manuals, brochures, corporate annual reports, technical bulletins, and other documents that are printed and distributed in large quantities. They are often overlooked, however, in reports, proposals, and other documents reproduced and distributed in limited numbers. Historically, this was because of the

FIG. NO.	PHOTO OF	SHOWING	COMMENTS
3-10	Clarifier	Bridge Controls	
3-11	RAS Pump Control Panel	All Controls	
5-X	Transfer Pump Station #1	Model View	
5-X	Transfer Pump Station #1	Motor Control Centers	
5-X	Transfer Pump Station #1	Pumps and Valves	Take as many shots
5-2	AMCC and BMCC	All Controls	as necessary
5-3	CMCC (TPS #2)	All Controls	
5-X	Transfer Pump Station #2	Model View	
5-X	Transfer Pump Station #2	Motor Control Centers	
5-X	Transfer Pump Station #2	Pumps and Valves	
5-X	TPS #1 Water Seal Units	Model View	
5-X	TPS #2 Water Seal Units	Model View	
5-X	TPS #2 Transformer		
5-X	Wet well drain sluice gate		
6-X	Lagoon and Lagoon Service Bldg.	Model View	Telephoto?
6-3	Main Diversion Box—MH	Handwheel Valve	
6-4	Splitter Box	Model	
6-5	Cell One Outlet Box	Three Sluice Gates	Show Control Wheels
6-6	Cell Two Outlet Box	Three Sluice Gates	Show Control Wheels
6-7	Cell Two Overflow Box	Sluice Gate	

Figure 8–8 Specimen List of Photos to Be Taken

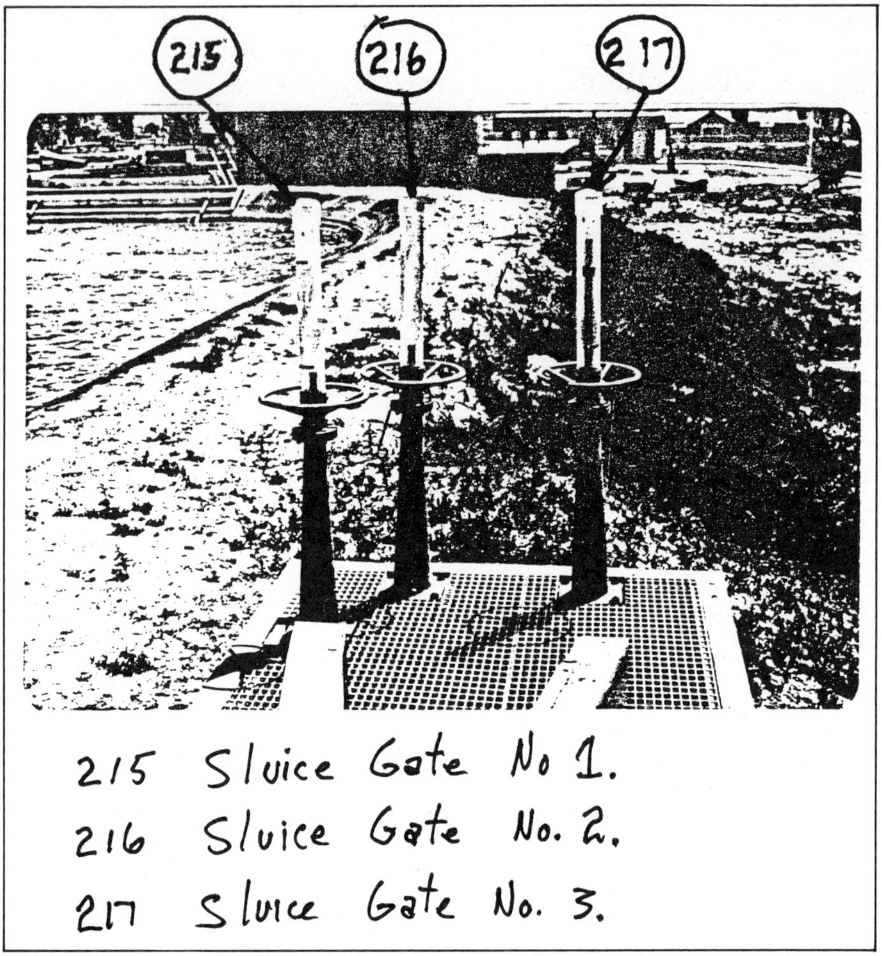

Figure 8–9 Marked-up Xerox Copy of a Photograph (Interwest Engineering, Inc.)

expense and difficulty in reproducing photographs, but for the last twenty years or so techniques of offset printing and electrostatic copying have been such that good clear reproductions of photographs can be produced cheaply and easily. It's high time photos were used more extensively in limited-distribution documents.

Most firms or agencies that make frequent use of photographs in their publications maintain files of photographs that are available for use. Check with the advertising or public relations departments of your organization or with the publications group or duplicating center. If nothing is available, you might suggest that a photo file be established. The file should contain the original negative or slide plus a 4-by-5-inch (or better yet, an 8-by-10) glossy black and white print. Each photo should be identified by a file number. In order to get a good reproduction, you need to start with a sharp, contrasty print that is neither too light nor too dark. Also, of course, each photo should exhibit good composition, with a single center of interest and an uncluttered background. Actually, the best way to get good photographs is with a good photographer; consult your phone directory for commercial or industrial photographers in your area.

Often, though, you will have to have photographs taken expressly for a particular report you are writing. Start by preparing a list (see Figure 8–8, page 121) of all the photos you will need; for each photo on the list include the complete figure caption and a brief description of what the photo should show. Gather all the items that have to be photographed and invite your photographer in (or if the items aren't portable, arrange to meet the photographer at the site). Have the photographer shoot several takes of each photo from different angles and with different lighting. Film is cheap when you consider the time and travel costs involved. When you get your prints back, select the ones you want to use and have extra prints made. Then number and file all negatives and extra prints for possible future use. Run Xerox copies of each of the prints you want to use and mark the Xerox copies for cropping (if you don't want to use the full area of the print) and callouts, if any (see Figure 8–9). Also, write the figure number of each photograph at the bottom of its corresponding Xerox copy. Then send the original prints plus Xerox copies to your art department for processing (or to an advertising agency or commercial printer, if your organization doesn't have an art department).

□ Success Strategy for Student Writers □

Obviously, it's time-consuming and expensive to have special photos taken for a classroom writing assignment. What you can do, though (with permission from your instructor), is set up your photo list as suggested above and leave space for each photo in the final draft of your paper. Type in each caption, and then in the space you leave include your brief description and a list of all items that you would call out in the photo. What you wind up with is a page that looks like the one shown in Figure 8–10.

□ Line Drawings

Line drawings have some advantages over photographs. Artists can eliminate needless detail and concentrate on what's important, and they can use various techniques to show internal parts in their working relationships to one another. Although a good line drawing costs about as much to prepare as a photograph does to take, line drawings can be reproduced slightly less expensively than photos. Cost, however, shouldn't be the major factor in deciding whether to use a photograph or a

This report recommends that Jim Jensen purchase the Miller Dialarc 250 AC/DC arc welder for his farm shop. Jime came to me to ask about a welder that would suit his purposes. He wanted something that would be economical and durable but would also give him a wide range of amperage settings and would weld most types of metals. In general, he uses a welder once or twice a week for an hour or so each time. Jim was also concerned about availability of parts and service.

Photo of Dialarc Model 250 AC/DC

Figure 1--Dialarc 250 AC/DC

The alternative models I looked at were the Dialarc Model 250 AC/DC, the Dialarc Model 250 AC, and the Model CP-250-75 DC. All of these are manufactured by Miller, as this is the line of arc welders handled by the local Cenex dealer in Trenary. Criteria in order of importance were 1) purchase price, 2) duty cycle, 3) warranty, 4) operational costs, and 5) range of amperage output. I did not consider availability of parts and service as a criterion because the service on all the models was essentially the same.

Photo of Model CP-250-75 DC

Figure 2--Model CP-250-75 DC

PURCHASE PRICE

The initial price of the Dialarc 250 AC/DC (see Figure 1) was $771.00--well within Jim's money range. The Model CP-250-75 DC (see Figure 2) was $1199.00--a bit high. The third alternative--the Dialarc 250 AC (see Figure 3, page 2) had a purchase price of $610.00. In terms of the length of time Jim would have the welder, I didn't see a significant difference in price between the Dialarc 250 AC/DC and the Dialarc 250 AC.

Figure 8–10 Page from Classroom Writing Assignment with Spaces Left for Photographs

drawing: if a photo shows it better, use a photo; if a drawing makes for more effective communication, use a drawing. Here are some different varieties of line drawings you will probably have occasion to use in reports and other documents:

Two-dimensional drawing (see Figure 8–11) These are based on the common plane view of drafting or mechanical drawing. In fact, often you

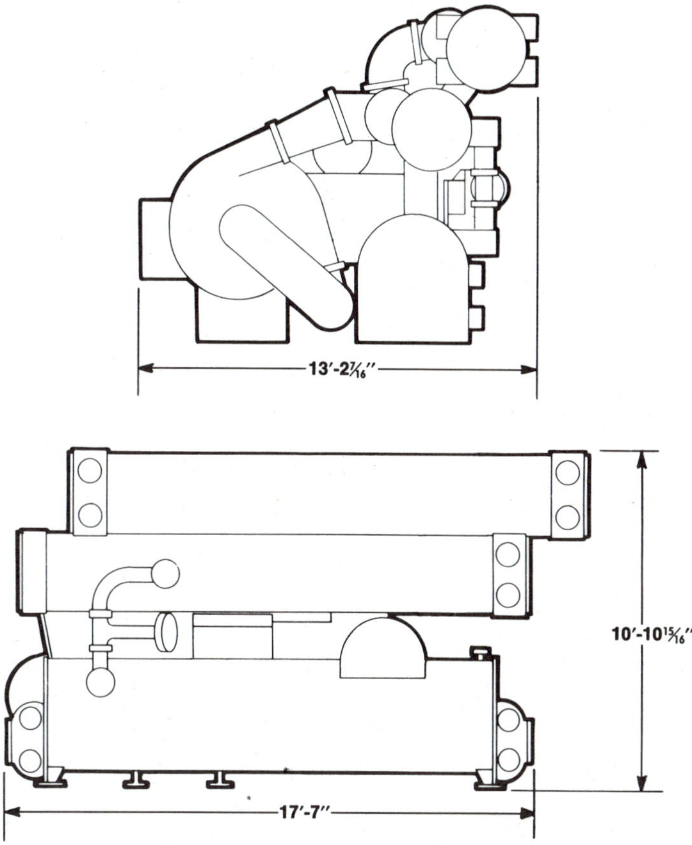

Figure 8–11 Two-Dimensional Line Drawing (The Trane Company)

can lift two-dimensional drawings from the plans or blueprints of the object or mechanism you are working with.

Three-dimensional (perspective) drawing (see Figure 8–12) Three-dimensional drawings are a little more realistic and allow most people to visualize an object a little more easily. (Purists distinguish between isometric and true perspective drawings, but most people call all three-dimensional drawings perspective drawings.)

Cross section (see Figure 8–13) A cross section is a special variety of two-dimensional drawing that shows an object as if it had been sliced or sawed through the middle. By convention, the "cut" portions of the object are shown cross-hatched. The cross section is particularly useful in showing internal components in their assembled relationships to one another.

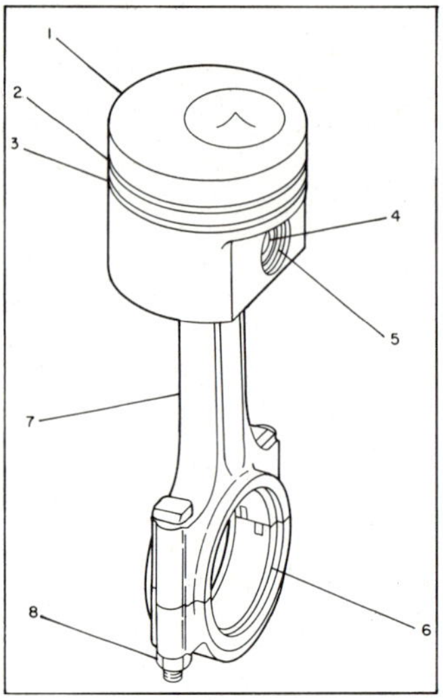

Figure 8–12 Three-Dimensional Line Drawing (Caterpillar Tractor Co.)

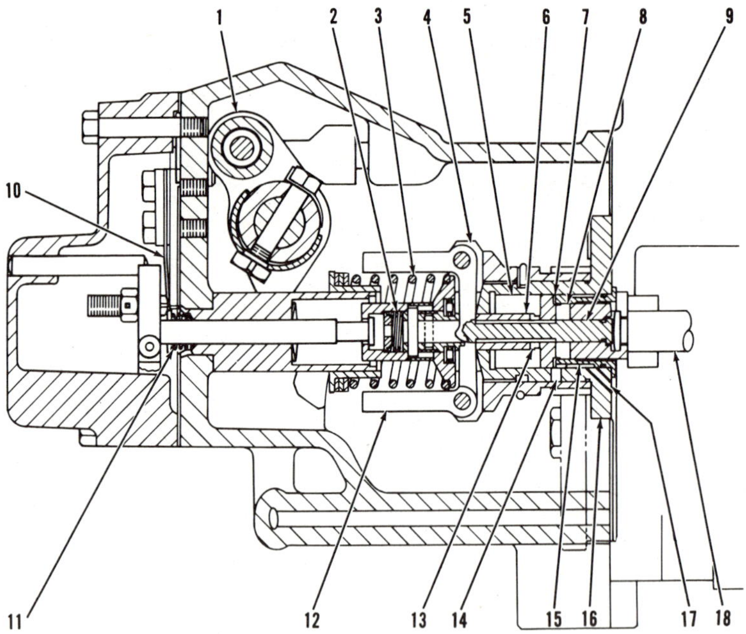

Figure 8–13 Cross Section (Caterpillar Tractor Co.)

Exploded view (see Figure 8–14) An exploded view shows the components of a mechanism strung out along a common axis. Exploded views are often used with parts lists, and mechanics like them for seeing how a mechanism is to be assembled.

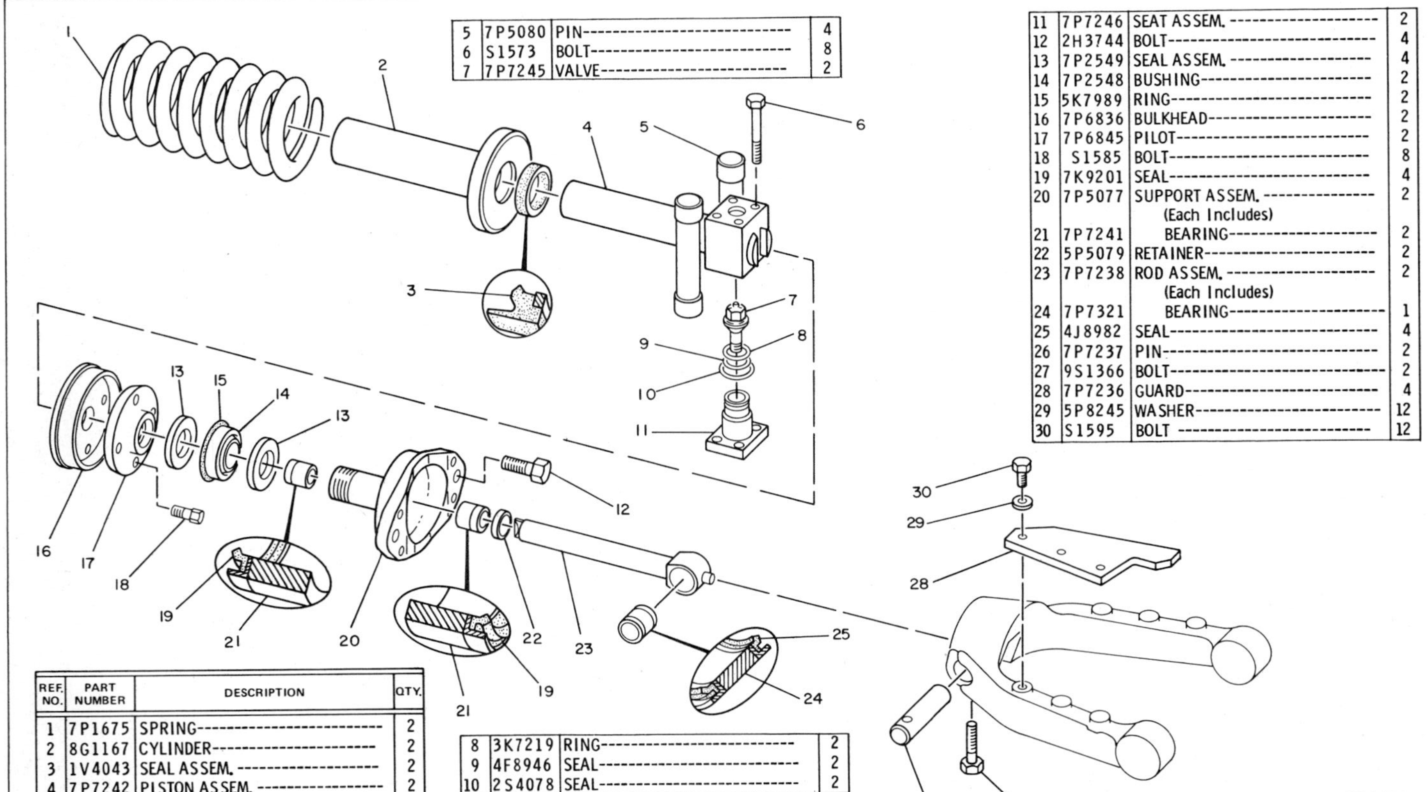

REF. NO.	PART NUMBER	DESCRIPTION	QTY.
1	7P1675	SPRING	2
2	8G1167	CYLINDER	2
3	1V4043	SEAL ASSEM.	2
4	7P7242	PISTON ASSEM.	2
5	7P5080	PIN	4
6	S1573	BOLT	8
7	7P7245	VALVE	2
8	3K7219	RING	2
9	4F8946	SEAL	2
10	2S4078	SEAL	2
11	7P7246	SEAT ASSEM.	2
12	2H3744	BOLT	4
13	7P2549	SEAL ASSEM.	4
14	7P2548	BUSHING	2
15	5K7989	RING	2
16	7P6836	BULKHEAD	2
17	7P6845	PILOT	2
18	S1585	BOLT	8
19	7K9201	SEAL	4
20	7P5077	SUPPORT ASSEM. (Each Includes)	2
21	7P7241	BEARING	2
22	5P5079	RETAINER	2
23	7P7238	ROD ASSEM. (Each Includes)	2
24	7P7321	BEARING	1
25	4J8982	SEAL	4
26	7P7237	PIN	2
27	9S1366	BOLT	2
28	7P7236	GUARD	4
29	5P8245	WASHER	12
30	S1595	BOLT	12

NOTE: QUANTITIES SHOWN ARE THOSE REQUIRED FOR BOTH R.H. AND L.H. SIDES

226670

Figure 8–14 Exploded View (Caterpillar Tractor Co.)

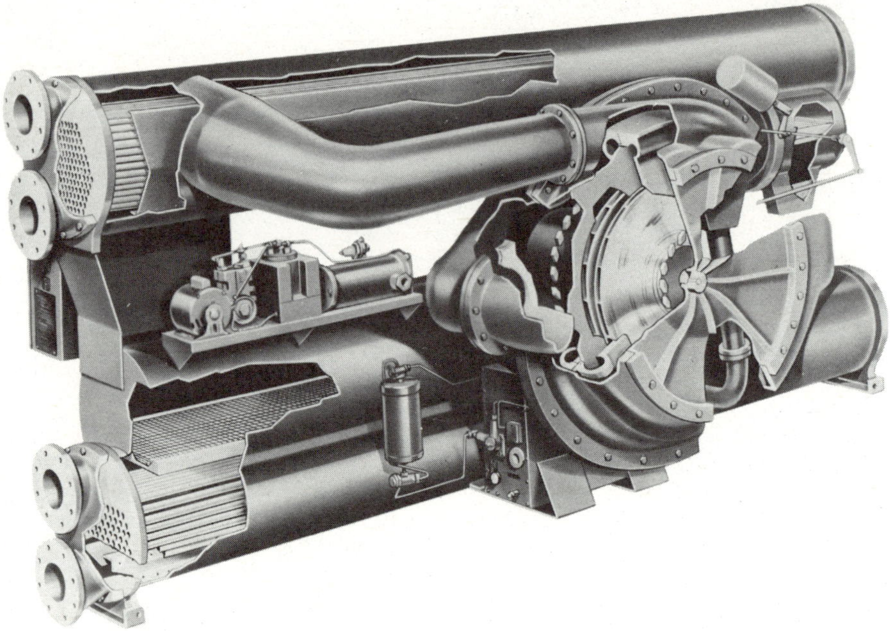

Figure 8–15 Cutaway View (The Trane Company)

Cutaway view (see Figure 8–15) A cutaway view is a special variety of perspective drawing in which portions of the external housing of a mechanism are graphically "cut away" to show internal components. Cutaway views are particularly useful with functional descriptions.

Techniques for Generating Line Drawings

The techniques suggested here are oriented toward the on-the-job situation where you have access to a drafting or graphic arts department. Obviously, if you're on your own you'll have to cut some corners or do some of the things yourself.

- □ Have all line drawings made originally at least twice final size and then photo-reduced to fit your page. This tends to sharpen everything and covers up minor drafting mistakes. Keep in mind, though, that as you reduce the overall drawing, line widths and any lettering or type will be reduced proportionally. If you start with light-weight lines, they will tend to drop out; similarly, if you start with regular-sized lettering or type, it will be too small to read after reduction.
- □ If you lift two-dimensional drawings or cross sections from plans or blueprints, have your artist eliminate all dimension lines, special symbols, and other clutter. Also, have him or her trace from the original print and then photo-reduce. It's cheaper and quicker than starting from scratch.

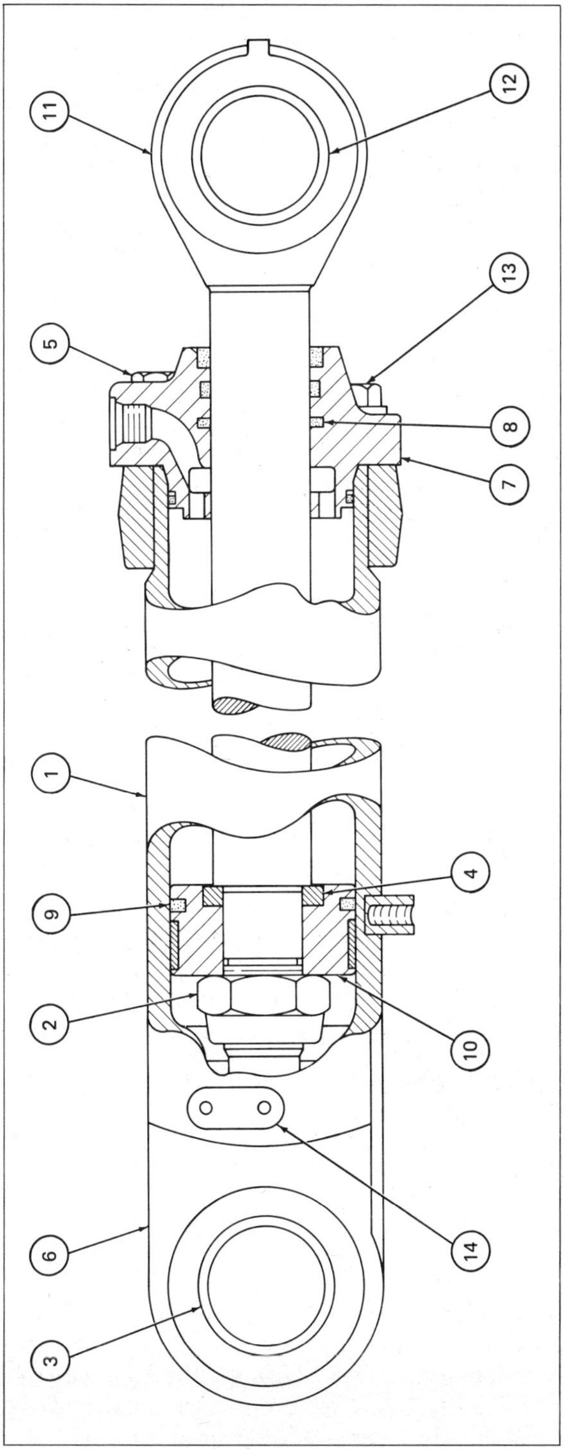

Figure 8–16 Drawing with Poorly Located Numerical Callouts

- ☐ Have your line drawings inked on tracing cloth or mylar without caption and without callouts or callout arrows. Save the original; you may want to use it again elsewhere with different callouts and a different caption. Have the callouts and callout arrows put on a photocopy of the original, and then have the whole works photo-reduced.
- ☐ Inked arrows and template lettering are time-consuming and expensive. Sheets of preprinted arrows, callout numbers, and lettering are available from art supply or office supply stores. Use them. Or have verbal callouts set in an appropriate type size and typeface and rubber-cemented to the photocopy of your original artwork.
- ☐ When using numerical callouts, run the numbers in order clockwise around the periphery of your drawing or in rows across your drawing. If you just throw them on anywhere, as in Figure 8–16 on page 129, it becomes hard for the reader to find a number that he or she is looking for.

☐ Success Strategies for Student Writers ☐

Pencil drawings and neat hand lettering are usually considered adequate for classroom writing assignments—check with your instructor. Don't draw directly on your typing paper—make your drawings on separate sheets of paper and fasten them in place with transparent mending tape. That way you don't have to redo your drawings if you retype your paper. Not all drawings have to be full-page; consider using half-page and quarter-page drawings. If you limit half-page drawings to 4 by 6 inches and quarter-page drawings to 4 by 3 inches, you'll have enough white space around them to avoid crowding. Also, if you have access to a reducing copy machine, try making your drawings twice size and giving them a 50 percent reduction.

☐ Tables and Graphs

All of us have read long lists of numbers; none of us likes to. So the next time you have a bunch of numerical data to present, take pity on your readers and put it in the form of a table or graph. When should you use a table and when should you use a graph? Tables are better for giving discrete data; graphs are better for continuous data. Tables enable you to be more precise, but graphs are better for showing trends and proportions. Tables lend themselves to data consisting of a single independent variable and several dependent variables; with graphs you can show several independent variables (by using different horizontal scales), but if you have more than two or three dependent variables you wind up with a mess. Sometimes what you need are the same data presented in both tabular and graphic form. Or if your data are extensive or complex, you may need a set of tables (one for each independent variable) or a set of graphs (one for each dependent variable). Whatever your decision, take the time to make your table or graph neat and attractive and easy to read and understand.

Techniques for Building Tables

A table can be anything from a two-column list to a complex array with several horizontal and vertical dimensions. Figure 8–17 shows a table of moderate complexity with the key components labeled. Here are some points to keep in mind when presenting data in tabular form:

- □ Many people distinguish between a simple list and a table (or an informal table and a formal table) on the basis of whether or not the array in question has row heads. Those who make the distinction would not call a two-column array a table, and they would not give it a table caption or enter it in the list of tables and figures at the front of the report. Others feel this is a distinction without a difference. If you feel an array of data warrants a title of its own, give it a full caption and call it a table. If not, don't.
- □ People read tables from left to right in each successive row from top to bottom, unless they're consulting a table for reference purposes—in which case they read from top down in the left-most column and then to the right. So arrange your data the way people are going to read it (or try to read it); you'll save every-

STUB HEAD

TABLE CAPTION

COLUMN HEADS

TABLE 16-2 CCKA 100E With CAUA 1000 Air-Cooled Condenser

LEAVING CHILLED WATER TEMP (F)	AMBIENT TEMPERATURE (F)													
	85		90		95		100		105		110		115	
	TONS	KW	TONS	KW	TONS	KW	TONS	KW	TONS	KW	TONS	KW	TONS	KW
40	87.3	87.4	84.8	89.8	82.2	92.1	79.5	94.4	76.7	96.6	73.7	98.8	70.8	100.9
42	90.3	88.8	87.6	91.2	84.9	93.7	82.2	96.0	79.3	98.4	76.4	100.7	73.2	103.0
44	93.2	90.2	90.5	92.7	87.7	95.3	84.9	97.8	82.0	100.3	78.9	102.7	75.8	105.1
46	96.3	91.5	93.4	94.2	90.6	96.8	87.6	99.5	84.6	102.0	81.6	104.7	78.4	107.2
48	99.2	92.9	96.4	95.6	93.4	98.4	90.4	101.1	87.3	103.9	84.2	106.6	80.9	109.5
50	102.3	94.2	99.3	97.1	96.4	100.0	93.2	102.9	90.1	105.7	86.8	108.7	83.5	111.7

ROW HEADS

BODY

AIR-COOLED PERFORMANCE TABLE NOTES

NOTES

1. Ratings based on 10 F water temperature difference in evaporator. Ratings are applicable to temperature differences of 8 to 12 F.
2. Ratings based on 0.0005 fouling factor in evaporator.
3. Interpolation between points is permissible.
4. Extrapolation is not permitted.
5. Ratings are in accordance with ARI 590-76.
6. 2 F pressure drop in suction line is assumed.

Figure 8–17 Table with Components Labeled (The Trane Company)

body a lot of trouble. In particular, put your independent variable in the leftmost column and put your dependent variables in successive columns to the right.

- □ Make sure every column in your table has a head or subhead and every row has a head or subhead. In general, avoid going to a tertiary breakdown with either columns or rows; go to a separate table instead.
- □ If you use a stubhead, see that it pertains to rows rather than columns (downward rather than across).
- □ Keep the number of horizontal and vertical lines, or rules, in your table to a minimum. A few rules will help your readers track across and down your table, but too many just clutter up the works. Use a bold rule (or double line) to separate column and row heads from the body of your table, and use a light rule (or single line) under each fifth or so row of data in the body of your table. Avoid vertical rules in the body unless you feel they are absolutely necessary for clarity.
- □ Use lowercase letters (rather than numbers) for table notes, and put the notes right under the table rather than at the bottom of the page. By doing this, you avoid confusion with footnotes in your text.
- □ If you need to set a table up the long way on a page, arrange it so that your reader rotates your document clockwise to read the table. That is, put the bottom of the table at the right margin.
- □ Make your headings clear and complete. If you use abbreviations or symbols, clarify all but the simplest with notes. Make sure you have identified all units of measure clearly, either in the headings, in the body of your table, or in a note.
- □ Align numerical data vertically in your columns on the decimal point. Align nonnumerical data vertically on the first letter or leftmost symbol.
- □ For tables or lists giving financial data in dollars, use a dollar sign only on the first entry in a column and the last entry in the column (which is usually the total).
- □ Be generous with your white space within and around a table. Skip a line or two rather than using a horizontal rule to separate rows. Leave plenty of space between columns. And leave an inch of white space above and below your tables to clearly set them off from the body of your text.
- □ Use parallel grammatical form in all headings and all subheads of equal rank. Arrange your headings in some logical order, if at all possible. Make your tables clear, logical, and easy to read. You're including tables to help your readers, not confuse them.
- □ Be sure you have a reference in the text to every table, and be sure the first reference falls ahead of the table itself. It's all right to make subsequent references to a table; but if more than five

pages or so intervene, it's more convenient for your readers if you repeat the table, giving it a caption like the following:

Table 6 (=Table 2)—Head Bolt Diameters and Torques

- Don't assume that the first way you set your table up is the best way. Try switching rows for columns (rotating about the orthogonal axis, for you math jocks) or try switching main heads for subheads in both columns and rows. You may well find that an alternate arrangement for your tabular data is easier to read, more pleasing to the eye, or both.

Techniques for Building Graphs

Graphs (or charts, as many people call them) come in all sorts of shapes and descriptions, but they're fundamentally the same. Basically, a graph is a curve (in the mathematical sense) drawn through a set of points and projected against a rectangular coordinate grid, as in Figure 8–18. If instead of connecting the points with a curve you draw a fat line from each point to the horizontal axis, you have a bar graph (or if you draw fat lines to the vertical axis you have a horizontal bar graph). If you make your fat lines out of little men or little houses or little factories, you have a pictograph. And if you draw several curves through several sets of points, you have a multiline graph. Avoid pie graphs. They're all right for *Newsweek* magazine and the United Way campaign posters, but they

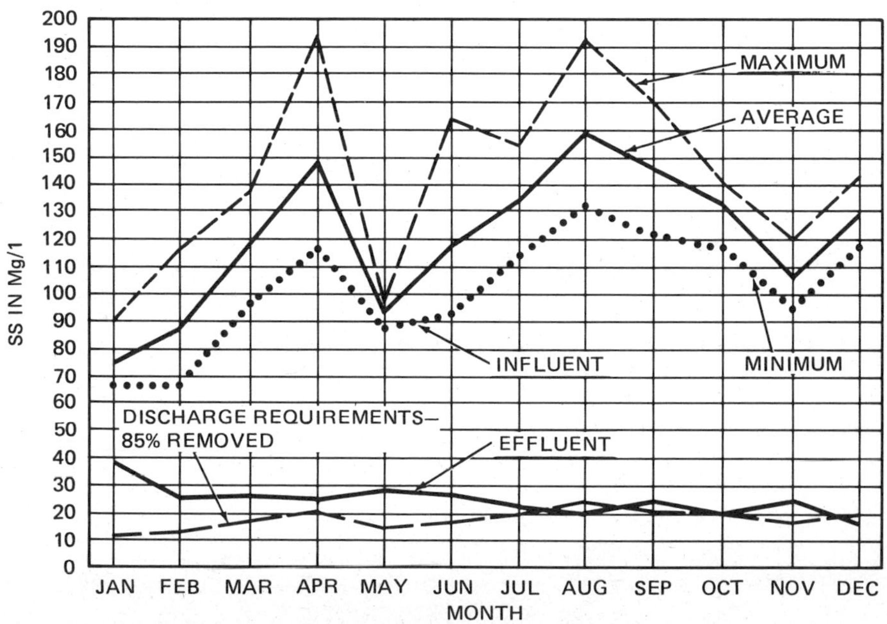

Figure 8–18 Basic Graph (Interwest Engineering, Inc.)

don't have much of a place in scientific or technical reports. Here are some tips for presenting data in graphs:

- Keep your graphs simple and uncluttered. Include no more data than a reader can absorb in a minute or so of study. If you find yourself building a complicated graph, eliminate some points on the curve, use several graphs, or present the data in a table instead.
- Plot your dependent variable on the vertical axis and your independent variable(s) on the horizontal axis. This enables your readers to enter the graph from the bottom, read up to the curve, and then read over to get a corresponding value on the vertical axis. (One exception: *time* is conventionally plotted on the horizontal axis regardless of whether it is independent or dependent.)
- In general, use a real zero for both horizontal and vertical scales, and increase both scales in regular, common intervals (2, 5, 10, 50, and 100). If your value range is so high that you can't conveniently do this, indicate clearly with a jagged line between your zero and your first scale label that some numbers have been left out (see Figure 8–19) or add the note "Zero Suppressed" under (or alongside) the scale. A particularly noxious habit of second-rate statisticians is to exaggerate minute differences by showing

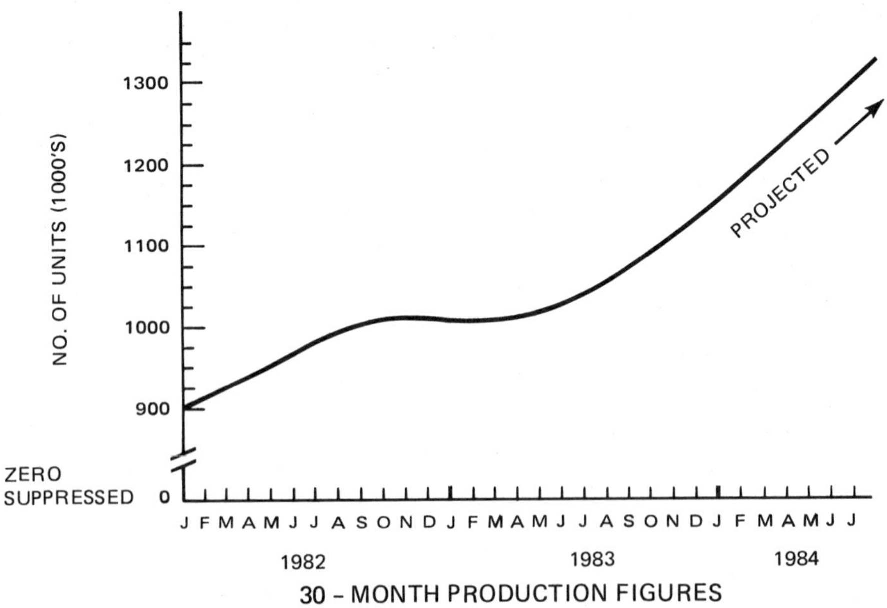

Figure 8–19 Graph with Zero Suppressed

a graph with a drastically suppressed zero. Typical of this is the newspaper chart showing the past week's Dow-Jones performance. The chart looks like the Grand Teton skyline, but in fact all of the week's variation took place in less than eight-tenths of one percent.

- Choose your scales wisely, so that your main data plot is more or less centered and you have a reasonable amount of vertical distance between plotted points. If your data range is much more than an order of magnitude, though, use a logorithmic scale on your vertical axis, your horizontal axis, or both.
- Remember that the ratio of line width to scale implies the precision of your data. (The same is true, to a degree, of the size of your plotted points with respect to your scale.) Don't hide the precision of your data with a fat line and a small scale, but more importantly don't imply more precision than is inherent in your data by using a fine line and a large scale. In general, your line width should be about equal to the confidence interval (level of precision) of your data.
- To avoid clutter, leave the grid lines off your graph and use ticks (hash marks) instead (see Figure 8–20). Put the ticks on both sides of your graph, though, so that those readers who wish to can lay a straightedge across it. (An exception to the rule of

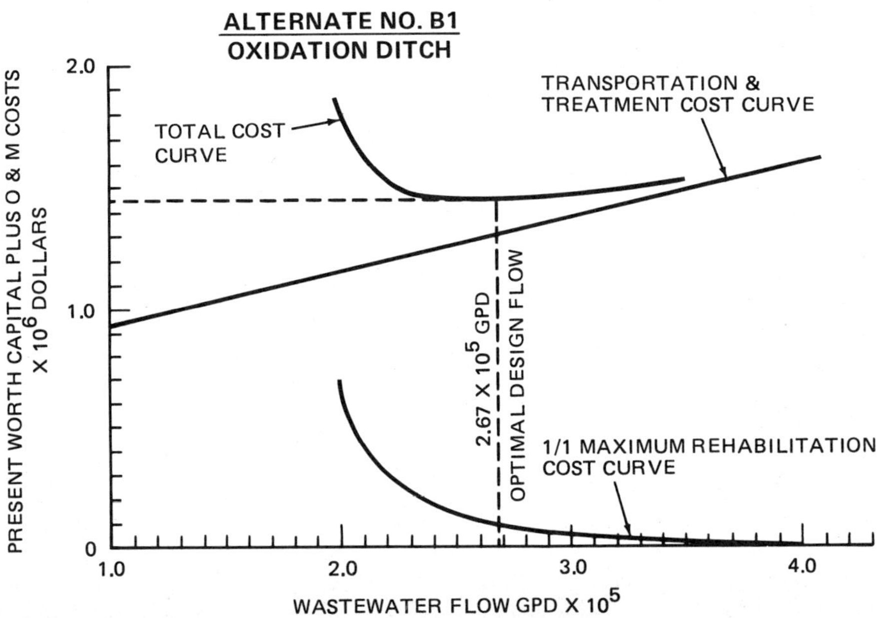

Figure 8–20 Graph with Ticks (Hash Marks) Instead of Grid Lines (Interwest Engineering, Inc.)

omitting grid lines is graphs you include for reference purposes, like pump performance curves or heating unit capacities. If you want precision, though, use a table instead.)

- Draw your graphs (or have them drawn) two to four times oversize and then photo-reduce them to fit your report page, just as with line drawings. Remember to use appropriate line widths and lettering sizes.

Maps and Schematics

Two categories of useful artwork that don't quite fit anywhere else are maps and schematics. Often, with reports or proposals that deal with a particular geographic area, it's useful to include a site map (Figure 8–21) and/or a vicinity map (Figure 8–22). As with two-dimensional line

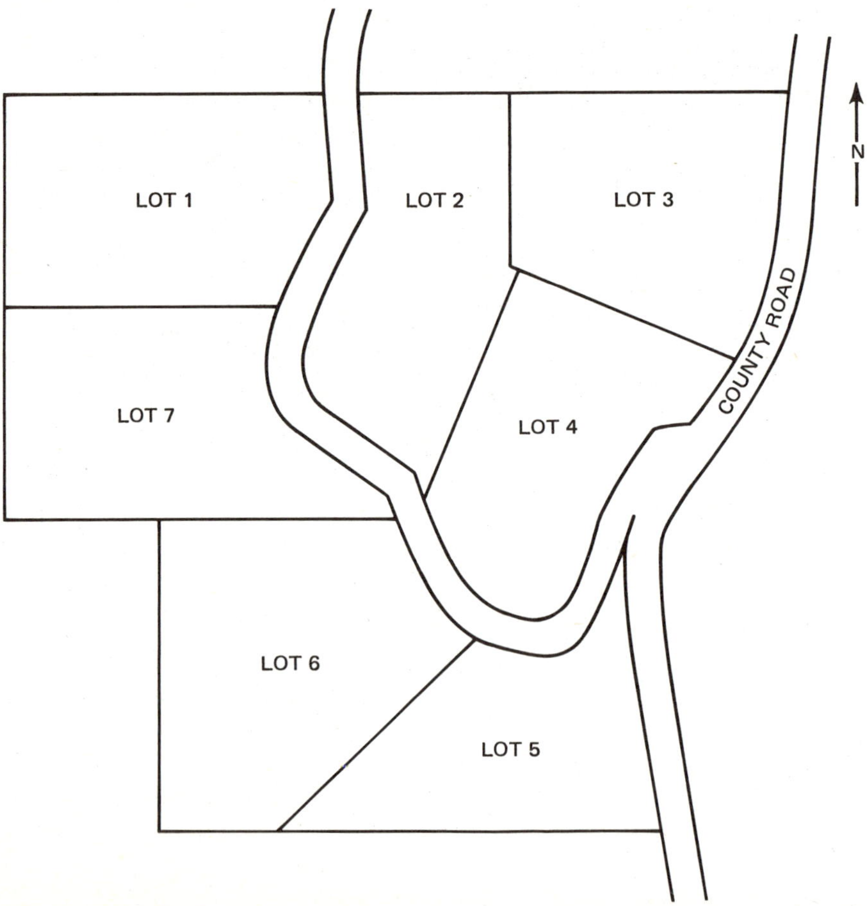

Figure 8–21 Site Map (Interwest Engineering, Inc.)

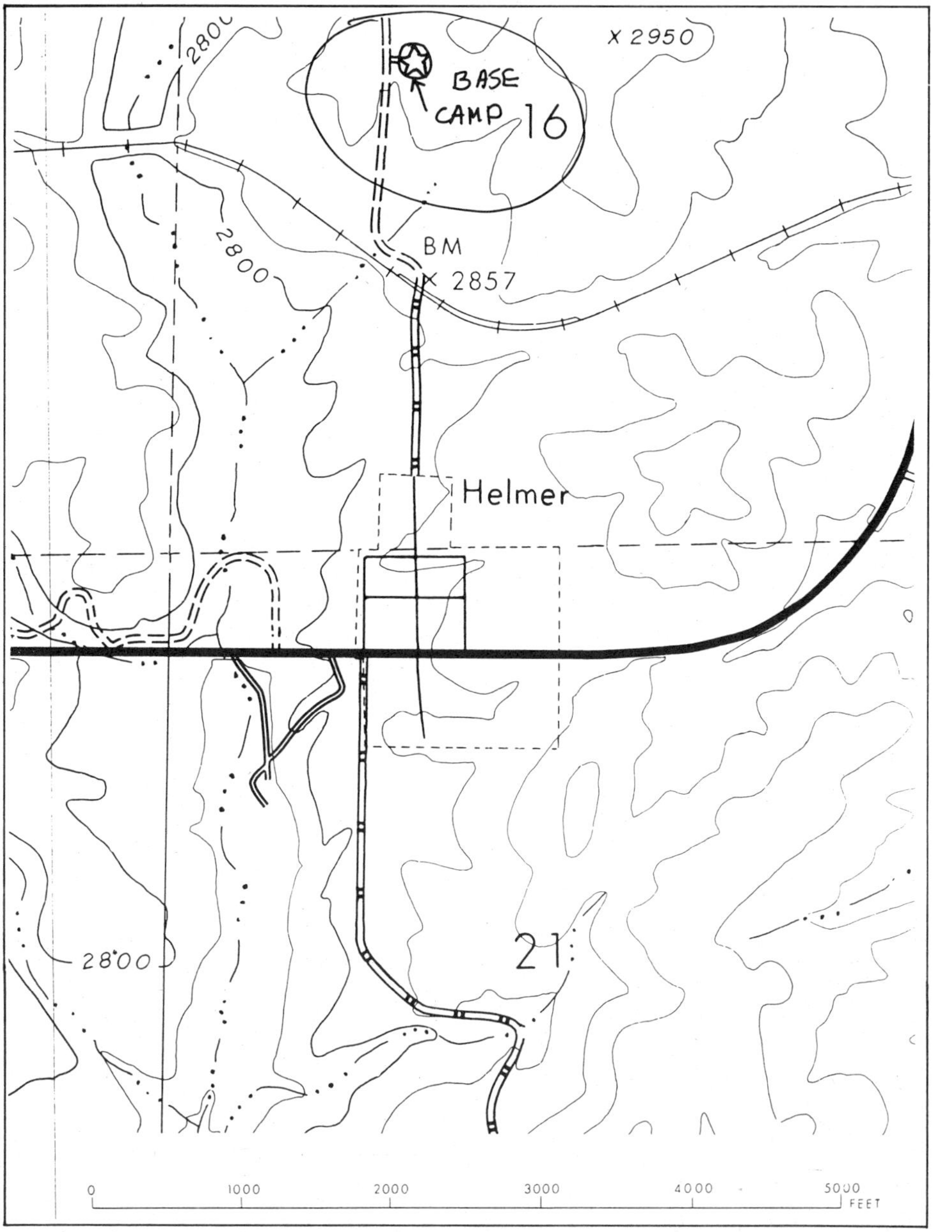

Figure 8–22 Vicinity Map (Palouse-Clearwater Search and Rescue, Inc.)

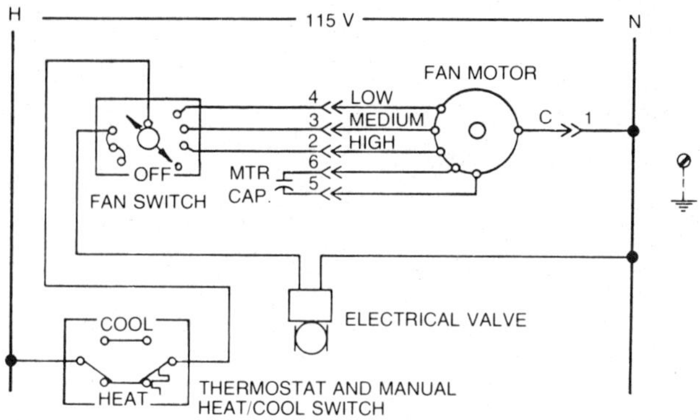

Figure 8–23 Simplified Schematic (The Trane Company)

drawings lifted from plans or blueprints, you're usually better off having a new map drawn rather than just copying an existing map. Most existing maps show far more detail than you need or want.

Schematic drawings are conceptually related to maps. A map, in effect, is a schematic of roads, topographic features, and property boundaries. A schematic drawing, on the other hand, is a "road map" of electrical pathways, fluid flow, lines of authority, process flow, or what have you. A simplified schematic like the one in Figure 8–23 is most useful for a functional description of a system or process, while a more detailed field diagram like the one in Figure 8–24 is more useful to accompany a physical description or maintenance instructions. Often in a manual you need both types of schematic. Don't try to make one serve both purposes. For fluid flow schematics, use double lines for pipes and use stylized pumps, valves, and fittings, as in Figure 8–25. These are clearer to most people than the single-line diagrams with special symbols that are common to engineering drawings (Figure 8–26).

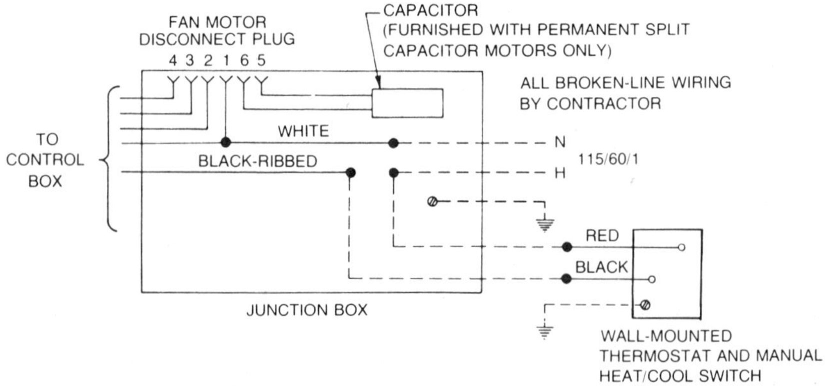

Figure 8–24 Field Writing Diagram (The Trane Company)

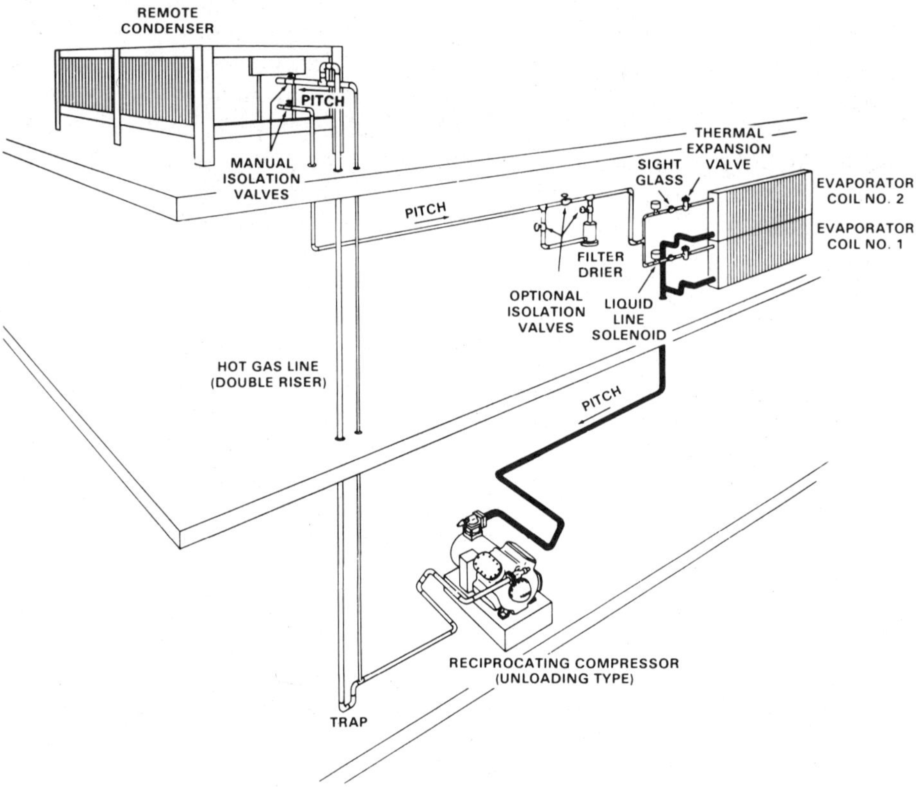

Figure 8–25 Piping Schematic with Stylized Components (The Trane Company)

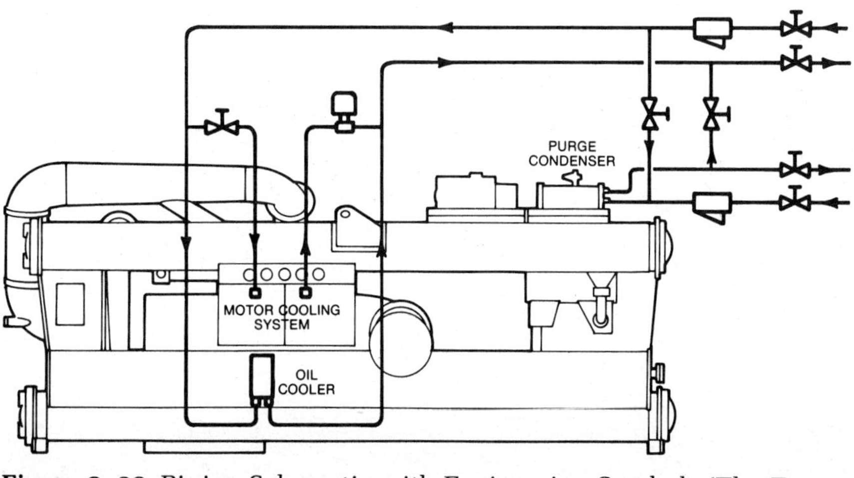

Figure 8–26 Piping Schematic with Engineering Symbols (The Trane Company)

□ Handling Artwork

For all but the simplest of tables, you will want to handle and process your artwork separately from the main text of your report. They both start with you, but as Figure 8–27 indicates they follow separate paths, coming together again only at the printing or duplicating center. As you write your initial draft and make your preliminary revisions, put figure references and table references in at appropriate places (see below) but leave the figure or table number blank "(see Figure)." This allows you to add illustrations later without having to change all your figure or table numbers. Toward the end of your revision stage, generate a complete list of all your figures and a complete list of all your tables. Separate your list of figures into photographs and line artwork (drawings, graphs, schematics). For the list of photos, check off the ones you have available in photo files and provide a brief description of what is to be shown in each of the others (see Figure 8–28). For a list of line artwork, prepare rough sketches of what you want to show or indicate sources (blueprints, maps, existing drawings) from which the line artwork can be made. If your organization has an art coordinator, turn the lists and roughs over to him or her. If you have to coordinate artwork production yourself, arrange for a photographer to take the photos you need and arrange for a draftsman or graphic illustrator to produce the line artwork.

As a final step in revision for the text of your report, plug in all your figure numbers and table numbers in the blanks you left. Have the review draft of your report text typed manuscript style (8½-by-11-inch sheets, double-spaced), keying in the approximate locations of your figures and tables as shown in Figure 8–29 (page 142). Have each of your tables typed one to a sheet of paper; don't attempt to integrate the tables into your text at this point. Once your photographs and line artwork are completed, run a Xerox copy of each piece of artwork and carefully proof-check figure numbers, callout numbers, and nomenclature against the review draft of your text. This is a particularly important step, because more typographi-

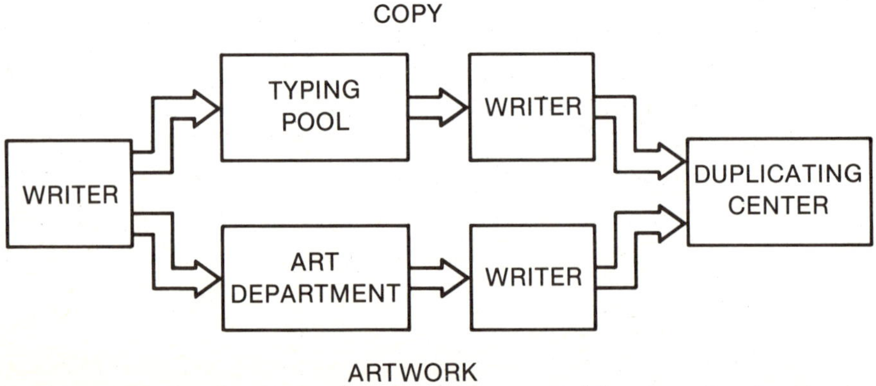

Figure 8–27 The Flow of Components in Publications Production

RUPERT PHOTO LIST

FIG. NO.	PHOTO OF	SHOWING	COMMENTS
1-3	MD Pump Control Panel	All Controls	
1-X	Main Drain Pump Station	Model View	
2-3	Pretreatment System	Model View	
2-5	Bar Screen	Model View	Combine, if possible
2-6	Bar Screen	ON-OFF Pushbutton Station	
2-8	Grit Collector	Model View	Combine, if possible
2-9	Grit Collector	ON-OFF Pushbutton Station	
2-10	Pretreat Inlet Box	Sluice Gate & 2 Stop Gates	
2-11	Pretreat Outlet Box	3 Sluice Gates	Show Control Wheels
3-2	Aeration Cell Parting Wall	Both Distribution Boxes	
3-5	Helixor	Model View (Installed)	
3-6	Blower Air Line along Basin	Air Line Plug Valve	
3-7	Blower Room	Blower, Blast Gate, Check Valve, Butterfly Valve	
3-8	Blower	Pressure Gauge, Ammeter, ON-OFF Switch	
3-8A	Blower Selector Switch		
3-9	Clarifier	Scum Box	

Figure 8–28 Photo List with Description of Contents (Interwest Engineering, Inc.)

cal errors and inconsistencies creep in between text and artwork than anywhere else in a technical document. Using Xerox copies of artwork for proof-checking (and review) has three advantages:

- □ It minimizes handling of the artwork itself. (Your original artwork is both fragile and valuable.)
- □ It prevents both you and your reviewers from marking changes and corrections directly on the artwork.
- □ It cuts down considerably on the bulk of the package you circulate for review and approval.

If the text of your report is short (say, under ten pages) and you have no more than a dozen or so pieces of artwork, just put the Xerox copies of your figures and tables all together at the back of your text. If your text is longish or if you have more than a dozen pieces of artwork, you may want to interweave the copies of your artwork into the text manuscript as a convenience for your reviewers.

CHAPTER 4--TECHNICAL CORRESPONDENCE

Just as we define technical writing as the job-related writing a technical person does, so we can define technical correspondence as the job-related correspondence a technical person might be asked to produce. We can identify three varieties of technical correspondence:

..... Memos

..... Letterhead Letters

..... Non-Letterhead Letters

Some people make a great deal of the distinction between memos and letters; in point of fact the difference is minimal. A *memo* (see Figure 4-1) is nothing but an in-house letter, a document sent from one individual in a firm or organization to one or more other individuals in the same firm or organization. The format of a memo--particularly the opening and closing--is different from that of a letter, but the body organization, the tone, and the style of writing is essentially the same for both memos and letters.

// FIGURE 4-1 //

A *letterhead letter* (see Figure 4-2) is a document that you write as a representative of the firm or organization identified in the letterhead. It is official correspondence: company business. It is sent from you as a representative of your firm or organization to one or more other people, often as representatives of their firms or organizations. If you have occasion to write a business letter in a role as private citizen, representing no one but yourself, don't use company stationary but rather prepare a *nonletterhead letter* on plain white paper (see Figure 4-3).

// FIGURE 4-2; FIGURE 4-3 //

The best that can come of using letterhead stationary for personal business is to be perceived as insensitive to the distinction between organizational and private business. At worst, you will be percieved as the sort of person who steals stationary and stamps--and probably steals pencils, notepads and calendars. The most glaring offense in this department is using letter-

Figure 8–29 Manuscript Page with Figure Locators Keyed in

When your package of manuscript and artwork comes back from review (see Figure 8–27) and you have made all necessary changes and corrections, send the corrected copy of your manuscript and your original artwork to the printing or duplicating center. Here, the text of your report will be set in type and individual pages will be laid out to leave room for the tables and figures. Your photographs will be cropped, reduced, and put in place, your tables and line art will be integrated into the text, and your whole report will be put into its final form.

Figure References

Handling figure references and callout numbers in the text of a report is a fairly straightforward operation, but it causes so many problems for inexperienced writers that it warrants a section of its own.

First, the location for your figure references. The general rule is to put the reference to a figure or table into your text at the point at which you want your readers to refer to the figure. If a figure is pertinent to an entire section or subsection and you want your readers to look at the figure before they begin reading your text, you can put the figure reference in right after the heading or subhead, as in Figure 8–30. Alternatively, you can put your figure reference in an early sentence of the paragraph, either medially in the sentence (see Figure 8–31, page 144) or at the end of the sentence (see Figure 8–32). Still another technique is to put your figure reference between two sentences. (See Figure 8–33, page 145.) This is useful when the illustration is pertinent both to the sentence that precedes the reference and the one that follows the reference. In general, avoid a stand-alone figure reference after the last sentence of a paragraph; it looks as if you just threw it in as an afterthought. Second and subsequent references to the same figure are quite in order, particularly if

EVACUATION AND DEHYDRATION

(See Figure 16, page 6)

It is imperative that all moisture and air be removed before refrigerant is charged into the system. The importance of evacuation and dehydration cannot be overemphasized.

Moisture in a refrigerant system can cause the formation of highly corrosive acids or copper plating of compressor parts. Further, moisture and/or air results in oil breakdown and sludging.

Never, under any circumstances, use the compressor as a vacuum pump. It is not designed for this purpose and if so used, may be seriously damaged. A vacuum pump capable of producing 1 mm Hg absolute must be used.

1. Connect an accurate high vacuum gage, such as a Stokes or Zimmerli gage, to the vacuum line. Zimmerli gages are available through the local Trane sales office.

Figure 8–30 Figure Reference Following a Heading

```
     When your package of manuscript and artwork comes back from
review (see Figure 11-35) and you have made all necessary changes
and corrections, send the corrected copy of your manuscript to-
gether with your original artwork to the printing or duplicating
center.  Here, the text of your report will be set in type and
```

Figure 8–31 Figure Reference Medially in a Sentence

you want to remind your readers of the existence of the illustration or call their attention to another aspect of the illustration. You definitely should include an additional reference if a half-page or so of text, a paragraph break, or a section heading or subhead intervenes between your first (or most recent) reference to the figure and the subsequent point you want your readers to look at the illustration again.

Next, let's consider the format for figure references. Unless your figure reference is grammatically part of a sentence, like the reference to Figure 8–30 on page 143, put parentheses around it. If the reference is in the middle of a sentence (see Figure 8–31), use a lowercase s on *see* and put any normal sentence punctuation *after* the closing parenthesis. Avoid setting parenthetical figure references off with commas rather than parentheses, and in no circumstances use both commas and parentheses. If your figure reference comes at the end of a sentence, it's still structurally part of the sentence (see Figure 8–32). Again, use a lowercase s on *see* and put the period after the end parenthesis. The only time you use a capital S and put the period inside the right parenthesis is when you have a stand-alone figure reference that is punctuated as a separate sentence in its own right. (See Figure 8–33.) If you follow these simple conventions, people will think you've been using figures and figure references all your life.

> Install the outdoor thermostat box on the inside of the rear panel on the Outdoor Unit, using the mounting studs (see Figure 26).
>
> Using the bulb holder which is provided, mount the temperature sensing bulb(s) on the right side panel of the Outdoor Unit (see Figure 26).

Figure 8–32 Figure References at the Ends of Sentences

> 1. Check the compressor suction and discharge pressures while the compressor is operating. (See Figure 151.) Be sure the pressures are within the limitations of the compressor. If pressures are excessive, it may be necessary to clean the condenser, purge air from the system, add a crankcase pressure regulating valve, modify the system control, or take such other action as may be necessary to avoid excessive pressures.

Figure 8–33 Figure Reference Between Two Sentences

Writing Exercises

1. Consult two issues each of a half-dozen scholarly journals in your field. Calculate the average number of illustrations per article and per page. Classify them as to line drawing, photograph, chart, table, and the like. Write a report of your findings for the class.
2. Do the same for trade journals in your field.
3. Do the same for general science journals such as *Scientific American, Nature,* or *Science.*

Writing about Writing

1. Interview the editors of several publications groups on your campus. (Most of the research centers or institutions will have an editor.) Determine the costs of reproducing various types of illustrations. More important, get their views on including illustrations in technical documents. Summarize your findings in a report to the class.
2. Interview the graphic artists or art directors of several publications groups on your campus. Determine their recommendations for handling artwork for technical documents and publications. Write a report for the class.

9

Documentation

As a student writer you documented or "footnoted" your term papers both to avoid allegations of plagiarism and to provide your instructor with a handy list of your sources for checking to see if you were thinking for yourself or just stringing together other people's ideas. There was mention of giving credit where it was due and perhaps some mention as well of providing your readers with a list of sources in case they wanted to pursue your subject matter in greater detail. All in all, though, documenting your papers was a time-consuming chore that seemed to have little relevance to your actual writing.

As a writer on the job, however, you will find all of this turned around. First, you won't do all that much documentation; most of the things you write will not require it. When you do document, you will do so primarily to help your readers find additional information on the subject you are writing about. Questions of proper credit and formal recognition will be present, but they will be secondary. You include references to other sources because you are convinced that it is in your readers' best interests to consult these other sources and you want to make it as easy as possible for them to find these sources. In short, on the job you document as an act of communication.

□ Situations Where Documentation Is Needed

The most common situation where formal documentation is needed is the scientific or technical paper you prepare for a conference or for submission to a scholarly or professional journal. Chances are, though, that you will not write many of these. Literature searches and annotated bibliographies are, of course, built around formal documentation, but again these are not common forms of technical writing for most people. The state-of-the-art section of a proposal—particularly a research grant request—almost always needs formal documentation to cite references to work done by others. Similarly, instruction manuals often need references to other related manuals and to sources of background information. A laboratory procedure manual might make reference to operation and maintenance instructions for specific pieces of laboratory apparatus as well as to standard works on basic analysis and measurement practices. A progress report might make reference to other project documents: the proposal, other progress reports, special data reports, and the like. Even in your correspondence you may have occasion to cite a specific publication or two. Finally, environmental impact statements should incorporate other environmental studies and assessments by reference rather than present a rehash of the original data. You will have a few occasions to document your writing, and as long as you will have to do it, you might as well do it right.

□ Some Preliminaries

Documentation, as we will use the term, refers to incorporating in-text references and bibliographic references, or citations, into a piece of technical writing. An *in-text reference* is the device you use in the body of your text to refer your readers to your bibliography. A *bibliographic reference* is the actual entry for a given source that you incorporate in your bibliography. The *bibliography* itself is nothing more than a list of books or references. In fact, some authors and publications use "List of References" or "References" as headings for their bibliographies. We could say, then, that documentation consists of referring your readers to certain entries in your bibliography and providing in each of these entries sufficient data for a reader to locate a copy of each work you cite.

A comprehensive bibliographic reference consists of seven *groups* of information: Authorship, Title, Edition, Imprint, Physical Description, Series Statement, and Notes. Of these, the Authorship group, the Title group, and the Imprint group are the ones that appear most often, with the others appearing when needed for the complete description of multiple edition works, microfilmed or other nonprint works, or special works like patent documents, legal materials, unpublished works, maps, films, recordings, and the like. Each information group consists of one or more bibliographic *elements;* for example:

Authorship Group	Primary Author Element
Title Group	Analytic Title Element Collective Title Element
Imprint Group	City of Publication Element Publisher's Name Element Date of Publication Element Volume-Identification Element Pagination Element

We'll look at examples of groups and elements later in this chapter.

□ Style Guides for Documentation

Each major field of scholarship and research has its professional society, and each society seems to have its own style guide. Moreover, individual journals, individual corporations and government agencies, and individual publishing houses often have style guides of their own. And up until a few years ago it seemed every style guide had different specifications for bibliographic references. In the early 1970s, however, representatives of all major scientific and technical societies gathered together under the aegis of the American National Standards Institute to work out a framework for the standardizing of bibliographic references. In 1977, ANSI published the *American National Standard for Bibliographic References*, which has become very much the bibliographic bible for U.S. scientific and technical publications. Since then, most societies and organizations represented have published new editions of their style guides that reflect the ANSI standard. The ANSI publication is the ultimate authority for bibliographic style and format, but it is a highly technical and involved document that is of limited use for the general writer. Your best bet for most of your writing is to pick the most appropriate style guide for your field from the list below and follow its recommendations.

General

American National Standards Institute. *American National Standard for Bibliographic References* (Z39.29–1977). New York: American National Standards Institute, 1977.

U.S. Government Printing Office. *Style Manual*, rev. ed. Washington, DC: GPO, 1973.

University of Chicago Press. *The Chicago Manual of Style*, 13th ed. Chicago: University of Chicago Press, 1982.

Biology

Council of Biology Editors, Committee on Form and Style. *CBE Style Manual*, 4th ed. Washington, DC: American Institute of Biological Sciences, 1978.

Chemistry

American Chemical Society. *Handbook for Authors of Papers in the Journals of the American Chemical Society*. Washington, DC: American Chemical Society, 1978.

Engineering

Engineers Joint Council, Committee of Engineering Society Editors. *Recommended Practice for Style of References in Engineering Publications*. New York: Engineers Joint Council, 1977.

Geology

U.S. Geological Survey. *Suggestions to Authors of Reports of the United States Geological Survey*, 6th ed. Washington, DC: GPO, 1978.

Mathematics

American Mathematical Society. *Manual for Authors of Mathematical Papers*, 5th ed. Providence, RI: American Mathematical Society, 1980.

Language and Literature

Modern Language Association. *MLA Handbook*. New York: Modern Language Association, 1977.

Linguistic Society of America. *LSA Bulletin*, No. 71 (Dec. 1976), pp. 43–45.

Medicine

American Medical Association, Division of Scientific Publications. *Style Book and Editorial Manual*, 6th ed. Chicago: American Medical Association, 1976.

Physics

American Institute of Physics, Publication Board. *Style Manual*, rev. ed. New York: American Institute of Physics, 1978.

Psychology and Other Behavioral Sciences

American Psychological Association. *Publication manual of the American Psychological Association*, 3rd ed. Washington, DC: American Psychological Association, in press.

The ANSI Standard is a broad one, allowing for individual variation. As you thumb through the style guides for various fields, though, you will find variations in ordering of elements, abbreviations used, punctuation and types of information included. We cannot cover all the variations here, of course, but we can look at basic formats for the major bibliographic groups and we can examine sample references for each of the three major variations of in-text and bibliographic references that are most widely used in scientific and technical publications. That should take care of most of your documentation needs. Keep in mind, though, that if a specific style guide conflicts with (or expands on) the general guidelines given here, follow the specific rules. Also, if you encounter a situation not covered here, don't guess—consult a more comprehensive source.

□ Basic Reference Formats

The variations allowed by the ANSI standard tend to group themselves into three basic formats: the footnote format, the author-date format, and the bracketed number format. These differ chiefly in the form of the in-text reference, but there are some differences in punctuation and order of elements in the bibliographic references as well.

The Footnote Format

The footnote format is characterized by the use of a superscripted numeral as in the in-text reference:

> Recent studies[12] confirm the use of this technique.

or

> Recent studies confirm the use of this technique.[12]

The oldest of the three basic formats, this is probably the one you are most familiar with, as it is the one generally taught in high school and college English classes. Although its popularity is dwindling in scientific and technical publications, it still enjoys widespread use in the arts and humanities and in those scientific journals which have been in print for a good many years. The most comprehensive style guide for the footnote format is the *MLA Handbook*.

The superscripted numeral does not refer the reader directly to the bibliography but rather to a note at the foot of the page (hence the term *footnote*). Footnotes are not limited to bibliographic citations; they may contain author's comments, definitions of terms, explanations, and a host of other items. We are concerned here, however, only with those footnotes that provide bibliographic information. The footnote for the first reference to a work cited contains complete bibliographic information:

6. Frank A. Ilardi, *Computer Circuit Analysis: Theory and Applications* (Englewood Cliffs: Prentice-Hall, Inc., 1976). p. 265.

Notice that the author's name is in normal order—first name first—and that the reference includes a page number. Footnotes for subsequent references to the same work contain the author's last name and a shortened version of the title:

9. Ilardi, *Circuit Analysis*, p. 325.

Ibid. and *op. cit.* are not used any more.

Since footnotes are scattered throughout the text, a separate bibliography is usually also employed. The bibliographic entries are similar to first-reference footnotes but are arranged and punctuated somewhat differently. Here are some examples:

Ilardi, Frank A. *Computer Circuit Analysis: Theory and Applications.* Englewood Cliffs: Prentice-Hall, Inc., 1976.

Atkinson, Richard C. and Mary C. Harlow. "Activated Bio-Filter Secondary Systems." *Journal of the Sanitary Engineering Division, American Society of Civil Engineers*, Vol. 101, 1981, pp. 44–56.

Johnson, Harold J. "Short Term Memory and SAT Tests." *Psychology Today*, December 1980, pp. 47–52.

Notice first of all that the author's names are inverted—last name first—and that specific page number references are omitted. (The page numbers given in the last two examples are the pages of the entire article.) Also, the three main groups—authorship, title, and imprint—are separated by periods rather than commas, and the city of publication and publisher are not enclosed by parentheses. Finally, second and subsequent lines of each entry are indented, whereas the first line of a footnote is indented and second and subsequent lines are flush left.

Notice also that titles of books, journals, and magazines are underlined (italicized in print), titles of articles are enclosed in quotes, and each of the major words in article titles is capitalized. Another peculiarity of the footnote format is that authors' first names and words such as *journal*, *association*, and *engineering* are spelled out rather than abbreviated.

An increasingly common practice—especially with publications in scientific and technical fields—is to gather footnotes at the end of an article or chapter rather than place them at the bottom of individual pages throughout the text. This makes for easier (and hence less expensive) typing and typesetting. Such a list of notes (now called "endnotes" rather than "footnotes") looks much like a bibliography and is a source of confusion for many. There are three giveaways, however: notes are numbered, while bibliographic entries are not; notes are listed in the order they appear in the text, while bibliographic entries are alphabetical; and notes are punctuated as single sentences, while bibliographic entries follow the arrangement and punctuation conventions outlined above. Figure 9–1 illustrates a set of endnotes for a technical article.

The Author-Date Format

The author-date format is characterized by the use of the author's last name and the date of publication in parentheses as in the in-text reference:

> Recent studies (Johnson, 1981) confirm the use of this technique.

or

> Recent studies confirm the use of this technique (Johnson, 1981).

This reference format is used principally in the social sciences and in business and economics publications, but it is gaining popularity in the biological sciences and some of the newer disciplines such as environmental studies, urban planning, and family relations. It is also used widely in education journals. It has the advantage of providing the reader useful information directly in the text. For people familiar with the literature on a given subject, often the author's name and date of publication are sufficient for them to identify the work cited completely. The author-date format also eliminates the difficulties of typing footnotes and avoids the confusion between endnotes and bibliographies. Of the three basic reference formats, it is the one that is gaining in popularity the

[1]E. A. Fagen, J. Appl. Phys. **50,** 6505 (1979).
[2]A. Catalano and M. Bhushan, Appl. Phys. Lett. **37,** 567 (1980).
[3]M. Bhushan and A. Catalano, Appl. Phys. Lett. **38,** 39 (1981).
[4]P. S. Nayar and A. Catalano, Appl. Phys. Lett. **39,** 105 (1981).
[5]A. Catalano and R. B. Hall, J. Phys. Chem. Solids **41,** 635 (1980).
[6]J. P. Donnelly and A. G. Milnes, Proc. IEEE **113,** 1468 (1966).
[7]R. L. Anderson, Solid State Electron. **5,** 341 (1962).
[8]G. A. Zubina, V. P. Toropeev, Yu. V. Toropeev, and O. S. Schukin, Neorg. Mater. **13,** 355 (1977) in Russian.
[9]B. Ray, *II-VI Compounds,* edited by B. R. Pamplin (Pergamon, New York, 1969), Vol. 2, p. 85.
[10]W. G. Oldham and A. G. Milnes, Solid State Electron. **7,** 153 (1964).
[11]A. R. Riben, J. P. Donnelly, and D. L. Feucht, IEEE Trans. Electron Devices **ED-12,** 511 (1965).
[12]J. P. Donnelly and A. G. Milnes, IEEE Trans Electron. Devices **ED-14,** 63 (1967).
[13]R. L. Anderson, Appl. Phys. Lett. **27,** 691 (1975).
[14]L. E. Terman, Solid State Electron, **5,** 285 (1962).

Figure 9–1 Endnotes for a Technical Article

fastest. The most comprehensive style guide for the author-date format is the *Publication Manual* of the American Psychological Association.

The parenthetical author-date in the text refers the reader to the corresponding bibliographic reference in the bibliography. Bibliographic references in the author-date format are similar to those of the footnote format, but there are some significant differences. Here are some examples:

Ilardi, F. A. *Computer Circuit Analysis.* Englewood Cliffs: Prentice-Hall, 1976.

or

Ilardi, F. A. (1976). *Computer Circuit Analysis.* Englewood Cliffs: Prentice-Hall.
Atkinson, R. C. & Harlow, M. C. Activated bio-filter secondary systems. *Journal of the Sanitary Engineering Division, American Society of Civil Engineers,* 1981, 101, 44–56.
Johnson, H. J. Short term memory and SAT tests. *Psychology Today,* December 1980, 47–52.

As with the footnote format, authors' names are inverted, entries are alphabetical, and the three main groups are separated by periods. The date of publication is sometimes moved from the end of the imprint group to a position immediately following the author's name. The author-date format employs several conventions that shorten entries and simplify the typing or typesetting: authors' first names are often omitted in favor of initials, subtitles are often omitted, article titles are not enclosed in quotes and have only the first word and proper nouns capitalized, and the abbreviations *Vol.* and *pp.* are omitted. Figure 9–2 (page 154) shows several in-text references in the author-date format together with the corresponding bibliography.

The Bracketed Number Format

The bracketed number format uses an in-text reference consisting of a one- or two-digit number enclosed in square brackets or (in a typewritten manuscript) slashes or parentheses:

> Recent studies [16] confirm the use of this technique.

or

> Recent studies confirm the use of this technique [16].

or

> Recent studies /16/ confirm the use of this technique.

or

> Recent studies confirm the use of this technique (16).

The bracketed number format is used mainly in engineering, computer science, mining and geology, the physical sciences, and other hardware-oriented technological disciplines. The most comprehensive style guide for the bracketed number format is the Engineers Joint Council's *Recommended Practice for Style of References in Engineering Publications.*

tabulate T-units (Larsen-Freeman and Strom 1977, Vann 1978). In other words, whereas an index like the mean number of morphemes per utterance (MLU) seems to be useful only *below* a certain level (approximately five morphemes per utterance) of language development (Schumann 1976), T-unit analysis seems to be useful only *beyond* a certain level of development in the target language.

The second limitation has to do with the discriminating power of the T-unit. Length of error-free T-unit (and related measures) will distinguish learners with low proficiency from those with a high degree of proficiency, but it is not as sensitive an indicator of second language development as might be desired. In a number of studies (Monroe 1975, Cooper 1976; Larsen-Freeman and Strom 1977, Larsen-Freeman 1978), considerable overlapping between adjacent groups was noted. This is not the case with data in first language development. The difference may be due to the fact that while the ability to subordinate and embed sentences develops gradually over a number of years in first language acquisition, the process is by comparison far more compressed in adult second language acquisition (Cooper 1976).

REFERENCES

Chaudron, Craig. 1978. English as the medium of instruction in ESL classes: An initial report of a pilot study of the complexity of teachers' speech. Unpublished paper, Modern Language Centre, Department of Curriculum, Ontario Institute for Studies in Education.

Cooper, Thomas C. 1976. Measuring written syntactic patterns of second language learners of German. *Journal of Educational Research* 69:176-83.

Endicott, A. L. 1973. A proposed scale for syntactic density. *Research in the Teaching of English* 7:5-12.

Gaies, Stephen J. 1976. Sentence-combining: A technique for assessing proficiency in a second language. Paper read at the Conference on Perspectives on Language, University of Louisville, Louisville, Kentucky, May 6-8. (ERIC ED 130 512)

———. 1977a. A comparison of the classroom language of ESL teachers and their speech among peers: An exploratory syntactic analysis. Unpublished Ph.D. thesis, Indiana University.

———. 1977b. The nature of linguistic input in formal second language learning: Linguistic and communicative strategies in ESL teachers' classroom language. In *On TESOL '77: Teaching and learning English as a second language.* (Brown, H. D., R. H. Crymes, and C. A. Yorio, Eds.) Washington, D.C.: Teachers of English to Speakers of Other Languages, 204-12.

———. 1979. Linguistic input in formal second language learning: The issues of syntactic gradation and readability in ESL materials. *TESOL Quarterly 13,* 1: 41-50.

Golub, Lester and Carole Kidder. 1974. Syntactic density and the computer. *Elementary English 51*:1128-31.

Hunt, Kellogg W. 1965. *Grammatical structures written at three grade levels.* Research Report No. 3. Urbana, Illinois: National Council of Teachers of English.

———. 1970. Syntactic maturity in school children and adults. *Monographs of the Society for Research in Child Development 53,* 1 (Serial No. 134).

Larsen-Freeman, Diane, and Virginia Strom. 1977. The construction of a second language acquisition index of development. *Language Learning* 27:123-34.

Larsen-Freeman, Diane. 1978. An ESL index of development. Paper read at the Twelfth Annual TESOL Convention, Mexico (D.F.), April 4-9.

Loban, Walter. 1976. *Language development: Kindergarten through grade twelve.* Research Report No. 18. Urbana, Illinois: National Council of Teachers of English.

Moffett, James. 1968. *Teaching the universe of discourse.* Boston: Houghton Mifflin Company.

Monroe, James H. 1975. Measuring and enhancing syntactic fluency in French. *The French Review 48*:1023-31.

Figure 9–2 In-text References in the Author-Date Format and Corresponding Bibliography

> selective ion implantation to localize active devices in the semi-insulating GaAs substrate, and gate densities over 100 000/cm^2 have been reported on very dense circuit structures [5].
>
> The development of planar fabrication methods and high speed logic gate structures with modest power dissipation (<1 mW/gate) and high density has created interest in the extension of GaAs digital IC's into the large or very large scale of integration range. The achievement of LSI or VLSI is important if very high speed components for signal processors, computers, and memory are to be feasible in GaAs IC's. The high complexity of such systems makes it very desirable to include large numbers of components on a single chip to reduce the large quantity of high speed interconnections. Trends in this direction are also quite evident in silicon IC implemented signal processing and computational hardware [6], [7].
>
> Very recent demonstration of high speed LSI parallel multiplier chips in GaAs MESFET logic [8] has proven that such chip complexities (over 1000 gates) are feasible in a well-controlled IC process. While further development and optimization

Figure 9–3 In-text References in the Bracketed Number Format

The bracket number refers the reader to a numbered reference in the bibliography. The bracketed number format moves even farther than the author-date format in the direction of abbreviation and simplification. Here are some examples:

14. Ilardi, F. A. *Computer Circuit Analysis: Theory and Applications.* Englewood Cliffs: Prentice-Hall Inc., 1976.
6. Atkinson, R. C. & Harlow, M. C. Activated bio-filter secondary systems. *J. Sanitary Engr. Div., ASCE,* 1981, 101:44–56.
10. Johnson, H. J. Short term memory and SAT tests. *Psych. Today,* Dec. 1980, 47–52.

Entries are arranged alphabetically and the entire bibliography is numbered; in-text reference numbers correspond to bibliographic reference numbers; and words such as *journal, association, division,* and *engineering* are abbreviated. Figure 9–3 (above) shows representative in-text references in the bracket number format, while Figure 9–4 (page 156) shows the corresponding bibliography.

□ Which Reference Format Should You Use?

The question of which reference format to use is both simple and complex. It is simple in the sense that if you have format requirements imposed on you, you follow them. Period. No choices and no decisions. Many firms and organizations have their own style guides. Others use a standard style guide such as the Chicago *Manual of Style* or the *CBE Style*

REFERENCES

[1] T. Mizutani, N. Kato, S. Ishida, K. Osafune, and M. Ohmori, "GaAs gigabit logic circuits using normally-OFF MESFET's," *Electron. Lett.*, vol. 16, pp. 315–316, Apr. 24, 1980.

[2] M. Cathelin, M. Gavant, and M. Rocchi, "A 3.5 GHz single-clocked binary frequency divider on GaAs," *Proc. Inst. Elec. Eng.*, vol. 127, pt. 1, no. 5, pp. 270–277, Oct. 1980.

[3] M. Cathelin and G. Durand, "Logic IC's using GaAs FET's in a planar technology," *L'onde Electrique*, vol. 58, pp. 218–221, Mar. 1978.

[4] G. L. Troeger, A. F. Behle, P. E. Friebertshauser, K. L. Hu, and S. H. Watanabe, "Fully ion implanted planar GaAs *E*-JFET process," *1979 Int. Electron Devices Meeting, Tech. Dig.*, pp. 497–500, Dec. 1979.

[5] B. M. Welch *et al.*, "LSI processing technology for planar GaAs integrated circuits," *IEEE Trans. Electron Dev.*, vol. ED-27, pp. 1116–1123, June 1980.

[6] H. Horikoshi *et al.*, "An example of LSI-oriented logic implementation in a large-scale computer, the HITAC M-200H," *COMPCON Tech. Papers*, pp. 62–65, Spring 1980.

[7] R. J. Blumberg and S. Brenner, "A 1500 gate, random logic, large scale integrated (LSI) masterslice," *IEEE J. Solid-State Circuits*, vol. SC-14, pp. 818–822, Oct. 1979.

[8] F. S. Lee *et al.*, "High speed LSI GaAs digital integrated circuits," GaAs IC Symp. Abstracts, Las Vegas, NV, Nov. 1980.

[9] R. C. Eden *et al.*, "The prospects for ultra-high speed VLSI GaAs digital logic," *IEEE J. Solid-State Circuit*, vol. SC-14, pp. 221–239, Apr. 1979.

Figure 9–4 Bibliography Corresponding to the References in Figure 9-3

Manual or the USGS *Suggestions to Authors*. (See the lists on pages 148, 149, and 150). Of course, if you are writing an article or paper for a scholarly journal or trade magazine, you are obliged to follow its requirements. On the other hand, the question can be complex. Frequently, there are no format requirements and you, as writer, must decide for yourself which format to follow. Further, there is the question of which format to use in your initial draft. Let's look at the strengths and weaknesses of each basic format.

First, the matter of in-text references. From the standpoint of reader convenience, the author-date format is best because it gives many readers the information they need without having to check a note or a list of references. The footnote format runs a close second, especially if the notes are at the bottom of the page rather than gathered at the end. The bracketed number format is least useful; every reader has to refer to the bibliography for every reference. From the standpoint of writer convenience, the author-date format again is best. You can put references in, take them out, and rearrange them pretty much at will. Further, you don't have to stop to prepare a complete bibliographic reference in the midst of writing your initial draft. With the footnote format, adding or deleting a reference means renumbering all subsequent notes. Moreover, you must keep good notes for your footnotes if you want to avoid quoting something and attributing it to the wrong source—or forgetting completely which source you wanted your readers to refer to. The bracket number format is even worse; you have to leave all your references blank until your bibliography is complete and then come back and fill in the numbers. Again, if you add or delete a reference after you have filled in the numbers, you have to go back and change all subsequent references. From the standpoint of typesetter and printer convenience, the bracketed number format is easiest and cheapest, with the footnote format with endnotes (if that makes sense) a close second. The author-date in-text references are not terribly inconvenient, but they take up more space than either of the other two. Full footnotes at the bottoms of pages are—as you well know—the most time-consuming and troublesome to generate.

Second is the matter of bibliographic references. The traditional footnote format is the best one for readers because it is the most complete. Abbreviations and initials are kept to a minimum, and with a full complement of capital letters, quotes, and underlining, there is no confusion as to what part of the reference is article title, what part is journal title, and so forth. The abbreviated bibliographic entries common with the author-date and bracketed number formats are less convenient for readers because they are less complete—the information is there, but it is quite dense. There is little difference among the three as far as writer convenience goes. It is just a matter of getting used to the conventions of one format or the other. As for printer convenience, there is no doubt that the bracketed number format is the most economical—in terms both of space and time required for typesetting.

My recommendation for your initial drafts and for those situations where you have the choice of format for final draft is a combination of author-date in-text references and bibliographic references in the full footnote format. This gives you a good combination of reader and writer convenience and complete bibliographic references which can easily be converted to other formats (see below). Of course, if all your writing is done within the guideline of a single basic format, you might as well practice all the time what you have to do most. But if you have some latitude, try author-date in-text references and complete bibliographic references.

□ Sample Bibliographic References

This section presents samples of bibliographic references for some common and some not-so-common sources. I have included not only the usual print media sources but also some nonprint media, such as motion pictures, slides and transparencies, video recordings, microfilm, and computer data bases. The sample references are in the comprehensive format recommended by ANSI in the *American National Standard for Bibliographic References.**

1. Journal Article
   Dunlap, Connie R. Cataloging in publication. Library Journal. 99(18): 2573–2578; 1974 October 15.
2. Journal Article with Subordinate Title
   Saarikoski, Pirkko-Liisa; Suomalainen, Paavo. Studies on the physiology of the hibernating hedgehog. 15: Effects of seasonal and temperature changes on the in vitro glycerol release from brown adipose tissue. Ann. Acad. Sci. Fenn., Ser. A4.1972; 187: 1–4. CODEN: AAFBAU; ISSN: 0066–1996.
3. A Single Page of an Article
   O'Connell, Daniel P.; Shore, Harvey H. The Hungarian method in personnel work. Personnel. 51(4): 60–68; 1974 July–August (p. 64, Figure 3).
4. Book with Single Author
   Heilbrun, James. Urban economics and public policy. New York: St. Martin's Press; 1974. 380 p.
5. Book with Two Authors
   Gruber, Edward C.; Willdorf, Barry S. Law school admission test (LSAT). New York: Simon & Schuster; 1974. 432 p.
6. Book with More Than Two Authors
   Janick, Jules [and others]. Plant science: an introduction to world crops. 2d ed. San Francisco: W. H. Freeman; [1974]. 740 p. ISBN: 0–7167–0713–6.
7. Book with an Editor
   Evans, R. R., ed. Social movements: a reader and source book. Chicago: Rand McNally College Pub. Co.; [1973]. 605 p. ISBN: 0–528-68038–2.
8. Book with a Corporate Author
   International African Institute Library. Cumulative bibliography of African studies: author catalog. Boston: G. K. Hall; 1973. 2 v. ISBN: 0–8161-1045–X.
9. Chapter in a Collection
   Barny, R. Sur la situation idéologique du Neveu de Rameau. Hommage à Georges Fourrier. Paris: Belles lettres; 1973: 37–70. 379 p. (Centre de recherche d'histoire et littérature au XVIIIe et XIXe siècles; [Publication] v. 4). (Annales littéraires de l'université de Besançon; 142).

*This material (except entries 27, 28, 31, 32, and 45) is reproduced with permission from American National Standard Z39.29-1977 for Bibliographic References, copyright 1977 by the American National Standards Institute. Copies of this standard may be purchased from the American National Standards Institute at 1430 Broadway, New York, NY 10018.

10. Reprint
    Smith, Joseph. Bibliotheca anti-Quakeriana: a catalog of books adverse to the Society of Friends, alphabetically arranged with biographical notices of the authors together with the answers which have been given to some of them by Friends and others. London: Joseph Smith; 1873. New York: Kraus Reprint; 1968. 32 p.

11. Paper in a Proceeding
    Miro, P.; Sanchez, J. The influence of surface-active agents on the reactivity of wool in alkaline solutions. Jakab, E. ed. Proceedings of the 18th Hungarian textile conference: textile chemistry, clothing industry, material testing; 1970 September 14–20; Budapest. Budapest: Hungarian Technical and Scientific Society of the Textile Industry; 1971; 2: 105–110.

12. Page in a Proceeding
    Gaylord, Norman G. Effects of polymers on grafting. Burke, John J.; Weiss, Volker, eds. Block and graft copolymers. Proceedings of the 19th Sagamore Army materials research conference; 1972 September 5–8; Racquette Lake, NY. Syracuse, NY: Syracuse University Press; c1973; 19–35 (p. 20).

13. Conference Proceedings—Unpaged Paper
    Merker, H. J.; Körtge, P. Elektronenmikroskopische Untersuchungen an menschlichen Lebern bei Virushepatitis. Beck, K., ed. Ikterus: International symposium; 1967 October 27–29; Freiburg i. Br. Stuttgart: Schattauer; c1968: [Section] 12.

14. Paper Presented at a Conference or Symposium
    Lüllmann, H. Metabolism of glycosides. Szekeres, L.; Papp, J. Gy, eds. Symposium on drugs and heart metabolism; 1971; Budapest, Hungary. Budapest: Akademiai Kiado; c1973: 107–110. (First congress of the Hungarian Pharmacological Society; v. 2).

15. Conference Proceedings in a Journal
    Singhal, G. N. Temperature measurement in industries. Proceedings of the seminar on instruments and control in process industries; Bombay, India. In: Indian Chemical Journal. 8(10): Supplement 17–35; 1974 April.

16. Routine Report
    Roberts, D. A. Review of recent developments in the technology of nickel-base and cobalt-base alloys. Columbus, OH: Battelle Memorial Institute, Defense Metals Information Center; 1961 August 4; DMIC memorandum 122. 2 p. Available from: NTIS, Springfield, VA; AD 261292.

17. Routine Report Treated as a Serial
    Langdon, Lawrence E.; Gabriel, Richard R.; Marsh, Alan H. Investigation of DC-8 nacelle modification to reduce fan-compressor noise in airport communities, pt. VI, psychoacoustic evaluation. NASA Contractor Report. 1970; NASA CR-1710. CODEN: NSCRAQ; Avail-

able from: National Aeronautics and Space Administration, Washington, DC.

18. Report with STRN (Standard Technical Report Number)
    Davis, Richard A.; Tepolilli, Ralph N.; Back, Kenneth C. Toxic point determination of selected hazardous materials. Washington, DC: U.S. Department of Transportation, Office of Hazardous Materials Operations; 1975 Dec.; DOT/MTB/OHMO-7514. 12 p. Available from: NTIS; PB 250450.

19. Newspaper Article, Unsigned
    Nassau Trust to open branches. Glen Cover (NY) Record-Pilot. 1974 July 18: 3 (col. 2).

20. Signed Newspaper Article in a Special Section
    Russell, John. Persistent pop: the paintings are outliving their subjects. The New York Times. 1974 July 21; Sect. 6: 6–7, 25, 27, 30, 32–34.

21. Newspaper Article with a Dateline
    Weinraub, Bernard. Shah offers oil but bans U.S. 'dictation' of terms. The New York Times. 1974 Oct. 4: 3 (col. 1). New Delhi, Oct. 3.

22. Separate Catalog or Brochure
    Gestetner Corporation. From original to printed copies . . . in minutes. Yonkers, NY; [n.d.]. [2 p.].

23. Numbered (Serial) Catalog
    Trigger Electronics. Trigger value catalog for the man in electronics. River Forest, IL; 1973; Catalog 140. 76 p.

24. Unpublished Paper
    Feast, M. W.; Garton, W. R. S. The Schumann-Runge bands of O_2 in emission and absorption in the quartz ultraviolet region. 1950. Unpublished draft supplied to author by W. R. S. Garton.

25. Manuscript
    MS D. c. 2–392; Novum organum botanicum. 1 leaf. Located at: University Library, Edinburgh, Scotland.

26. Dissertations and Theses
    Cairns, R. B. Infrared spectroscopic studies of solid oxygen. Berkeley. CA: Univ. of California; 1965. 156 p. Dissertation.

    Allen, David Yahling. From romanticism to realpolitik: studies in nineteenth-century German conservatism. New York: Columbia University; 1972. Available from: University Microfilms, Ann Arbor, MI; Publication number 72–15,559. 176 p. Dissertation.

27. Translations
    Coulson, Jessie, trans. Crime and punishment. By Fedor Dostoevsky. George Gibian, ed. New York: Norton; 1964.

    or:

    Dostoevsky, Fedor. Crime and punishment. Trans. Jessie Coulson. George Gibian, ed. New York: Norton; 1964

Sastre, Alfonso. Sad are the eyes of William Tell. Trans. Leonard C. Pronko. In The new wave Spanish drama. George E. Wellwarth, ed. New York: New York Univ. Press; 1970, pp. 265–321.

28. Dictionary

Webster's seventh new collegiate dictionary. Springfield, Mass.: G. & C. Merriam Company; 1970.

29. Letters

DeRussy, R. E., Lieut. Col. [Letter to Brevet Gen. J. G. Totten]. 1853 May 15. 1 leaf. Located at: National Archives, Washington, DC; Record Group 77.

30. Maps

World [Political]. Washington, DC: National Geographic Society; 1957 March. 1: 39, 283, 200; Van der Grinten projection; 41 x 29 in.; colored. Prepared for the National Geographic Magazine.

Antarctica [Topographic]. Robert McDonald, cartog. Revised [ed]. Washington, DC: U.S. Geological Survey; 1972. 1: 250,000; 80 x 44.5 cm; colored. English. (Reconnaissance Series; [sheet] ST 57–60/6.)

31. Personal or Telephone Interview

Poussaint, Alvin F. Telephone interview. 1975 May 7.

32. Radio or Television Programs

The first Americans. Narr. Hugh Downs. Writ. and prod. Craig Fisher. NBC News Special. 1968 March 21.

The joy ride. Writ. Alfred Shaughnessy. Upstairs, downstairs. Created by Eileen Atkins and Jean Marsh. Dir. Bill Bain. Prod. John Hawkesworth. Masterpiece Theatre. Introd. Alistair Cooke. PBS; 1977 Feb. 6.

33. Motion Pictures (Special Packaging Method)

Convex mirrors [Motion picture]. Educational Development Corp. New York: Universal Education and Visual Arts; 1969. 2 min.; 1 loop film cartridge; si.; b&w; super 8mm. Accompanied by: teacher's notes on cartridge.

34. Motion Picture (with Accompanying Material)

Ripple [Motion picture]. Jim Lawless Motion Picture Consultants. Seattle: Lawless; 1971. 29 min.; magnetic sd.; color; 16 mm. Accompanied by: leader's guide, 12 p.

35. Slide

Rouse, Irving. Areas and periods of culture in the Greater Antilles [Slide]. Madison: Bureau of Audio-Visual Instruction, University of Wisconsin Extension; 1951. 35 slides, color, 2 x 2 in. Accompanied by: notes, 11 p.

36. Slide with Sound Recording

Center for Humanities. Man and his values: an inquiry into good and evil [Slide]. Middle school edition. White Plains, NY: Center for Humanities; c1972. 160 slides; color; 2 x 2 in. Accompanied by: sound recording; 2 discs; 12 in.; 33-1/3 rpm.

37. Transparency—Reference to One in a Set
    Rickard, Judy. Information and communication theory [Transparency]. San Jose, CA: Lansford Pub. Co.; 1972: transparency 6. 10 transparencies; color; 10 x 12 in. Accompanied by: overlays; teacher's guide, 12 p.
38. Filmstrips with Accompany Material
    The era of black slavery [Filmstrip]. New York: Educational Resources; 1971. 72 fr.; color; 35mm (400 years: black history in America; 1). Accompanied by: sound recording; disc; 12 in.; 33-1/3 rpm; teacher's guide, 35 p.
39. Videorecordings
    Evolving toward women [Videorecording]. Chicago: University of Illinois at Chicago Circle; 1972. 1 tape cassette; 55 min.; sd.; b&w; 1/2 in.

    Mammals of Australia [Videorecording]. Hamburg, Ger.: Teldec; 1972. 1 disc; 5 min.; sd.; b&w; 1500 rpm; 9 in.
40. Sound Recording
    Beethoven [Ludwig van]. Concerto no. 1 in C major for piano and orchestra: op. 15 [Sound recording]. Rudolf Serkin, pianist; Philadelphia Orchestra; Eugene Ormandy, conductor. [New York]: Columbia; [1954]; ML 4914; (x"LP"30829-2B; x"LP"21341-2C). 1 disc; 12 in./30cm; 33-1/3 rpm.
41. Book as a Sound Recording
    Mitchell, Margaret. Gone with the wind [Sound recording]. Claire, Helen, reader. Washington, DC: Library of Congress; [1956]; AFB (American Foundation for the Blind) 55-995-55-1066. 72 discs; 12 in.; 33-1/3 rpm. Edition used for recording: New York: Macmillan; 1936. Talking book, solely for the use of the blind.
42. Computer Program and Documentation
    Theoretical prediction of interference loading on aircraft stores: Part I—subsonic speeds and Part III—programmer's manual [Computer program]. Fernandes, F. Dan. Pomona, CA: General Dynamics, Electro Dynamics Div.; 1972 June. 1650 card images; Fortran IV, for CDC-6000. National Aeronautics and Space Administration. NASA CR-112065-1. Available from: University of Georgia, COSMIC, 112 Barrow Hall, Athens, GA, 30602; LAR-11249.
43. Magnetic Tape, Reference to an Entire Tape Service
    SPIN, Searchable Physics Information Notices [Data base]. New York: American Institute of Physics. Magnetic-tape reels; 7- or 9-track, 800 bpi, EBCDIC, odd parity.
44. Computer Data Base, Reference to a Specific Portion
    Laporte has stated that its naphthalene-based . . . year, will not be considered . . . until 1973. Eur. Chem. News. 1971 December 10; 20: 4. From: Chemical Industry Notes [Data base]. Columbus, OH: Chemical Abstracts Service; [1971]; CIN Citation CINTAW 00105 A00452. Magnetic-tape reels; 9 track; 800 bpi, ASCII–8 and binary, NRZ1.

45. Reference to a Single Item on a Computer-Readable Magnetic Tape
Bacteriophage T_4D-induced proteinase. Poglazov, B. T.; Levshenko, M. T. J. Mol. Biol. 1974; 83(3): 463–467. In: Chemical Abstracts Condensates [Data base]. Columbus, OH: Chemical Abstracts Service. 1974 September 30 (81)13: abstract no 74715f. Magnetic tape; 9 track, 800 and 1600 bpi, CAS standard distribution format, ASCII.
46. Book in Microform
Shirley, James. The gentlemen of Venice: a tragi-comedie presented at the private house in Salisbury Court by Her Majesties servants [Microform]. London: H. Moseley, 1655. 78 p. New York: Readex Microprint; 1953. 1 microprint card; 23 x 15 cm. (Three centuries of drama: English, 1642–1700).
Battles, Ford Lewis; Miller, Charles. A computerized concordance to Institutio Christianae religionis 1559 of Joannes Calvinus: Based on the critical text of Petrus Barth and Guilemus Niesel, corrected from the original text of 1559 [Microform]. Pittsburgh, PA: Clifford E. Barbour Library, Pittsburgh Theological Seminary; 1972. 3 microfilm reels; 16mm.
47. Patent Documents
Harred, John F.; Knight, Allan R.; McIntyre, John S., inventors; Dow Chemical Co., assignee. Epoxidation process. U.S. patent 3,654,317. 1972 April 4. 2 p. Int. Cl.[2] C 07D 1/08,1/12.
48. A Section of an Article in the U.S. Constitution
U.S. Const. art. I, §9.
49. A Section of a Federal Statute (Session Law)
Act of Sept. 8, 1950. Stat. 64-798. (Ch. 924, §2).
50. A Section of a Federal Statue (Code)
U.S.C. 1964. 28: §105 (a)(3)—(b)(1).
51. A State Statute in a Code
Blue Sky Law. N.Y. Gen. Bus. Law. §352. (McKinney 1962).
52. A Portion of a State Statute (Code)
D.C. Code Ann. 1951. §§1-243.

□ Techniques for Documenting a Report

This section presents some tips and suggestions for making your documentation easier and more effective. I have flagged some of the common problems novice writers have with documentation and offered suggestions for overcoming these difficulties. The earlier subsections here deal with documentation in general, while the later subsections deal with aspects of individual basic reference formats.

What Needs to Be Cited?

- □ First, give complete bibliographic citations for all outside references you want your readers to consult or feel might be useful for them to consult.

- □ Cite sources for any direct quotations you include in your report.
- □ Cite sources for any data you are paraphrasing or summarizing—especially if you feel any of your readers might want to consult the original to check the accuracy of your paraphrase or summary.
- □ Do not cite sources for statements of fact that are common knowledge—statements that can be found in many different sources.
- □ Give source citations for all statements of opinion or interpretation unless such opinions or interpretations are uniquely your own.
- □ Use blanket acknowledgements or citations when it would be cumbersome to document individual statements. Just include a statement like "Much of the material in this section is based on discussions by Smith (1978) and Johnson (1981)."
- □ If you are in doubt, go ahead and include a citation. It is easier for a reviewer or editor to delete excessive documentation than it is to supply missing documentation. Err in the direction of too much rather than too little.

Preparing a Working Bibliography

As you gather data necessary for your report (see "Data-Gathering" in Chapter 5), keep a record of complete bibliographic information for each source you consult—including personal interviews. It is much easier to capture this information while you are gathering your data than it is to attempt to reconstruct it after you have written your preliminary draft. For each entry in your working bibliography include both a complete bibliographic reference in the full footnote format (complete punctuation, all words spelled out, and so on) and an in-text reference in the author-date format. Figure 9–5 shows a sample entry for your working bibliography.

If you have access to a computer terminal and a permanent data file, keyboard your entries as you acquire them. That way you can add to and delete from your list periodically and rearrange your list alphabetically

Horan, John D., Charles B. Rubenstein, and Richard W. Werner. "Density of Atomic Hydrogen in the Trans-Pluto Sector," Journal of Intersellar Astronomy, Vol. 3, pp. 426-455, Third Quarter, 1984.

(Horan, et. al., 1984)

Figure 9–5 Sample Entry for a Working Bibliography

with a simple sorting routine any time you want. Also, if you have your working bibliography in machine-readable form you will be able to use a text-editing program to generate your final bibliography in whatever basic reference format you choose. If you don't have access to a computer, put each entry for your working bibliography on a separate 3-by-5-inch index card so you can hand-sort your list from time to time.

When your working bibliography is complete, generate a printout in alphabetical order by author (or arrange your cards alphabetically). If you plan to use the bracketed number format, number each entry sequentially once you have the entries in alphabetical order.

Handling In-Text References

- □ As you write your preliminary draft, include an author-date in-text reference for each source you wish to cite. This is the most convenient manner for keeping your in-text references and bibliographic references straight.
- □ If you include the author's name as part of your sentence, put only the date of publication in parentheses: "As Smith (1982) reports, . . ." "On the other hand, Johnson (1978) argues that . . ."
- □ If you have two or more publications by the same author published in the same year, use lowercase letters to distinguish them: (Brown, 1980a) (Brown, 1980b). Make sure, of course, that the letters in your in-text references correspond to the order of entries in your bibliography.
- □ If you wish to specify a particular page number or page numbers, do so as follows: (Jones, 1975, p. 57) (Smith, 1979, pp. 42–48) or (Jones, 1975:57) (Smith, 1979:42–48).
- □ Each time you cite a reference, put an × (or a dot or a red mark) alongside the appropriate entry in your working bibliography. When your preliminary draft is complete, purge all entries with no marks from your working bibliography, renumbering if necessary.
- □ Each time you cite a reference in your preliminary draft, put an × (or a dot or a red mark) in the margin alongside the line with the citation in it. That way, you will be able to find all of your in-text references on a subsequent pass through your draft.
- □ To convert your in-text references to the footnote format, simply go through your draft and substitute a superscripted number for each author-date entry. (For direct quotations, put the superscripted number at the end of the quotation rather than immediately after the author's name. Also, keep in mind that footnote references appear at the end of a sentence more often than do either author-date or bracketed number references.) Be sure to write the appropriate note (see "The Footnote Format," above) at the time you insert the superscripted number.
- □ To convert from author-date to bracketed number format, make

sure your working bibliography is complete and numbered sequentially and then just substitute the appropriate entry number for each author-date reference in your preliminary draft. For example, suppose (Johnson, 1981) is entry number 12 in your working bibliography. Change the sentence "According to Johnson (1981), . . ." to "According to Johnson [12], . . ." Or if you have an in-text reference like "Another report (Johnson, 1981) states that . . ." change it to read "Another report [12] states that . . ."

- Regardless of the format you use, make a final revision pass through your entire preliminary draft to make sure that all of your in-text references correspond to appropriate and correct bibliographic references. Double-check all publication dates, spellings, page numbers, and so on with your earliest (preferably your original) records.

Alphabetizing and Ordering Your Bibliography

Putting bibliographic entries in alphabetical order is generally not too much of a problem, but here are a few things to watch for:

- Use *Anonymous* or *Anon.* as an author element only if the source is signed "Anonymous." Otherwise, use the corporate author or the title for purposes of alphabetizing.
- Alphabetize corporate authors just like individual authors:

  Ford, Gerald R.
  Ford, Thomas C.
  Ford Motor Co.

- If you have a source without an individual or corporate author, alphabetize by title, ignoring *a, an,* and *the:*

  Carlson, John C.
  "City Council Rejects Airport Proposal,"
  Crandon, Alice H.
  "The Creative Writer in America,"

- Remember the rule that "nothing precedes something." For example:

  Meyer, Robert C.
  Meyer, Thomas J.
  Meyers, George R.
  Meyerson, Anita H.

- For authors with identical last names, alphabetize by first name—or if necessary by middle initial.
- For multiple works by the same author, arrange the entries alphabetically by title *unless* you are using the author-date reference format. In that case, arrange entries chronologically by publication date. (In the case of two or more publications by the same author in the same year, arrange them chronologically by month, if possible, or arrange them alphabetically by title.)

- For multiple works by the same author, repeat the author's name for both the author-date and the bracketed number formats. For the footnote format, use a series of ten hyphens in place of the author's name for second and subsequent entries.
- If you have a mixed bag of single-author entries and multiple-author entries with the same first author, list all single-author entries first and then list multiple-author entries alphabetically by the second author:

  Johnson, Richard C.
  Johnson, Richard C. and Mary K. Foeling.
  Johnson, Richard C.; John C. Thomas; and Mary K. Foeling.

- The alphabetizing of foreign names involves a number of problems we need not go into here. If you have a number of foreign authors in your bibliography, consult an editor or refer to the bibliographical section of *Webster's Eighth New Collegiate Dictionary*.

Full Format versus Reduced-Format Bibliographic Entries

Here are several bibliographic entries in their full footnote format and their corresponding reduced format that is used with some versions of the author-date and bracketed number formats.

Peters, Harold C.; Jane W. Summers; and Richard K. Johnson. "Isolating Interferon: Some Analytic Techniques," *Journal of Organic Chemistry*, Vol. 32 (1981), pp. 156–167.
Peters, H. C.; Summers, J. S.; Johnson, R. K. Isolating interferon: some analytic techniques. *J. Organ. Chem.* 32:156–67 (1981).

Hanson, Robert J. "High-Altitude Airfoils for Re-entry Vehicles," *Journal of the Aeronautical Sciences*, Vol. 39, No. 10 (October 1978), pp. 605–613.
Hanson, R. J. High-altitude airfoils for re-entry vehicles. *J. Aero. Sci.* 39–10:605–13 (Oct. 1978).

Ilardi, Frank A. *Computer Circuit Analysis: Theory and Applications*. Englewood Cliffs: Prentice-Hall, Inc., 1976.
Ilardi, F. A. *Computer Circuit Analysis*. Englewood Cliffs: Prentice-Hall, 1976.

Notice that the full-format entries are more complete and are, in the opinion of many, easier to read. On the other hand, notice that most reduced-format entries are a line shorter than the corresponding full-format entries. To convert from a full-format entry to a reduced-format entry:

1. Eliminate authors' first names in favor of initials.
2. For multiple-author works, put *all* authors' names in inverted order—last name first.
3. Delete the quotation marks from around magazine or journal titles and make all words in the title except the first word and proper nouns lower case.

4. Abbreviate common words in journal titles according to the style guide you are following.
5. Eliminate the words *Vol.*, *No.*, and the like, in favor of the following code: volume-number:pages (date).

Keep in mind, however, that not all journals and style guides use the fully reduced format. The American Psychological Association journals, for instance, use initials for authors' names and eliminate the quotation marks and conventional capitalization from titles of articles, but they spell out all journal titles in full. The Engineers Joint Council recommends initials for authors' names but employs the full footnote format for everything else in the bibliographic entry—even though their publications use the bracketed number form of in-text reference.

Writing Exercises

1. Consult ten periodicals in your field or discipline to see which of the three basic reference formats are used. Does any one predominate?
2. Interview several experts in your field or discipline to see which style guides or standards they use for documentation. Prepare a list.
3. Prepare ten bibliographic entries in the footnote format. Convert each to author-date format and to bracketed number format.
4. Prepare a list of five documentation endnotes. Convert the list to an author-date bibliography.
5. Find actual references in your field or discipline to ten or more nonprint sources (maps, microfilm, microfiche, films, videotapes).

Writing about Writing

1. Write an essay explaining which basic reference format you feel is best and why.
2. Write an essay exploring the advantages and disadvantages of a single standardized documentation format for all fields of science and technology (or all fields of study, for that matter).

□□□ THREE □□□

Types of Written Communication

The eight chapters of Part Three form the core of *Technical Writing: Process and Product.* It is here that you get a chance to look in some detail at the organization, content, and approach strategies that are appropriate for the major categories of technical writing. Before you proceed with Part Three, however, you would profit from reviewing Chapter 1, "Writing in Business, Industry, and Government." (In fact, you would profit from reviewing all three chapters of Part One.) Then as you study each particular type of written communication, give some thought to these questions:

- □ How does each type of writing discussed fit into the overall scheme of writing in business, industry, and government?
- □ How do the demands of adequacy and excellence affect the amount of time and effort you should devote to each type of writing?
- □ In what ways do particular communication contexts modify the organization of, content of, and approach to each type of writing?

In general, the chapters of this part move from the more common and less demanding types of writing to the less frequent but more demanding varieties. What you will notice very quickly as well, Part Three also moves from the shorter forms of writing—letters and memos—to the longer forms of writing—articles, proposals, and major reports. What may not be immediately apparent, however, is that the forms of writing discussed earlier tend to be those where adequacy is sufficient, while those discussed later tend to be those where excellence is often demanded. There are exceptions, of course. Some of your correspondence may be exceedingly important and demand the utmost of your writing skills. Likewise, some reports and proposals are more or less routine and perfunctory. On the whole, though, the movement is from simple and routine to complex and demanding.

Chapter 10, "Technical Correspondence," and Chapter 11, "Employment Applications," cover letters and memos—the most common and frequent kinds of things you will be asked to write when you get on the job. Chapter 12, "Project Reports," covers the common internal documents that technicians and technical experts write—activity reports, progress and status reports, and common data reports. Chapter 13, "Abstracts, Summaries, and the Like," presents information on a variety of types of writing that involve summarizing and condensing material written by others. The next four chapters deal with more substantial forms of writing. Chapter 14, "Instructions and Manuals," gets into instruction writing, mechanism description, process explanation, and procedural presentations. Chapter 15, "Recommendation Reports," and Chapter 16, "Proposals," deal with what I feel is the most important category of writing for technical people: management decision-making documents. If you can do well with these, you can do well with any other form of technical writing. Finally, Chapter 17, "Technical Articles," puts it all together, with guidelines and suggestions for presenting any of a wide range of technical information to any of a wide range of audiences with broad and diverse reasons for reading what you write. Chapter 17 also serves as a transition into the advanced writing techniques presented in Part Four: Revising for Excellence.

10

Technical Correspondence

Just as we define technical writing as the job-related writing a technical person does, so we can define technical correspondence as the job-related correspondence a technical person might be asked to produce. We can identify three varieties of technical correspondence:

- □ Memos
- □ Letterhead letters
- □ Nonletterhead letters

Some people make a great deal of the distinction between memos and letters; in fact the difference is minimal. A *memo* (see Figure 10–1) is little more than an in-house letter, a document sent from one individual in a firm or organization to one or more other individuals in the same firm or organization. The format of a memo—particularly the opening and closing—is different from that of a letter, but the body organization, the tone, and the style of writing are essentially the same for both.

A *letterhead letter* (see Figure 10–2, page 173) is a document you write as a representative of the firm or organization identified in the letterhead. It is official correspondence: company business. It is sent from you as a representative of your firm or organization to other people, often as representatives of their firms or organizations. If you have occasion to write a business letter as a private citizen, representing no one but

ARCHIBALD ENTERPRISES, LTD., AKRON, OHIO 44307

Interoffice Memorandum

TO: Peter Darrington, Manager
Mid-Central Sales

FROM: Harold P. Miller, Vice President
Sales Division

SUBJECT: Second Quarter Sales Report DATE: July 22, 1983

I was quite pleased to note that sales in the Mid-Central section were up 22 percent from the First Quarter, 1983. No doubt, the Salem Quarry contract accounted for a good bit of the increase, but at the same time I see that all of your people have made more contacts and increased their sales volumes over the previous quarter. This comes from dynamic leadership, Peter! You're doing a fine job with the Mid-Central section, and you have my full support with your new programs. Congratulations! And keep up the good work!

And please relay my congratulations and encouragement to every one of your sales staff. They have done a fine job this past quarter, too.

Figure 10–1 Representative Memo

yourself, don't use company stationery but rather prepare a *nonletterhead letter* on plain white paper (see Figure 10–3, page 174).

The best that can come of using letterhead stationery for personal business is to be perceived as insensitive to the distinction between organizational and private business. At worst, you will be perceived as the sort of person who steals stationery and stamps—and probably pencils, notepads, and calendars. The most glaring offense in this connection is using letterhead from company A when writing an employment application letter to company B.

□ Memo Format

Most firms and organizations have pads of memo blanks available, so the format for memos you write offers no problem. Just fill in the blanks in the

SCIENCE ASSOCIATES PUBLISHERS

425 Madison Avenue
New York, N.Y. 10017

January 30, 1983

Professor Edwin V. Anderson
Deparment of English
Bradley University
Peoria, Illinois 61600

Dear Professor Anderson:

I refer to our very pleasant telephone conversation of January 26th, and I would like to thank you for agreeing to review and critique Harold Adams' new book *Essays in Word Processing and Text Editing*.

I am enclosing the manuscript, and we would very much appreciate having your review and comments by March 15th. This will enable us to come to a prompt decision about publishing and be ready for marketing by the beginning of 1984.

An honorarium of $500 (plus expenses) has been set aside for you.

You are at liberty to mark up the pages of the manuscript with suggestions, corrections, and annotations. Additionally, a letter giving an overview, your ideas about pricing and marketing, and any other suggestions you might have would be quite welcome.

Should you have any questions, please call me at Science Associates during business hours or at home (496) 768-3329.

Very truly yours,
SCIENCE ASSOCIATES PUBLISHERS

Steven R. Abels

Steven R. Abels
Editor

Encl.

Figure 10–2 Representative Letterhead Letter

4479 Citrus Drive
Crystal River, FL 32629
January 17, 1984

Heart of Dixie Arms
425 South Halsted Street
Albany, GA 31701

Gentlemen:

Last fall, I purchased one of your new Litearm .36 calibre muzzle loaders for turkey hunting. I had excellent service from the gun last fall, but now I am experiencing some difficulties with the set trigger mechanism. About half the time the rear trigger won't set the front one, and even when it does, the front trigger won't always release the hammer.

Have you had similar complaints from other Litearm .36 owners? Is there something I can do myself to correct the problem? If not, should I return the trigger and lock mechanism to your factory, or should I return the whole rifle?

I don't have a pressing need to get the problem taken care of immediately, but I would like to be able to use my rifle for some varmint shooting in March. Can you let me know how best to proceed?

Sincerely yours,

Elmer K. Roberts

Elmer K. Roberts

Figure 10–3 Representative Nonletterhead Letter

To:

From:

Subject: Date:

Figure 10–4 Memo Constructed on Plain Paper

heading. If you have to construct a memo on plain paper, set it up as shown in Figure 10–4. (The Latin *re* or *in re* is sometimes used in place of *subject*, but this is considered obsolete and somewhat pompous usage by most businesspeople today.) For an informal memo, it is quite proper to use first names or even nicknames in the TO: and FROM: blanks, as in Figure 10–5. For a somewhat more formal situation, use a last name with single initial (N. Abbott, D. Pearlman), and for a formal memo, use full name and title, as in Figure 10–6. For the SUBJECT: blank, fill in a short statement of the memo's purpose: "Request to take night courses," "Confirmation of your parts order," "Attached monthly report," and so on.

Practices concerning when and where to sign a memo vary considerably from organization to organization, so the best thing to do is follow the local ground rules. More often than not, a memo isn't signed; it is official even without a signature. In some groups, the writer of the memo will

To: Dan

From: Skip

Subject: Travel to IEEE Conference   Date: April 15, 1983

Figure 10–5 Heading for an Informal Memo

To: Daniel Pearlman, Manager

From: Wayne Hamilton, Design Supervisor

Subject: Travel to IEEE Conference   Date: April 15, 1983

Figure 10–6 Heading for a Formal Memo

initial the original right after his or her name in the FROM: blank (see Figure 10–7). In other cases, the writer of the memo will include both a typed and a handwritten signature at the close of a memo, similar to those in a regular business-letter (see Figure 10–8). A memo, however, should not include a complimentary close.

To: All Engineering Staff

From: Daniel Pearlman, Manager D.R.P.

Subject: Annual Christmas Party   Date: December 12, 1983

You are all invited to the annual Christmas party at the home of Mr. and Mrs. Weldon Harrington. Cocktails and other refreshments will be provided; dress is semi-formal; and the festivities will begin at 8:00 p.m. December 17th. Hope to see all of you there.

Figure 10–7 Memo with Initials in the Heading

University of Idaho
Inter-Office Memorandum

To Limnology Department Faculty and Staff

From Wm D. Meyers, Chairman

Subject Use of Computer Terminals   Date October 22, 1983

Computer Services has made effective a new rate schedule for the time-shared computer terminals and other peripheral equipment which has much lower charges for evening and weekend use. I would urge all of you who can to do as much computer work as possible during the low-charge periods. Of course, computing time for key research projects may continue to be done during the regular working day.

Wm. D. Meyers

Wm. D. Meyers, Chairman

Figure 10–8 Signed Memo

□ Letterhead Letter Format

In general, you don't have to worry about format for letterhead letters because you will probably have a secretary or assistant to type your letters who will be familiar with the preferred company format. It might be useful, however, for you to be familiar with standard nomenclature for the basic parts of a business letter (see Figure 10–9).

Heading. Tells the reader where the letter came from and when it was written. Consists typically of a printed letterhead and typed date.

HEADING

ITASCA TOOL AND MANUFACTURING

425 Sherman Blvd. St. Cloud, Minn. 56301

February 5, 1983

INSIDE ADDRESS

Janice Trebbi Richards
Marketing Manager
NORTHWEST RIVER SUPPLY
1525 Madison Street
Minneapolis, MN 55412

SALUTATION

Dear Ms. Richards:

BODY

Here is the quotation for fabricating 2000 insulated aluminum coolers for use in rafts and canoes.

Also enclosed is a set of modified drawings and specifications. As you will see, we made a few changes in the overall design so that the coolers would fit between the thwarts on a wider range of canoes, and we added several more D-rings to provide both support and tie-downs when the coolers are used in rubber rafts.

If you find the modifications and the quotation to be acceptable, we can order materials and begin fabrication just as soon as you give us authorization to proceed.

COMPLIMENTARY CLOSE Sincerely yours,

SIGNATURE David S. Barber
David S. Barber,
Chief Engineer

Figure 10–9 Basic Parts of a Business Letter

Inside Address. Identifies the person or people to whom the letter is sent. Should include name, title, organization, and mailing address.

Salutation. A letter-writing convention, not a term of endearment. (See below.)

Body. What the letter is all about. (See "The Letter as a Minireport," below.)

Complimentary Close. Another letter-writing convention. (See below.)

Signature. Identifies the sender of the letter by name and title. Also indicates the sender's preferred form of address for return correspondence.

There are at least a half-dozen additional recognizable components of a business letter and as many variations for each component, but these are enough to get you by in most situations. If you need more information on business-letter format, consult any up-to-date business writing textbook or secretary's handbook.

□ Nonletterhead Letter Format

The nonletterhead letter is appropriate whenever you have business correspondence where you are representing yourself as a private individual. Don't confuse the nonletterhead business letter with the personal letter, though; the personal letter is sent only to relatives or good friends and deals primarily with nonbusiness matters. It is highly informal in style and tone and is formatted somewhat differently. For your nonletterhead letters follow the format shown in Figure 10–10; it is as good as any and better than most. (Or, if you don't like this one, consult some other source; the important thing is to follow some accepted format consistently rather than invent your own.) Here are some suggestions to help you avoid common pitfalls with nonletterhead letters:

- □ Do not include your name in the heading; it belongs in the signature. Type in the date your letter is likely to be mailed rather than the date you begin the letter.
- □ Control the vertical centering of your letter by following the guidelines in Table 10-1.
- □ Accuracy is important with the inside address: there is no quicker way to turn people off than to misspell their names or misuse the official corporate name of their firms. Consult previous correspondence from the firm or organization for proper

422 Ekton Drive East
Batesville, MS 38606
April 13, 1983

Mr. John C. Crouch,
Public Relations Director
BURGDORF CONSTRUCTION, INC.
5522 Ludington Street
Biloxi, MS 39531

Dear Mr. Crouch:

I am writing on behalf of the Batesville High School Track Committee to ask if Burgdorf Construction can help out with some earthmoving in the construction of Batesville High's new 400-meter track facility. We have several bulldozers, front-end loaders, and dump trucks available on each of the next three weekends, but we are desperately in need of one or two scrapers.

Bill Hermenson talked to the operators of the scrapers that you have working on the Highway 26 bypass, and they agreed to donate their labor for three weekends. Apex Producers is donating diesel fuel for all heavy equipment, so all we need is permission from your company to proceed. There will be no direct expense to Burgdorf Construction, Inc.

The construction of the new Batesville High track has been entirely a volunteer effort. Residents of the school district and interested firms and organizations in the area have donated their time and money to allow the track committee to accomplish approximately $135,000 worth of construction for an actual cash outlay of less than $8,000. In these times of financial difficulty for the school district, we feel this is a significant step toward keeping property taxes down.

If you find our project worthwhile and wish Burgdorf Construction to be a part of it, can you ask an appropriate officer of Burgdorf to authorize the use of one or two scrapers for each of the next three weekends?

Sincerely yours,

Donald R. Turner

Donald R. Turner,
Batesville Track Committee

Figure 10–10 Suggested Format for a Nonletterhead Letter

TABLE 10-1. Spacing Guidelines for Nonletterhead Business Letters

Letter Length	Lines Between Heading and Inside Address	Margin Settings
Short (under 100 words)	6–10	26–76 (Elite) 22–62 (Pica)
Medium (100 to 200 words)	4–8	21–81 (E) 17–67 (P)
Long (200 to 300 words)	2–6	16–86 (E) 12–72 (P)

spellings and usages. If you have none, consult the group's advertising or promotional literature or an agency or organizational directory (available in most libraries).

- □ Use the conventional "Dear ______" in your salutation, and follow your salutation with a colon (:), not a comma (,) or semicolon (;). Address your correspondent by first name if you call that person by first name on the phone; otherwise use a courtesy title (Dr., Mr., Ms.) and the person's last name.
- □ Single-space the body of your letter, and double-space between paragraphs. You may indent paragraphs or not, as you choose, but whatever you do, be consistent.
- □ Capitalize only the first word in your complimentary close (Sincerely yours, Very truly yours) and follow your complimentary close with a comma.
- □ Leave three blank lines (quadruple space) between your complimentary close and your typed signature. And don't forget to sign your letter after you pull it out of the typewriter!
- □ Avoid the temptation to put a cluster of initials at the bottom of your letter to make it look like someone else typed it. Unless you're a professional typist, you won't fool a single soul.

□ The Letter as a Minireport

A memo or letter is typically a short piece of writing—less than 300 words. Hence, it is tempting to regard it as relatively unimportant. Don't. If a communication context calls for written correspondence, the situation warrants all of your basic writing skills and a full writing procedure (see Chapter 5). Plan your correspondence, gather all your data before you begin, prepare an initial draft, revise your draft until you are satisfied with it, and then (and only then) type it or have it typed in final format. More business is lost, more time is consumed, and more ill will is generated from the sending of badly conceived and poorly written letters and memos than from any other single factor in the entire arena of business, industry, and government.

See that your letter has a beginning, a middle, and an end—an introduction, a body, and a conclusion. Make sure each of your letters and memos has a single, consistent theme. Don't write multipurpose, shotgun letters that read like you're trying to beat the U.S. Postal Service out of a couple extra quarters. You will write many more letters and memos than reports, so use every letter-writing situation as a chance to practice your report-writing techniques. If your letters are good, your reports will also be sound. And the converse is equally true. No one I have ever known writes ho-hum letters and excellent reports.

What Letters Say and What They Mean

Your letters and memos convey as much information about you and your firm or organization through their appearance, tone, style, and approach as they do through their content. A letter with typographical errors says the writer is careless. A letter that fails to follow conventions of form says the writer is indifferent. A snotty memo says the writer is snotty. A stuffy and pompous letter says the writer is pompous and stuffy. A neat, correct, cordial, brief, and direct letter or memo says the writer is the kind of person it is a pleasure to do business with. The choice is yours.

Appearance

Even though someone else types it, the appearance of your letter is solely your responsibility—that's your name at the bottom of it. You can learn a great deal about neat, well-formatted letters from a competent secretary, but don't let yourself be intimidated by a pushy typist. If it's not right, insist that it be done over. Don't settle for messy corrections when it takes only five minutes to retype a letter. Even if retyping takes another day, the good will gained is worth the time spent.

Tone and Style

The tone of a letter is every bit as important as tone of voice; and as any three-year-old knows, tone of voice is everything. Most novice letter writers err in the direction of too stuffy a tone and too formal a style in their technical corespondence. Modern business and governmental agency practice calls for more conversational letters. A good rule of thumb is this: when corresponding with a given individual, use the same tone and style in your letter or memo that you would use when talking to that individual on the telephone. Don't be overly chummy, but by all means don't be pompous. (Look back at Figure 3–4, Adjectives for Tone and Approach.)

A special aspect of tone in business and technical correspondence is taking care to avoid what is called in the trade "red-flag" words—words that most people find irritating or offensive when they're on the receiving end of them. Using red-flag words in your correspondence is like waving the proverbial red flag in front of the bull. Here is a list of common red-flag words and phrases:

against company policy
apparently
as per
defective
erroneous
failure (or failed)

incompetent
irresponsible
mistaken
must
no excuse for
nonsense

not in compliance	you forgot
obviously	you neglected
overlooked	you should know
pursuant to	your complaint
shoddy	your insinuation that
we must insist	

Of course, the list could go on and on, but you get the idea. In fact, something to consider is this: when you start to receive correspondence, make a list of the words and phrases that irritate *you*. Then make a point of never using such words and phrases in your own correspondence.

Use a personal writing style in your correspondence. Mention the person's name once or twice in the body of your letter, and, if appropriate, mention the name of his firm or organization at least once. Use an ample stock of first- and second-person pronouns. If you know your correspondent well, a sentence or two about family, personal affairs, or an upcoming visit is not at all out of place in a business letter.

□ Success Strategy for Student Writers □

Take the classroom writing situation as an occasion to push yourself as far as you dare in the direction of personal, conversational, informal writing. Rely on your instructor to let you know when you go too far, and then pull back just a bit. This should put you right on the mark as far as style and tone goes.

Approach

The approach you use in a letter or memo depends on your assessment of the communication situation—the type of letter or memo you are writing; your knowledge of your correspondent's beliefs, attitudes, and preferences; and the ends you wish to achieve with your correspondence. If you wish to alienate a client, write a surly, vindictive letter. If you wish to insult your boss or a fellow worker, write a harsh, condescending memo. If you wish to make yourself look big by making others look little, fill your letter or memo with pompous phrases, technical jargon, and convoluted syntax. On the other hand, if you wish to appear knowledgeable, write in a clear and informative manner. If you wish to establish or maintain good will, be polite and cordial. If you wish to save your correspondent some time, be straightforward.

Most of your on-the-job letters and memos will fall into one or another of the following categories:

- □ Requests (for information, favors, merchandise, etc.)
- □ Replies (to requests, invitations, direct questions, etc.)
- □ Announcements (of meetings, new products or services, new offices, etc.)

- □ Documentation (of phone calls, conversations, etc.)
- □ Reports in the form of letters or memos
- □ Cover letters or memos (for reports, publications, packages, etc.)
- □ Good-will letters
- □ Complaints and reprimands (about poor service, shoddy goods, etc.)

Of these, announcements and good-will letters are generally positive communications, complaints and reprimands are generally negative communications, and the rest are more or less neutral communications. For positive and neutral communications, use a direct approach. State the purpose of the letter in the first paragraph; in fact, get right down to business in your opening sentence. Don't beat around the bush.

If you have to write a letter of complaint or reprimand, try to find a positive aspect of the situation and open with that; then ease into the main matter. Don't be wishy-washy, but don't back your correspondent so far into a corner that all he or she can do is lash back in fear or anger. Always give your correspondents the benefit of the doubt, and try to leave a way for them to extricate themselves from an uncomfortable situation with a certain degree of grace. In short, for negative communications exercise a little empathy: put yourself in the other person's shoes and write the kind of letter you would want to receive. You catch more flies with honey than you do with vinegar.

Good-will letters are things like thank yous, congratulations, commendations, haven't-heard-from-you-latelys, and the like—letters that you don't *have* to write but sure make people feel good when you do write them. Never pass up a chance to write a good-will letter. The world is far too short of good will, anyway.

□ Specific Types of Correspondence

Most of the letters and memos you write will be instances of routine, straightforward correspondence, and many of these fit into fairly standard categories. There is a certain economy in writing standard letters for standard occasions, but there is a cost as well. No one likes to receive an impersonal form letter—no matter how routine the situation is. Your job as a technical correspondent is to balance the economies of standard approaches to standard situations with the enhanced communication that comes from thoughtful, hand-tailored, and well-written letters and memos. You must be both efficient and effective.

Requests for Information

The letter or memo requesting information is one of the more common forms of correspondence written by technicians and technical experts. In general, such requests are neutral or positive in nature; hence a direct approach is in order. Make clear in your opening sentence that your letter or memo is a request for information. Ask specific questions and identify your questions clearly—preferably by numbered list. If it's not immediately obvious why you need the information you are requesting, add a short paragraph explaining the nature of your problem or explaining how you intend to use the information. This may help your correspondent furnish additional information that may be useful to you.

Replies to Requests for Information

You have two goals with a reply to a request for information: to provide the information requested and to build good will for your firm or organization. To make sure you have answered all the questions, identify each specific question on the original letter, number the questions if they're not already numbered, and use the marked-up original as a checklist in drafting your reply. In the good-will department, offer to provide additional information, if appropriate, or suggest other sources of help. If you receive a request for information that you can't handle but someone else in your firm or organization can, forward the request yourself rather than returning it to the sender.

Other Kinds of Requests

Requests for literature, merchandise, parts, and so on are positive and routine; treat them accordingly. Be direct and be brief. Whenever possible, use an order form or a requisition form; such forms are designed to provide all necessary information in a convenient format. If you are requesting catalogs, bulletins, spec sheets, and the like, identify them fully by title, form number, and publication date. If you are requesting merchandise, parts, or components from a catalog or parts book, specify quantity, name, and part number or catalog number for each item requested. Also, identify the form number and publication date of the catalog or parts book you are ordering from and the make, model, serial number, and date of purchase of the device you are ordering parts or components for. A little extra information in your first letter may well save you considerable time and a second or third letter.

Transmittal Correspondence

Quite often—probably more often than not—copies of reports, proposals, and other documents are sent with a cover letter or cover memo attached. (These are discussed in some detail in Chapter 4.) Suffice it to say here that letters and memos of transmittal are positive and routine and as such should be direct and brief. Resist the temptation to summarize a report in its cover letter; save this kind of information for your abstract or executive summary (see Chapter 13, "Abstracts, Summaries, and the Like").

Documentation of Phone Calls and Conversations

An ever-increasing number of business transactions are being conducted by phone or in face-to-face meetings rather than by mail. It's cheaper and quicker. But it is nice to have something in writing—for the records—so you will often be expected to write a follow-up letter or memo. The key here is write an accurate, objective summary of whatever you agreed to orally. Don't use the written summary to bend things in your favor—what little you might gain in the short run will be more than offset if you earn the reputation as a "fudger" on phone-call memos.

□ Miscellaneous Techniques and Strategies

As best you can, avoid multipurpose letters; send separate letters instead. If you must write a multipurpose letter, get the good news up front—start on your most positive note. Then get down to the neutral or negative items, but save at least one positive item to close on.

Don't write when you can call. A phone call is quicker, cheaper, and usually more effective. If the situation demands written documentation, consider the phone call with a follow-up letter or memo.

Review a letter carefully before you sign it, and make your act of signing a letter an act of placing your stamp of approval on it. A corporate executive I know never signs a letter the same day he dictates it— he always waits until the next morning when he's fresh to review and sign his letters. If he finds changes that need to be made, he sends the letter back for retyping with the next day's date.

Draft an angry letter or memo in haste if you must (it's good catharsis), but *never* mail it in haste. Always give yourself a day or two to cool off and reconsider. You'll never regret it.

Insist on a written or typed rough draft of every letter you write. If you do your own typing, double-space to force a retyping. If you dictate, insist on rough drafts for your editing and revision before the final draft is typed. Your typist may complain, but your correspondence will be the better for it. And you'll save the company time in the long run. (There's never time to do it right, but there's always time to do it over!)

Avoid phrases like *above-referenced, as per, herewith, thereof,* and other quasi-legal jargon. Similarly, avoid parenthetic numerals following spelled-out numbers: *five (5) days.* Usage like this is of questionable value even in contracts and legal documents; it certainly has no place at all in your day-to-day correspondence.

End your letters on a strong positive note. Avoid weak and hackneyed closing sentences like "I look forward to seeing you," "If you have any questions please feel free to call,'' or ''Thank you in advance for your consideration.'' (If thanks are in order, send a separate thank-you letter after the favor you ask for has been granted.)

Be positive, be cordial, be courteous, and be personal in every letter or memo you write. For every technical-correspondence situation you encounter, send the kind of letter or memo you would like to receive if you were on the other side of the fence.

□ Specimen Letters and Memos

694 Lander Hall
Pepperdine University
Malibu, CA 90274
September 29, 1978

Institute for Certifying Secretaries
2440 Pershing Road
Crown Center G10
Kansas City, MO 64108

LOCATION OF THE CPS EXAMINATION

Where is the CPS examination offered in the Moscow, Idaho, area?

Currently, I am a junior in college majoring in Business Education. I am interested in taking the CPS examination during my senior year or shortly after. Next year I will be attending the University of Idaho and graduating in May, 1980.

Here are some further questions I have about the CPS examination:

1. What are the requirements for taking the CPS examination?

2. What courses in college would aid me in taking the examination?

3. Do many college students take the CPS or is it mainly taken by on-the-job secretaries?

Any other information you can send me concerning the CPS examination will be appreciated.

Carrie Lee

CARRIE LEE

Specimen Letter Requesting Information

427 Esmore Street
Muncie, IN 47305
October 15, 1983

Hi-Performance, Inc.
4255 San Rononnodo Drive
Golden, CO 80401

Gentlemen:

I am interested in installing one of your new four-barrel carb and manifold sets on my 1974 Toyota Land Cruiser. Can you tell me the following:

1. Are there any significant differences between the earlier and later '74s so far as mounting the manifold is concerned?

2. Can I use my existing throttle linkage?

3. Can I use my existing manual choke and throttle linkages?

4. What kind of increase in horsepower and gas mileage can I expect?

5. What should be my next investment to improve the performance of my Cruiser?

Thank you for your consideration.

Very truly yours,

Peter Boucher

Peter A. Boucher

Specimen Letter Requesting Information

HI-PERFORMANCE, INC.
4255 San Rononndo Drive
Golden, CO 80401

October 22, 1983

Peter A. Boucher
427 Esmore Street
Muncie, IN 47305

Dear Mr. Boucher:

Thank you for your interest in the Hi-Performance carb and manifold set for your 1974 Toyota Land Cruiser. In answer to your questions:

1. There are no differences at all between earlier and later '74s as far as manifold mounting is concerned.

2. We provide a new throttle linkage with the carb.

3. You can use your existing manual choke, but you must install an adapter provided with the carb kit to use your existing manual throttle.

4. Models vary, of course, but most customers have experienced a 30% increase in horsepower and a 15 to 20% increase in gas mileage.

5. You may want to consider next a set of Hi-Performance headers. With the four-barrel carb, this would let your Cruiser breath at its best.

I am enclosing our latest catalog and price list, with the items you are interested in circled. If you order the carb and manifold set and the headers together, we can offer you a 15% discount on the package.

Sincerely yours,

James R. Greeson

James R. Greeson
Sales Manager

Specimen Reply to a Request for Information

TO: Judy Tatko, President

FROM: Carrie Lee, Treasurer

SUBJECT: Information requested on Kappa Alpha Theta   DATE: March 22, 1981

Here is the information you requested regarding the current financial status of Kappa Alpha Theta. This information is complete only up to February 28th as not all of the bills have been paid for the month of March.

I hope this what you needed. Please contact me if you have any questions.

Specimen Cover Memo for a Report

TROY DRAY AND CARTAGE

TROY MAINE 04401

January 28, 1984

Mr. William P. Schwartz, Manager
Blue Diamond Battery Company
3448 Division Blvd.
Boston, Mass 02122

Dear Mr. Schwartz:

I am writing to ask for a warranty adjustment on one of your batteries. In July, 1979 we purchased twelve of your heavy-duty truck batteries for our fleet. We installed ten of them and kept two as spares. During the recent cold snap we needed one of the spares, but upon installing it found it wouldn't take a charge.

I asked your regional sales representative for a replacement or adjustment, but he pointed out that the warranty was from date of purchase--not date of use. I can understand your position, but I hope you can understand mine. Your sales representative inspected the battery and agreed that it was defective from the start; the other spare is fine.

We at Troy Dray and Cartage have enjoyed excellent service from Blue Diamond batteries, and we realize we should have inspected and tested the battery before we put it in storage. Can you make an exception to your warranty policy and give us a replacement battery or an adjustment on the purchase of a new one?

Sincerely yours,

Elmer Durheim

Elmer Durheim, President

EKD/bkr

Specimen Letter Requesting an Adjustment

INTERWEST ENGINEERING, INC.

engineers
planners
surveyors

October 25, 1983

City of Coeur d'Alene
City Hall
Fifth and Sherman Avenue
Coeur d'Alene, ID 83814

Re: Water Pollution Control Facilities
Engineering and Planning Proposal

Dear Council Members:

We appreciate the opportunity to present a proposal for preparing a water pollution control Facilities Plan under the U.S. EPA Step I grant program. This proposal is submitted as evidence of Interwest's qualifications and desire to provide engineering and planning services for this project.

If retained by the City, Interwest Engineering would accomplish the project in accordance with the enclosed proposal submitted for your review and consideration. We are confident that our well-qualified staff with recent experiences in projects of a similar nature can provide the aggressive approach required to formulate and expedite solutions to the facilities planning requirements of the City of Coeur d'Alene. Our staff and equipment are within 1.5 hours of your community and as such we can respond to your engineering service requirements rapidly, efficiently, and economically. If you have any questions or need additional information, please contact this office.

Sincerely,

INTERWEST ENGINEERING, INC.
Consulting Engineers

Jack S. Hammond

Jack S. Hammond, P.E.
President

Enc.
JSH/srs

333 PALOUSE RIVER DRIVE, P.O. BOX 9126, MOSCOW, IDAHO 83843 (208) 882-1510

Specimen Proposal Cover Letter

EVERGREEN ACCOUNTING SERVICE
St. Louis, MO
63104

May 22, 1984

Mr. Jay Martin, Manager
Halsted Farms, Inc.
Route One, Box 384
Uniontown, MO 63783

Dear Jay:

As you requested in our telephone conversation yesterday, we will prepare the following:

1. A statement of earnings for the 1st quarter, 1984

2. A profit-and-loss statement for calendar 1983

3. A request for delay in filing estimated income for 2nd quarter, 1984

In addition, we will run copies of all of your tax returns and related documents--both federal and state--for 1981, 1982, and 1983.

I know this situation is difficult for you, Jay, and largely unwarranted. Be assured, though, that both Evergreen and I, personally, will stand with you all the way.

Best regards,

Paul

Paul H. Dreyden,
General Manager

PHD:crs

Specimen Letter Documenting a Phone Call

Writing Exercises

1. Collect copies of correspondence from nearby firms and organizations. Note differences in format and briefly describe each significantly different one. Select the format you feel is most effective and write a brief paragraph justifying your selection.
2. Develop a revision checklist for routine correspondence. Working in small groups in class, compare checklists and develop a master revision checklist.

3. Revise the following sentences as necessary:

   Pursuant to your letter of April 16th, I beg to state that we did indeed complete the work as specified.

   I sure don't want to cause you boys any trouble, but that carburetor you put on my rig last week has gone henhouse.

   The concrete poured yesterday on the above-reference project failed standard 345.12. Further shoddy work of this type will not be tolerated.

   Please answer this letter as I want to hear from you soon on this matter.

   If the Board does not respond within five (5) days to this letter, the Association will within ten (10) days file for mediation pursuant to Section 30 (thirty) of the master Agreement.

4. Collect five examples each of (a) correspondence you judge to be excellent and (b) correspondence you judge to be unacceptable or minimally adequate.

Writing about Writing

1. Write a brief essay on the differences between tone and style as they relate to technical correspondence.
2. Write an arrogant demand for adjustment on some faulty product. Write an obsequious version of the same. Then split the difference. Write a brief essay on the relative effectiveness of the various approaches.
3. Write a detailed essay on the theme "Send the kind of letter or memo you would like to receive if you were on the other side of the fence."

Writing Assignments

1. Write six of the following types of letters or memos:

   A letter requesting specific technical information about a problem you have.

   A letter ordering parts for a camera, firearm, small appliance, power tool, etc.

   An internal memo requesting information on the status of a project.

   A letter inviting someone to speak at a meeting, luncheon, or the like.

   A letter accepting such an invitation.

   A letter declining an invitation to speak and suggesting someone else instead.

   A letter or memo of congratulation on receiving an award or distinction.

   A letter soliciting membership in an organization or group.

   A letter to a university requesting information on a graduate program you are interested in.

A cover memo for one of the specimen recommendation reports at the end of Chapter 15, "Recommendation Reports."

A cover letter for one of the specimen proposals at the end of Chapter 16, "Proposals."

A thank-you letter to someone who has helped you with a project.

A letter announcing new offices or new products or services.

A keep-in-touch letter to a customer or client you haven't heard from for six months or so.

2. Write a series of six collection letters that might be sent to customers with overdue accounts. For the early letters, remember to give the customer the benefit of the doubt. For all letters, remember the principle of empathy.

□ 11 □

Employment Applications

Perhaps the most important set of business letters you will ever send are those seeking full-time, permanent employment. The key components of your employment application package are your resume and your resume cover letter, although you should not neglect letters of inquiry and follow-up correspondence. Seeking employment after you graduate from college is a major undertaking; your job-hunting is the doorway between $50,000 worth of college education and $2,000,000 worth of lifetime earnings. Make sure that doorway is big enough.

□ The Employment Process

Job-hunting is not begging. When you seek full-time, permanent employment as a management-level professional you are not looking for a handout; you are marketing your knowledge, your skills, and your willingness to work. The employment process is a two-way street. A prospective employer has one or more job vacancies; he has needs that must be filled, and he has a certain number of dollars allocated to fill those needs. A prospective employee—you, in this case—has a need for dollars that must be met and a certain package of skills and knowledge that can be applied toward solving problems. Both employer and employee have a limited time frame within which to operate. Prospective employers want to find the best person available to fill their needs, and you as prospective employee want to find the best-possible job available. So you both begin.

The prospective employer starts by circulating vacancy notices and job descriptions in newspapers, trade journals, and college placement offices. As inquiries and applications come in, the employer does a preliminary screening to see if applicants meet basic qualifications. Those that do are usually sent a formal employment application form and are encouraged to fill it out and return it. Those that do not pass this initial screening are politely but firmly rejected. After the closing date for receipt of applications, the employer's personnel staff reviews the application packages and narrows the field down to the top candidates. Each of these then is invited for an on-site interview. Based on management's assessment of the interview performances, the finalists are rank-ordered, and an offer is tendered to the top-ranking candidate. If that candidate declines, an offer is tendered to the second-ranking candidate, and so on down the line. If all decline, it's back to the drawing board.

Note two things about this process. First, it is complicated, lengthy, and expensive. The prospective employer doesn't want to do it any more often than he has to. Second, the prospective employer is caught between the rock of wanting to delay selection as long as possible in hopes of finding a more suitable candidate and the hard place of waiting too long and seeing all of the top candidates accepting employment elsewhere.

How about the prospective employee? You start by scanning vacancy notices and talking to people about job prospects. You identify a number of organizations that employ people in the kinds of jobs you would like to work at. Then you send off a number of resumes and job application letters, or perhaps you make appointments for a number of on-campus interviews. If you are fortunate, you are invited for one or more plant-site or home office interviews. If you impress the prospective employer favorably during your interview, you receive a job offer. If you receive one, you take it; if you receive several, you pick the best and take that; if you receive none at all, you enroll in graduate school or take a job as a laborer. And begin your job-hunting all over again.

So why not do it right rather than do it over?

□ Employment Application File

Start early—in the fall of your junior year—to identify potential employers. Begin to build a list of corporations, businesses, and/or governmental agencies you might want to work for. (See your college or university placement center for basic information.) For each entry on your list, include the organizational name and mailing address and, if possible, the name, title, and telephone number of the personnel officer or contact person. Set up a table like the one shown in Figure 11–1 so that when you

	INQUIRY SENT	REPLY REC'D	RESUME SENT	REPLY REC'D	FOLLOW-UP SENT
Bureau of Communications Research Department of Journalism and Mass Communications South Dakota State University Brookings, South Dakota 57006 Dr. George Phillips, Director (605) 688-4171	2/3	2/12			
Bureau of Press and Broadcasting Research School of Journalism University of Montana Missoula, Montana 59801 Warren J. Brier, Director (406) 243-0211	2/3	2/18			
Center for Communication Studies College of Communication Ohio University Athens, Ohio 45701 Dr. Robert S. Goyer, Director (614) 594-5054	2/3	3/5	3/8	4/22	
Communication Research Center University of Denver Denver, Colorado 80210 Dr. Harold Mendelsohn, Director (303) 753-1964	2/15	2/27	2/27	3/18	
Communication Research Center School of Communication University of Washington Seattle, Washington 98105 Peter Clarke, Director (206) 543-2346	2/15	3/12	3/15	3/30	4/1
Institue for Communication Research Stanford University Stanford, California 94305 Dr. Wilbur Schramm, Director	2/20	3/1	3/2	3/8	3/21
Laborartory of Sensory Communication Syracuse University 821 University Avenue Syracuse, New York 13210 Dr. Jozef J. Zwislocki, Director (315) 476-5571	2/20	3/12			

Figure 11–1 Record of Employment Application Correspondence

get down to the actual letter-writing you will have a record of whom you have written to and when.

Begin to collect and document the details of your educational background and work experience. Get copies of your college transcripts. Include copies of your scores on special tests (SAT, GRE, and so forth). Identify those of your college professors who would be likely to recommend you favorably for employment and record their names and titles, business addresses, and office telephone numbers. (Consult the college catalog or bulletin for this information.) For each full-time job you have held and for each major part-time or summer job, record the name and address of the firm or organization, the dates (months and years) of your employment, and the name(s) and title(s) of your immediate supervisor(s). Document any special awards you have received, any stories or articles you have published, any foreign travel, and anything else you might want to include in an employment application package.

□ The Letter of Inquiry

Your initial contact with a prospective employer need not be a full-fledged employment application sent a month before you graduate. Consider sending letters of inquiry during your junior year and early in the fall of your senior year. Indicate your interest in working for the organization, request information about the organization, and inquire about the procedure of gaining employment with the organization. Figure 11–2 shows a sample letter of inquiry.

Keep your letter of inquiry brief and don't go into detail about your background and qualifications, as your letter will probably be handled by a clerk or secretary anyway. Make it clear in your first sentence that your

1010 Second Avenue South
Escanaba, Michigan 49829
February 9, 1983

Dr. James Lemert, Director
Division of Communication Research
School of Journalism
University of Oregon
Eugene, Oregon 97403

Dear Dr. Lemert:

This letter is to inquire into possibilities of employment with your communication research program as a communication research technician. I have a B.S. in Communication Theory from Michigan State University and three years' part-time experience in survey research and in-depth interview techniques.

Having been born and raised in the Pacific Northwest, I am eager to return. I am familiar with the Willamet Valley and I would like very much to make my home there.

If you feel that there might be career opportunities for me at the University of Oregon, may I send you a resume and transcripts?

Sincerely yours,

Steven W. Abels

Figure 11–2 Representative Letter of Inquiry

letter is just a request for information and not a formal job application. Such a letter nets you three things: (1) the name and title of the personnel officer or contact person, (2) some information about the organization and the kinds of jobs they have, and (3) an established contact that you can capitalize on when you are ready to send your resume and resume cover letter.

☐ Resumes

Your resume is a brief, well-organized, and attractive summary of your educational background, work experience, personal data, and other qualifications that a prospective employer will use to help him or her decide whether or not he wants to invite you for an interview. Notice that the purpose of your resume and application letter is not to get a job, but rather to get an interview. In this day and age, no one offers full-time, permanent professional employment on the basis of a letter and resume.

General Outline for Resumes

I. Personal Data
- A. Name, Present Address, Telephone Number
- B. Permanent Address
- C. (Optional) Birth Date, SS Number, Sex, Marital Status and Dependents, Citizenship (if other than U.S.), Handicaps (if any)
- D. Hobbies and Leisure-Time Interests

II. Educational Background
- A. High School Data
  1. Name and Town; Year Graduated
  2. Class Rank or GPA (if impressive)
  3. Major Activities
- B. College Data
  1. Name and City; Years Attended
  2. Degree, Major and Minor(s)
  3. Major Courses; Other Helpful Courses
  4. Scholarships; Awards
  5. Major Activities; Offices Held
- C. Special Schools or Training Programs

III. Employment Experience
- A. Major Jobs (for each:)
  1. Dates of Employment and Job Title
  2. Name and Location of Firm or Organization
  3. Brief Description of Duties and Responsibilities
  4. (Optional) Wages or Salary; Immediate Supervisor
- B. Minor Jobs (list)
- C. Pertinent Volunteer Work (list and describe)

IV. Additional Goodies (include as appropriate)
   A. Special Awards and Honors
   B. Significant Foreign Travel; Languages Spoken and Read
   C. Publications
V. References (for each of three to five)
   A. Name and Title
   B. Business Address
   C. Business Phone Number

Personal Data (see Figure 11–3)

In this section you state who you are and how someone can get in touch with you and tell a little bit about yourself. Avoid data on your height, weight, health, and hair color; you're not applying for a driver's license. Follow your personal preferences on age, marital status, and the like. If you have no objection to sharing such information with prospective employers, put it in. On the other hand, if you feel strongly about your privacy, you cannot be compelled to divulge such information in most employment situations. Give you birth date rather than your age; that way you won't have to reprint your resume after every birthday. (Similarly, don't date your resume; instead date your resume cover letter.)

Hobbies and interests don't constitute crucial information, but they can provide useful conversation-openers at the interview stage. The same can be said for a brief description of your background—where you were born, where you have lived, the size of your family, or similar data. In fact, some people like to develop all this information in a "Personal Background" section and include it toward the end of the resume, just ahead of "References."

```
                              RESUME

                         Casey Susan Schmidt

Permanent Address:                    School Address:
     3420 South Cole Road                    1119 King Road
     Boise, Idaho  83709                     Moscow, Idaho  83843

Phone:  (208) 362-3462                Phone:  (208) 882-2401

Date of Birth:  November 14, 1957     Marital Status:  Single
Age:  22                              Social Security Number:  518-78-0758

Hobbies:  Photography, backpacking, skiing, dance, disc-jockey.
```

Figure 11–3 Personal Data Section of a Resume

Educational Background (see Figure 11–4)

Start with the high school you graduated from; no one is crucially interested in anything earlier. If you were class valedictorian, or made the top 15 percent, or graduated with honors, say so. If not, keep silent. Don't brag about modest achievements. List major activities and achievements (class officer, all-state football, four-year letterman in track) but don't include your full yearbook entry. Ask yourself "Who cares?"

Include data on all colleges you have attended, but emphasize the college or university from which you receive a degree. In listing courses, focus on upper-division electives in your field and courses outside your field—the kind of courses that might set you apart from your fellow majors. More jobs are landed because of unique combinations of talents than because of excellence in a narrow field. (Not all of us are 4.0 students.) List enough extracurricular activities so you don't look like a bookworm but not so many that you look like a joiner. Stress service activities and organizations in which you held office.

If you have completed any noncollege schools or training programs, include a brief description of each. These might be such things as apprentice programs, military service schools, an EMT (Emergency Medical Technician) course, automotive mechanic or secretarial training schools—what have you. Increasingly, prospective employers are looking for people with breadth of background as well as depth in a special field.

```
Education

   1978 to Present--
        Universtiy of Idaho, Moscow, Idaho.
        Degree Objective:  Bachelor of Science, Computer Science.
        Graduation Date:  May 15, 1981.
        G.P.A.:  3.24 overall, 3.81 major.
        Subjects of Interest:  Basic, Cobol, Compiler Design, Data Structures,
           Fortran, HP21MX assembler, IBM 370 assemble and OS/JCL, Pascal,
           PL/I, and PL/M.
        Special Achievements:  Dean's List four out of six semesters.

   1974 to 1978--
        Moscow High School, Moscow, Idaho.
        Graduation Date:  June 2, 1978.
        G.P.A.:  3.76
        Organizations:  Junior Engineering Technical Society Summer Camp, June,
           1977; National Honor Society; Drama Club; D.E.C.A.; High School
           Bowl (televised quiz show) team member, champions 1978.
```

Figure 11–4 Educational Background Section of a Resume

Employment Experience (see Figure 11–5)

If you have held one or more full-time, permanent jobs, start with these. Treat military service as a full-time job; in fact, treat each of your duty

EXPERIENCE

September 1981- December 1981 (part time)	UNIVERSITY OF IDAHO, College of Forestry, Wildlife and Range Sciences, Moscow, ID 83843. Equipment room checkout--I was responsible for the maintenance and checkout of the college's forestry equipment.
May 1981 - August 1981 (full time)	BAIRD CONTRACTING & CONSULTING, P.O. Box 9322, Moscow, ID 83843. Forest technician--I was the assistant foreman for a 30-man planting crew, auger planting 270 acres. I conducted stand exam, determined variable plot size, identified habitat types and collected fuel data with the aid of aerial photos.
September 1980- October 1980 (part time)	BAIRD CONTRACTING & CONSULTING, P.O. Box 9322, Moscow, ID 83843. Stand exam foreman--I was responsible for a crew subcontracting stand exam inventory on the St. Joe National Forest.
May 1980- September 1980 (full time)	BOISE NATIONAL FOREST, Mountain Home, ID 83647. Timber sale administrator assistant--I was responsible for the reconnaissance, layout, sale, administration, and final closure of post and pole sales and commercial firewood sales. I also administered commercial thins and timber sales over 10 million board feet which included: utilization standards, erosion control, slash disposal, safety and fire inspections, and logging plan methods.
April 1979- September 1979	Same as above.
May 1978- November 1978 (full time)	BOISE NATIONAL FOREST, Mountain Home, ID 83647. Project crew person--I planted trees with an auger for one month and performed worthless tree removal and precommercial thinned with a power saw. I performed roadside rehabilitation, constructed erosion control structures and ignited slash piles, jackpots, and broadcast burns.
September 1977- May 1978 (part time)	NORTHEAST RESOURCE CONSULTANTS, Danvers, MA 01923. Survey crew member--I determined the total area of a watershed with the use of abney hand level and a transit.
June 1977- September 1977 (full time)	SAWTOOTH NATIONAL RECREATION AREA, Ketchum, ID 83340. Visitor information specialist--I was responsible for the Galena Overlook on Highway 75.
September 1976- May 1977 (part time)	NORTHEAST RESOURCE CONSULTANTS, Danvers, MA 01923. Resource management assistant--I conducted a line plot cruise with an abney hand level and a biltmore stick.

Figure 11–5 Employment Experience Section of a Resume,

assignments as a separate job. After you have listed full-time, permanent experience, then go on to list part-time or summer jobs that have some bearing on or relationship to the kind of permanent job you are seeking. If you have no full-time, permanent experience, list your part-time and summer jobs in chronological order. Start with significant part-time or summer jobs you held in high school; don't dip back to your baby-sitting or paper-route days.

Describe your duties and responsibilites in some detail. A summer or part-time job should be good for two or three lines, while a full-time job should be good for four or five lines at least. Don't just enter job labels like "chokerman," "stocker," "service writer," but describe in a few words just what you did. Stress any supervisory, financial, or decision-making authority that you had. If you directed the activities of a couple of other workers, say so. If you closed up the shop and cashed out the register, say so. If you made purchases, or kept your own time-sheets, or planned work assignments, say so. Prospective employers are interested in two things: (1) experience you have had related to the type of work you are applying for, and (2) evidence that you are a dependable, productive worker whose abilities have been recognized with steadily increasing responsibilities.

Don't overlook significant volunteer work. Include such activities as church mission work, camp counseling, political campaign work, Peace Corps or VISTA work, charitable fund-raising, and the like. Again, stress any supervisory, financial, or decision-making responsibilities you had in such jobs.

Additional Goodies

Include mention of special awards and honors, travel, publications, and so on, under separate headings, as appropriate. (Don't, for goodness sake, use "Additional Goodies" as a heading in your resume!) Strike a balance between being excessively modest on the one hand and bragging about trivia on the other. Such things as Eagle Scout, junior year abroad, and outstanding student awards are fair game. Hold back on runner-up in the Delta County Potato Queen contest or a three-day visit to Juarez.

References

In your employment application resume, you want professional references rather than character references. Omit neighbors and the pastor of your hometown church in favor of college professors and former employers. Include at least three but no more than five references, equally balanced between teachers and employers. Be sure to get permission before including someone's name on your reference list. Avoid the common "References will be supplied on request" entry. It makes you look either (1) lazy or (2) so undependable that you couldn't get anyone to recommend you.

Variations in Resume Content

If you hold (or are soon to earn) graduate degrees, you may wish to reverse the order of entries in your "Educational Background" section, starting with your latest (highest) degree and working backward. In this case, omit your high school entry.

If you have significant full-time, permanent experience (ten years or more), you should put your "Employment Experience" section ahead of your "Educational Background" section. Also, you should list your jobs in reverse chronological order—most recent first.

If you plan to seek employment in two or more widely divergent fields, consider preparing separate resumes, each stressing the aspects of your education and experience that are most pertinent.

If you spent four or more years in military service, you may wish to consolidate the training and work aspects of your service under a "Military Service" heading.

Resume Format

There is no single universally recommended resume format, but there are some guidelines you should follow. The main thing is to be neat. Come up with something that is pleasing to the eye. Avoid a wide-open format that you can throw a cat through anywhere (see Figure 11–6, page 206), but don't crowd things in so much that it looks like you're cheap with paper (see Figure 11–7, page 207). In general, set up your resume to conserve vertical space. Avoid long lists of items and excessive indentation. Arrange your section headings and your entries so that it is easy for a reader to select specific items of information; make your resume easy to follow. Finally, consider a typeset rather than typewritten resume (see Figure 11–8, page 208). These are fairly inexpensive and make for a very attractive package.

□ Resume Cover Letter

Some people refer to this document as an "employment application letter," but the term is a misnomer. For one thing, you're not really trying to get a *job* with your letter; you're trying to get an *interview*. It's the interview that will get you your job. Second, nowadays the trend is away from a long and detailed letter with supplementary data sheet and toward a full-developed resume with a cover letter that presents certain specific information not included in the resume. This is not to downgrade the importance of your letter, however; remember that your letter is what your prospective employer will see first.

Keep your resume cover letter to three or four paragraphs. In your opening paragraph, identify your letter as part of an employment applica-

RESUME

248 North Long Street
Williamsville, NY 14221
(716) 632-8263

Height: 6 feet
Weight: 177 lbs.
Marital Status:
Married 4-7-79

EDUCATION	1969 - 1972	Williamsville Central High -National Honor Society -Library Club Scholarship -PTA Scholarship -Regents Scholarship
	1972 - 1976	State University of New York at Buffalo -Graduated with highest distinction Linguistics Department -B.A. Magna Cum Laude
	1977 - 1982	State University of New York at Buffalo -M.A. (Linguistics) -Ph.D. Candidate (Sociolinguistics)
EMPLOYMENT	1974 - present	Sexton/Custodian Gethsemane Luthern Church
	1978	Tutor Empire State College
	1979 - present	Teaching Assistant SUNY Linguistics Dept. -Intro to Linguistics -Social Science of Lg -Acoustics Lab -Field Methods Lab
PUBLICATIONS	1979	"A Sociological Description of Religious Communication: A Case Study of an Evangelical Message," 1979 LACUS Fifth Forum

Figure 11–6 Resume with Excessively Open Format

tion package, identify the specific job or the general job category you are interested in, and present a one-sentence summary of your major qualifications. Use the body of your letter (1) to amplify on your special

RESUME of Harold J. Throcton

Present Address: 5224 Birch Street, Duluth, Minnesota 55804 Phone (903) 882-6793

Permanent Address: 3835 Westchester St., Mouthcard, Kentuckey 41548

Height: 5' 10" Weight: 190 Pounds SS No. 345-43-6789 Age: 22 Single

CAREER OBJECTIVE

Once I finish my educational program in forest resource management from the University of Minnesota, I desire to utilize my education and work experience in the field of forestry to secure a management position with a middle-sized or large timber-holding company or a lumber company or pulp and paper company. My main areas of interest are silviculture and/or timber acquisition.

EDUCATIONAL BACKGROUND

1979-1981 University of Minnesota at Duluth. Bachelor of Science in Forest Resource Management (management option). Cumulative GPA: 2.89. Courses included: Silviculture, Engineering and Harvesting, Mensuration, Intermediate Fire Behavior, Low Volume Roads, Land Resource Economics, Forest Protection, Dendrology, Fire Management, Forest Regulation and Finance, and Business English.

1976-1978 Essex Agricultural and Technical Institute, Hawthorne, Kentuckey. Associate Degree in Applied Science (Forest Technology). Cumulative GPA: 2.35. Courses included: Small Business Management, Pesticide Application, Tree and Shrub Pathology, Small Engine Operation, Natural Resource Management - I, Natural Resource Management - II, Soils and Soil Types, Aerial Photography, Entomology, and Dendrology.

EMPLOYMENT EXPERIENCE

8-80 to present University of Minnesota, Duluth, Minnesota. Tropical Plant greenhouse manager. Duties: responsible for operation and management of greenhouse for propagation, care, and maintenance of tropical, semi-tropical, and exotic plant species; teaching assistant for tropical dendrology class; care and maintenance of decorative houseplants for the School of Forestry. Supervised one other employee. My supervisor was: Dr. Frederick H. Johnston, Professor of Forestry.

6-80 to 8-80 Kimberley Clark Paper Company, Hayward, Wisconsin. Summer Intern in Silvicultural Engineering and Plant Propagation. Worked for the Department of Forestry Research and Development. Assignments included: work in cellular biology group doing research on the clonal propagation of Norway pine through tissue culture methods, Studied and research the feasibility of advanced propagation methods being applied to mass production of conifers for regeneration and genetic research. My Supervisor was: Dr. Abwestil El-Horrab, Senior Scientist.

6-79 to 8-79 United States Forest Service, Clearwater Ranger District, Grangeville, Idaho. If any of you have read this far, you have probably figured out that basically what I am doing here is typing in a bunch of junk just to fill up the page. All in all, makes for a pretty crowded format, don't you think? It's much better to use an extra sheet of paper if your background is rather extensive. Follow a format more like the one shown in Figure 11-8.

Figure 11–7 Resume with Excessively Crowded Format

qualifications and (2) to indicate your particular interest in the job situation in question. Close your letter with a positive request for action.

Cover-Letter Strategies

Keep the employment process in mind—particularly the initial screening process. In a tight job market such as exists today, an employer with a

Allan Briese

Permanent Address
W. 6401 Riverview Dr.
Post Falls, Idaho 83854
Telephone: (208) 773-5287

Temporary Address (until June 1982)
Rt. 3 Box 117 L-28
Moscow, Idaho 83843
Telephone: (208) 882-1455

EDUCATION

University of Idaho, Moscow, Idaho Candidate for B.S. degree in Metallurgical Engineering in May 1982. I have a 2.99 GPA, I also have taken Hydrometallurgy and Electrometallurgy and a directed study extractive metallurgy. Last fall I received a Ziegler Foundation scholarship.

Post Falls High School I graduated in 1978 with a 3.7 GPA and received the Post Falls Merchants scholarship (1978). I lettered in football, wrestling and track. In high school I was a state contender in wrestling and captain of the football team. I was also Vice-President of our 4-H youth group.

SUMMER AND PART-TIME WORK

Summer, 1981 — *Idaho State Highway Department,* Coeur d'Alene, Idaho. I drove dumptruck (diesel and gas), flagged and also drove roller. One of my other duties was to set up road signs and place flagmen.

Summer, 1980 — *Capp Homes Inc.*, Spokane, Washington. I was part of the yard crew and organized parts of shipments and drove forktruck. Unloading and chasing after parts was another of my duties.

Summer, 1979 — *Jacklin Seed Co.*, Post Falls, Idaho. I helped repair and set up seed mills. While the seed mills were operating I drove forktruck and kept the seed mills supplied.

Summer, 1978 — *Comet Corporation*, Spokane, Washington. My job was assemblyman and my duties were to assemble side-panels for semi-trailers. This included reading blueprints and setting up the assembly racks.

PERSONAL INTERESTS AND DATA:

Social Security Number 518-72-7807 Birth Date: April 7, 1960 Married - have no children. I am interested in cars, sports and the outdoors and enjoy working around machinery. Some of my spare time at college is spent weightlifting and running.

REFERENCES

Patrick Taylor, Associate Professor of Metallurgical Engineering, College of Mines,
University of Idaho, Moscow, Idaho 83843 Telephone: (208) 885-6769

Gene Bobeck, Ph.D., Metallurgy, College of Mines,
University of Idaho, Moscow, Idaho 83843 Telephone: (208) 885-7937

Sherrill Mills, Foreman, Idaho Department of Highways,
Coeur d'Alene, Idaho 83814

Figure 11–8 Typeset Resume

professional or managerial-level vacancy will receive an average of eighty applications for each job he has open. He (or his personnel manager) immediately sets up three lists:

- The A list—People we definitely want to invite for an interview
- The B list—Possibles
- The C list—Applicants we don't want to bother with

You have two goals: first, to stay off the C list; second, to make the A list.

Assuming you have the basic qualifications for the job, a neatly typed letter that is free from error and touches all the bases will keep you off the C list. On the other hand, making the A list is a little tougher. Here are some things you can do to increase your chances:

- □ Be direct and businesslike in your opening. Avoid being clever (unless you are exceedingly clever), but avoid such common and tired openings as "In response to your advertisement in the *Times-News* . . ." or "Professor Jones of the Civil Engineering Department has informed me that . . ."
- □ Personalize your letter. Address the personnel officer by name. Mention the name of the organization once or twice in the letter (instead of "your firm" or "your company").
- □ Give some indication in your letter that you know something about the firm or organization—its product lines, the locations of its major offices and facilities, its size, its reputation, its significant achievements. (This means study up on the organization before you write the letter. Your college placement center has extensive files on all employers who recruit on your campus.)
- □ Indicate your willingness (if not eagerness) to live permanently where the company has work. Express your desire to live in the East (or the West or the South), in a large city (or a small town or a rural area), to stay (or move) near the mountains (or the seashore).
- □ Avoid being overly egocentric in your letter; be confident but not cocky. Strive for a rough balance between the "I"s and the "you"s. Especially, avoid starting every paragraph with a first-person pronoun. Use phrases like "do you feel . . ." rather than "I think . . ."; "Pittsburgh is attractive" rather than "I like Pittsburgh"; "can you invite me . . ." rather than "I am available. . . ."
- □ Give the impression that you are likely to stay with the firm or organization for some length of time. Use words and phrases like *career, professional development, advancement opportunities, fringe benefits, employment stability*.
- □ Capitalize on any special circumstances that exist, like previous part-time or summer employment with the organization, relatives or close friends employed by the organization, or a keen interest you have in some aspect of the organization's activities.

Remember that out of the eighty applicants for the job you seek, fifty or so will make the B list and only a few will make the A list. Do whatever you can to stand out in the crowd, to be noticed. Only those who stand out—who are outstanding—get invited for on-site interviews.

□ Interview Techniques

A full discussion of interview strategies and tactics warrants a chapter (if not a whole book) in itself. Let me just suggest two vitally important and often-overlooked techniques. First, learn all you can about interviewing. Take a short course if your college offers one. Read a book or two on the subject. Talk to the people in your college's placement office. Second, practice. Never turn down an invitation to an on-site interview, even if you're not eager about the job. Start in your junior year with on-campus interviews. Get a half-dozen or so under your belt before you go after the jobs you really want. Make your mistakes (and you will make them) early, so that when you get to the interviews that really count, you will have some degree of confidence in your ability to handle yourself.

□ Follow-up Correspondence

It is unwise to burn your bridges behind you, but it is equally unwise to let them collapse through disrepair. The key here is "keep in touch." Consider these several varieties of follow-up letters:

Interview Thank You

On the day you return from an on-site interview send a brief note to the personnel officer who invited you, expressing your appreciation for the opportunity to visit the firm.

Request for Additional Information

Within a week after your return from an on-site interview, write to the person for whom you would likely be working and ask for further details on one or two aspects of the potential job. At the same time, express your pleasure with the division or department and your eagerness to become part of the team.

Follow-Up Inquiry

If you haven't heard from an organization after a month to six weeks following an on-site interview, write a letter inquiring about the status of your application. Suggest that you are being pressed to make a decision about employment and would like to be able to consider a career with the organization in question. Send this letter to the manager or supervisor of the department you would be working for; the personnel department may be dragging its feet.

Acknowledgment of Offer

Acknowledge receipt of a job offer as soon as you receive it. If you're ready to accept, do so. If not, at least let the organization know that you received the offer, express your appreciation for it, and indicate a time frame within which you will try to respond.

Acceptance of Offer

As soon as you are ready to accept an offer, post a letter of acceptance. Restate (for the record) salary and any conditions of employment, and acknowledge (or ask for) the date and place you show up for work.

Request for Extension of Time

Most often, any offer you receive will specify a date by which you have to either accept or decline. Sometimes that date approaches and you still have some prospects out. It is perfectly acceptable to write and ask for an extension. If they want you badly enough they will extend; if not, they will give you one last chance to accept. Either way, you can't lose.

Request for Withdrawal from Consideration

If you accept an offer and have applications pending (at any stage) with other organizations, it is just good manners to write to the other organizations and ask that your name be withdrawn from consideration. And such good manners can pay off in the future; who knows where you'll be looking for a job in several years?

Donna M.Sherman
086-58-5937
Birth:12/18/59

Permanent Address	Temporary Address
8 Elliot Ave E.Greenbush N.Y. 12061 (518) 477-6374	602 Carter Hall Moscow, Idaho 83843 (208) 885-7126

Education: University of Idaho, 1980 to present. canidate for B.S. degree in May,1983. Majoring in wildlife management. G.P.A. is a 3.40 out of a 4.00.

University of Vermont, 1978-1980. Majoring in wildlife biology with a minor in forestry. Overall G.P.A. was a 3.34 out of 4.00.

Educational Activities: University of Idaho; Program Director and member of the Student Chapter of the Wildlife Society. Member of the Xi Sigma Phi and Phi Sigma honor societies.

University of Vermont; Public Relations Officer for the Student Chapter of the Wildlife Society. Involved with hall government for two years. Sophmore Social Chairman.

Pertinent College Courses: Forest Mensuration, Computer Science, Arial Photographic Interpretation, Eastern Dendrology, Surveying.

High School: Columbia High School,E.Greenbush New York. Graduated with honors in 1978.

Work Experience:

8/81 to present Culinary Vender, University of Idaho, Moscow Idaho 83843.
Supervisor: Colleen Bumgarner
Responsibilities: distribution of prepared dishes, operation of fryers,ovens and microwaves.

9/80-5/81 Research and Lab Assistant for Spruce Budworm Project at the University of Idaho Moscow, Idaho
Supervisor: Oz Garton
Resposibilities: observation and collection of data on various species of birds, computer operations, trial run testing and literature search.

Specimen Resume

5/80-8/80 5/79-8/79	Jean Shoppe Clerk, T-Shirt Designer, Cashier. Joy Department Store, Rensselear N.Y. Supervisor: Mr. Nernee Responsibilities: monthly inventories, hot press operations,cashier and daily tillcounts, relation work with sales representatives.
9/79-5/80 9/78-5/79	Mailing Assistant; University of vermont, Burlington ,Vt. Supervisor: Lou Provincher Responsibilities: operation of mailing machines, collection and distribution of paychecks,cashier and daily accounts clerk.
9/76-6/78	Volunteer Office Help; 9th Grade English Dept. Columbia High School, E.Greenbush N.Y. Supervisor: Mr.J. MacArevey Responsibilities: input consultant for new teaching proceedures, test grading, file clerk, supervisor for three other office personnel.
Personal Interests	Photography, sewing, cooking, essay writing, experience working with harness horses and carriage driving.
References	
Oz Garton	University of Idaho, Moscow, Idaho 83843 Wildlife Department (208) 885-6389
James MacArevey	Columbia High School, E.Greenbush, N.Y. 12061 English Department (518) 477-2414
Lou Provincher	University of Vermont, Burlington, Vt. 05405 Mail Room (802) 882-3664

Specimen Resume, continued

RESUME for W. STEVEN MEYER

PERSONAL DATA

Present Address	G-17 Gault Hall University of Idaho Moscow, ID 83843 (208) 885-7181	Permanent Address	4125 Stratford Drive Boise, ID 83704 (208) 375-1005
Birth Date:	March 28, 1959	Soc. Sec. No.:	519-56-6592

Hobbies and Leisure-time Interests: Skiing, SCUBA Diving, Horseback Riding, Hunting and Fishing

EDUCATIONAL BACKGROUND

High School — Capital High School, Boise, Idaho. Graduated May, 1977 in the top ten percent of my class. Active in Army ROTC Program: rifle marksmanship team and honor drill guard.

College — Ricks College, Rexburg, Idaho. Studied in the general liberal arts program for two years: 1978 - 1979. GPA: 3.76/4.00.

University of Idaho, Moscow, Idaho. Graduated in May, 1980 with a B.S. degree in Civil Engineering. Advanced elective courses include: pavement design, highway construction, transportation planning, and engineering report writing. GPA: 3.54/4.00. Active member of the student chapter American Society of Civil Engineers.

EMPLOYMENT EXPERIENCE

Summer 1979 — U.S. Forest Service, Payette National Forest, McCall, Idaho. Worked as an Engineering Aide. Member of a four-man survey crew laying out logging roads. I kept notes, brushed line, and ran forward and rear chain.

1978-1979 — University Bookstore, University of Idaho, Moscow, Idaho. Stocked shelves, ran cash register, and assisted in shipping and receiving.

Summer 1978 — Wyeth Laboratories, Meridian, Idaho. Worked on assembly line, weighed vitamins and minerals for the baby formula, and painted the shop.

Summer 1977 — Luck Peak Tree Nursery, Boise, Idaho. Worked as a Forestry Aide. Thinned and weeded tree seedlings, set irrigation sprinklers, packed and sorted seedlings for shipment, and sprayed insecticide.

REFERENCES

Mr. Jeffery Kahler, Field Supervisor Payette National Forest McCall, ID 83638 (208) 634-2255	Mr. Ronald W. Lindquist University Bookstore 122 Line Street Moscow, ID 83843 (208) 885-6469	Dr. Frederick J. Watts Civil Engineering Department University of Idaho Moscow, ID 83843 (208) 885-6782

Specimen Resume

305 Lauder #403
Moscow, Idaho 83843
November 15, 1981

Mr. Gene Ray
Chicago Bridge and Iron, Inc.
76041 22 St.
Fremont, California 10328

Dear Mr. Ray:

This letter is to inquire into the possibility of employment with C.B.I. as an assistant construction engineer. Do you feel that I could be a good team player for C.B.I.?

As the enclosed resume indicates, my career objective is to become a project manager. I like Fremont, California and the challenges C.B.I. has to offer. C.B.I. is large enough to care about the development of its employees' careers. This is the sort of firm with which I would like to share my engineering career.

If you feel my qualifications are what C.B.I. needs, I will be available for an interview at your convenience.

Very truly yours,

Danny L. Forge

Danny L. Forge

Specimen Resume Cover Letter

12904 Hugh Graham Road
Albuquerque, NM 87111
April 11, 1983

Col. Mark E. Daily
Director of Laser Testing Program
Weapons Laboratory
Kirtland AFB
Albuquerque, NM 87108

Dear Col. Daily:

Recently, I read in *Physics Today* about the work your branch of the Weapons Lab has been doing with lasers. I am aware of the computer programming necessary in laser testing and would like to become involved in the development of software for computer aided design and testing procedures with lasers.

I have had the opportunity to use several different systems including an IBM 370 and HP 1000 at the University of Idaho and a VAX 11/780 and IBM 370 at the University of New Mexico. Since I have had a lot of experience programming at different institutions with scientifically oriented languages such as FORTRAN and PASCAL, I feel I can learn to program quickly and effectively for you at the Weapons Lab. I will be graduating from the University of Idaho in May of this year with a Bachelor's Degree in Computer Science and a 3.65 GPA (on a 4.0 scale).

If you feel there are career opportunities for me with the Weapons Lab please contact me as soon as possible.

Sincerely yours,

Mike Daily

Mike Daily

Specimen Resume Cover Letter

107 E. Fourth Street
Ann Arbor, MI 48103
March 17, 1983

Mr. Leon Gray
Director of Personnel
Michigan Bell Telephone Company
Lansing, MI 48900

Dear Mr. Gray:

Through the placement office of the University of Michigan, I have learned that your company has an opening for an electrical engineer in the field of research and development.

In December, I will graduate from the University of Michigan with a Bachelor of Science degree in electrical engineering. I have the necessary courses in advanced electronics, integrated design, modern solid state, which I believe will meet your requirements.

As indicated on the attached resume, I have worked two summers through the Internship Program, and as a result I have picked up a moderate amount of experience as an engineering technician. As for my education, I have always enjoyed working with analog analysis and digital design. Because of this practical experience and the theory learned in school, I can fill your vacant position with minimal training.

If you feel, as I do, that I can contribute to Michigan Bell's goals, I am available for an interview at your convenience. Please let me know if you need additional information. I am looking forward to discussing my qualifications with you in more detail.

Sincerely yours,

Dangngoc Tai

TAI NGOC DANG

Specimen Resume Cover Letter

Writing Exercises

1. Prepare a list of the names and addresses of the contact people in ten firms or organizations with which you might apply for full-time, permanent employment after graduation.
2. Write three letters of inquiry about summer or part-time employment. After reviewing them with your instructor, mail them (being sure to keep a copy of each letter you mail). Turn in copies of the letters you send together with copies of the replies you get.
3. Collect five resumes that you feel are excellent and five you judge unacceptable or minimally adequate.

Writing about Writing

1. Write to the personnel managers of five nearby firms or organizations inquiring what they feel is important in the resume and letter of application. Summarize your findings in a report to the class, with a copy going to your department chairman or major professor. Attach copies of your original letters and the responses to your report.
2. Consult a half-dozen or so technical writing textbooks, business writing textbooks, or other appropriate sources on the question of resume format, content, and organization. Summarize your findings in a report to the class.

Writing Assignments

1. Prepare a resume for yourself and then write two resume cover letters to two *different kinds* of prospective employers. Assume you are writing in the spring of the year you will graduate and are seeking a full-time permanent job. Tailor each letter to the particular employer and opening.
2. Team up with another student in the class and exchange playing roles of applicant and prospective employer. Prepare two complete exchanges of employment-related correspondence, including letter of inquiry, resume and cover letter, interview thank-you, request for additional information or follow-up inquiry, acknowledgement of offer, and acceptance of offer or request for extension of time and request for withdrawal from consideration.

□ 12 □

Project Reports

A *project*, as the term is used in business, industry, and government, is an extensive undertaking that requires the concerted effort of several people. Sure, we still see a few individual, solo projects, and occasionally someone wants to call a small task a project; but the norm is a set of activities that takes time, costs money, and involves a group of people. Projects take place within the framework of organizations—businesses, industries, and government agencies. And wherever you find projects, you find project reports. What we'll cover in this chapter are the routine, day-to-day records of data and activities and reports of status and progress that help keep people in organizations informed about projects. In order to give us a framework for discussion, let me suggest the following categories of project reports:

- □ Reports of individuals' activities
- □ Reports of data for archival purposes
- □ Progress and status reports

Since project reports are routine in nature, they tend to be informal in format (see Chapter 4), with most being either fill-in-the-blanks reports or memo reports. Also, they will be among the first kinds of reports you will be asked to write when you get on the job. In the following sections we will look at examples of various types of routine project reports and offer a few hints and suggestions for preparing them.

□ Activities Reports

In any organization, there are certain people (called bosses) who like to keep track of what other people (called workers) are up to. Hence the activities report. Perhaps the most common of the activities reports is the periodic activities report (weekly or monthly), in which you explain how you've spent your time during the preceding reporting period.

As you go about your day-to-day activities, keep some personal records, so that when the end of the month rolls around you can recall what you were doing on the first or the third or the fourth. Your desk calendar or a personal appointment calendar is handy for this. Keep a record of both "milestones" (dates on which you started and finished individual tasks) and the hours each day you devote to individual tasks. That way, you've got something to work with at the end of the month.

Reports are always after the fact. You prepare last week's activities report on Monday or Tuesday of this week; you fill out July's activities report on the fourth or fifth of August. This situation makes it tempting to fudge a bit. You really completed that Johnson report on the first, and those last two sections of guard rail went in on the second, but why not claim them for last month? What the heck—they're finished, aren't they? It's not as if you're claiming something is completed that really isn't. Resist the temptation. Sooner or later you're going to come up short. In fact, if you do anything at all, fudge a little bit in the other direction—save those tasks completed on the twenty-ninth or thirtieth for next month. That way you'll have a little cushion to fall back on if you have a lean month.

□ Archival Data Reports

Some of the data reports technicians and technical experts generate find immediate use for planning and decision-making or serve as input for larger aspects of a project. But many—perhaps most—just go into the files so that they are available at some later date for plotting trends or establishing benchmarks or serving some other historical or archival purpose. Myriad examples of such data reports exist: the meteorologist's reports of temperature, barometric pressure, humidity, and rainfall; the process engineer's reports of temperature, pressure, and flow; the paramedic's reports of pulse, respiration, and blood pressure; the lab technician's reports of pH, suspended solids, and concentrations; the librarian's reports of patronage and circulation; the branch manager's reports of sales.

The overwhelming majority of such reports are fill-in-the-blanks reports that contain primarily—if not exclusively—numerical data. Preparing them is no big deal; all you need is accuracy, neatness, and a certain amount of diligence so you don't get too far behind in your record-keeping.

☐ Laboratory Reports

The term *laboratory report* is used almost as widely and indiscriminately as the term *report* itself. Things called lab reports can range all the way from a cryptic fill-in-the-blanks report of measurements taken to a full-blown doctoral dissertation. Most on-the-job lab reports, however, tend to be more like the former than the latter. The working lab report tends to focus on results and discusses theory, apparatus, methodology, or calculations only when such considerations might have a bearing on results or their interpretation. The successful lab report—like the successful archival data report—is accurate, neat, and brief.

☐ Progress and Status Reports

Progress and status reports tend to be longer and more involved than activities reports and archival data reports. Moreover, they tend to be written by supervisors and managers rather than technicians and technical experts. (This is not to say that as a chemist or engineer or accountant you will never write a progress report; it is instead to say that you will write progress and status reports in your role as project team leader or supervisor rather than in your role as technical expert.) Progress reports and status reports are similar in that they both discuss the state of affairs with respect to the collective activities of a group of people. Progress reports, however, tend to be project-oriented, while status reports tend to be oriented toward a task force or work group (or department or division). Furthermore, progress reports tend to cover a period of time— a month or a quarter—while status reports tend to deal with a slice of time—the end of a month or the end of a quarter. Other names for reports that fall in this category include periodic reports (monthly reports, quarter reports, annual reports), interim reports, and just plain project reports.

Progress Reports

A good progress report (or series of progress reports) is based on a sound project proposal (see Chapter 16, "Proposals"). If the proposal outlines the projected time and work schedule, breaks the overall project down into individual tasks, and identifies the personnel working on the project and what each is accountable for, then the progress reports can be fairly straightforward and concise, referencing back to the initial proposal. On the other hand, if the project proposal was skimpy—or missing altogether—the initial progress report will have to supply the detail of tasks, schedules, authorities, and accountabilities. Here is a general outline for a progress report:

I. Introduction
  A. Purpose of the Report
  B. Time Period Covered
  C. References to Other Reports
II. Summary of Project Status
  A., B., Etc. Different Phases or Tasks of the Project
III. Work Completed During Report Period
  A., B., Etc. Different Phases or Tasks
IV. Work Anticipated for Next Period
  A., B., Etc. Different Phases or Tasks
V. Special Problems or Difficulties

In the introduction, give your readers whatever background information they might need to understand your report and to see where it fits in the scheme of things. In particular, refer by title, date, and identification number (or author) to the original project proposal and all earlier progress reports, as well as any background papers and data reports that might be pertinent. Next, summarize the status of the project:

> All phases are on schedule, the project is within budget, and no unusual problems or difficulties have been encountered.
>
> The pretreatment section is two weeks behind schedule, pending delivery of the communitor, and the interconnecting piping has been held up accordingly. Digester construction is catching up but is still one week behind schedule. All other phases are on schedule except for the control building, which is three weeks ahead of schedule.
>
> Because of the heavy rains last month, corn harvest is three weeks behind schedule, and more than half of the crop is still in the fields. Unless we have extra help or go to a second shift, we will not be able to meet our delivery commitments for November 1.

The summary is important because it lets your readers see immediately the salient aspects of your report. Don't make your boss wade through the details to see that the project is behind schedule or over budget; you won't win any points that way. He may not be too tickled to find out that things are fouled up, but you can be certain he'll be a lot more grumpy if he finds out from someone else.

In the "Work Completed" section, present the details of schedules, activities, and accomplishments that led you to the conclusions you reported in the summary. This accomplishes two things: first, it provides data to support your conclusions; second, it allows your readers to analyze the data and draw their own conclusions—in case they want to. In the "Work Anticipated" section, make reasonable and realistic projections so that higher-level supervisors and managers can make whatever changes may be necessary in work loads, personnel, and facilities and equipment. Finally, in the "Special Problems" section, discuss extenuating circumstances or unforeseen difficulties that might account for changes in schedule or budget. Don't just make excuses; point out special aspects of the project that might make planning for similar projects more realistic and effective. Similarly, if you anticipate special problems or difficulties during the coming weeks, say something about them now so that your boss can take steps to head them off or mitigate their impact. The best surprise is no surprises at all.

Status Reports

A status report is similar to a progress report but tends to have project phases or tasks as main headings and time periods or dates as subheads. Alternately, a status report may have divisions, departments, or task groups as main headings. Here is a general outline for a status report:

I. Introduction
- A. Purpose of the Report
- B. Date of the Report
- C. References to Other Reports

II. Summary of Project (or Department) Status

III. First Project Phase (or Department Section)
- A. Major Accomplishments Since Last Report
- B. Tasks Anticipated During Coming Period
- C. Special Problems or Difficulties

IV., V., Etc. Second (and Subsequent) Project Phases (or Department Sections)
- A. Major Accomplishments Since Last Report
- B. Tasks Anticipated During Coming Period
- C. Special Problems or Difficulties

Specimen Project Reports

01-812-10730-04

EMPLOYEE'S TIME RECORD

NAME: JOE BENNINGTON

MONTH OF: JUNE 19 82

JOB NO.	DESCRIPTION OF WORK	1	2	3	4	5	6	7	8	9	10	11	12	13	14	15	16	17	18	19	20	21	22	23	24	25	26	27	28	29	30	31	TOTAL
651	ENG. SET POINT - SM	3	6	7	8	X	X		2			4	X	X		2				X	X						X	X				X	32
1207	CAMSHAFT GEAR - 342	1	2					3		1					4	1																	12
3280	DIFF & BEVEL GEAR - 166	4						5	1						2	2																	14
1274	3208 FUEL LINES			1					4	7							2																14
1424	NEUTRAL START SWITCH								1		3						4																8
1307	TURBO LUBE VALVE										3							8						4	1								16
4313	TRACTOR DRAG LINK										2	4					1		4					1		1				1			14
4092	PLANET GEAR SPACER														2		1		1							2							6
1408	WIRING HARNESS															2			1				3	2		2			1				11
5057	983 HYD HOSE																		2				2	1		3			1				9
1286	FUEL SYSTEM																						3		7				6				16
1155	BRAKESAVER																													7	8		15
	SECTION MEETING															1																	1
ABSENCE																						8											8
TOTAL DAILY HOURS		8	8	8	8	X	X	8	8	8	8	8	X	X	8	8	8	8	8	X	X	8	8	8	8	8	X	X	8	8	8	X	176

LESS: TOTAL OVERTIME -0-

TOTAL WORK HOURS FOR ACCOUNTING MONTH JUNE 1 THRU JUNE 30 176

Specimen Monthly Activities Report

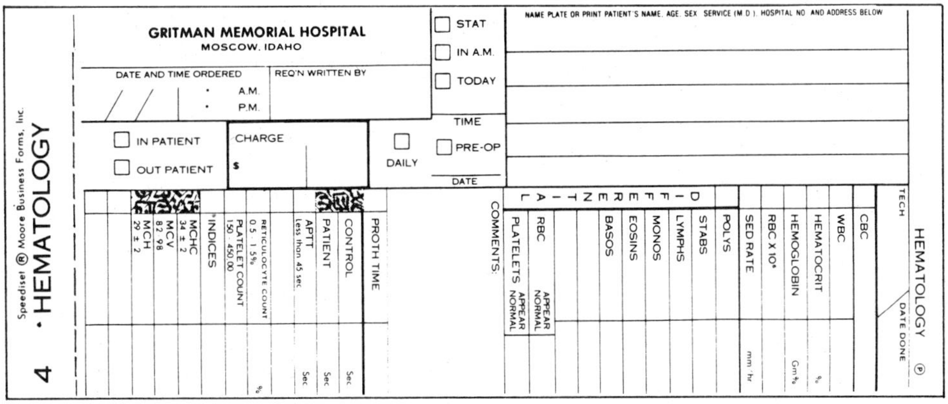

4 · HEMATOLOGY
Speediset ® Moore Business Forms, Inc.

GRITMAN MEMORIAL HOSPITAL
MOSCOW, IDAHO

DATE AND TIME ORDERED / / · A.M. · P.M.
REQ'N WRITTEN BY

☐ STAT
☐ IN A.M.
☐ TODAY
TIME
☐ PRE-OP
DATE
☐ DAILY

☐ IN PATIENT
☐ OUT PATIENT
CHARGE $

NAME PLATE OR PRINT PATIENT'S NAME, AGE, SEX, SERVICE (M.D.), HOSPITAL NO. AND ADDRESS BELOW

MCH 29 ± 2
MCV 82-98
MCHC 34 ± 2
INDICES
PLATELET COUNT 150 – 450.00
RETICULOCYTE COUNT 0.5 – 1.5% %
APTT Less than 45 sec. Sec
PATIENT Sec
CONTROL Sec
PROTH TIME

COMMENTS:
DIFFERENTIAL
PLATELETS APPEAR NORMAL
RBC APPEAR NORMAL
BASOS
EOSINS
MONOS
LYMPHS
STABS
POLYS
SED RATE mm/hr
RBC X 10⁶
HEMOGLOBIN Gm%
HEMATOCRIT %
WBC
CBC
TECH
HEMATOLOGY
DATE DONE
Ⓟ

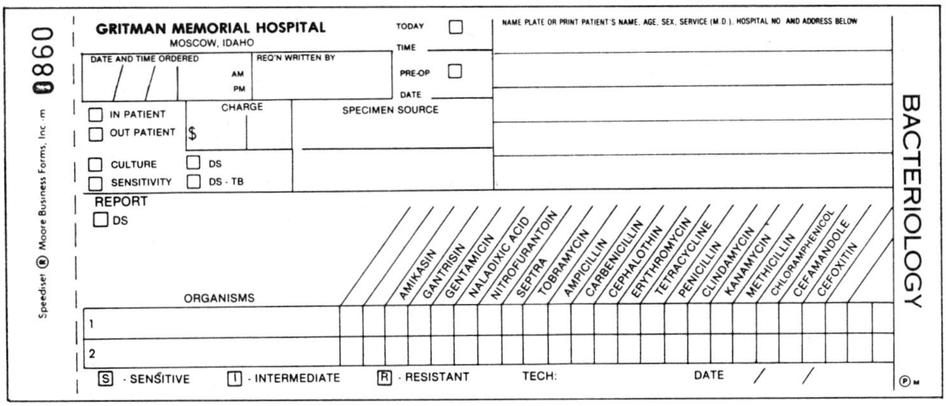

0860
Speediset ® Moore Business Forms, Inc.

GRITMAN MEMORIAL HOSPITAL
MOSCOW, IDAHO

DATE AND TIME ORDERED / / AM PM
REQ'N WRITTEN BY
TODAY ☐
TIME
PRE-OP ☐
DATE

☐ IN PATIENT
☐ OUT PATIENT
CHARGE $
SPECIMEN SOURCE

☐ CULTURE ☐ DS
☐ SENSITIVITY ☐ DS - TB

REPORT
☐ DS

NAME PLATE OR PRINT PATIENT'S NAME, AGE, SEX, SERVICE (M.D.), HOSPITAL NO. AND ADDRESS BELOW

BACTERIOLOGY

ORGANISMS
1
2

AMIKASIN, GANTRISIN, GENTAMICIN, NALADIXIC ACID, NITROFURANTOIN, SEPTRA, TOBRAMYCIN, AMPICILLIN, CARBENICILLIN, CEPHALOTHIN, ERYTHROMYCIN, TETRACYCLINE, PENICILLIN, CLINDAMYCIN, KANAMYCIN, METHICILLIN, CHLORAMPHENICOL, CEFAMANDOLE, CEFOXITIN

S - SENSITIVE I - INTERMEDIATE R - RESISTANT TECH: DATE / /
Ⓟ

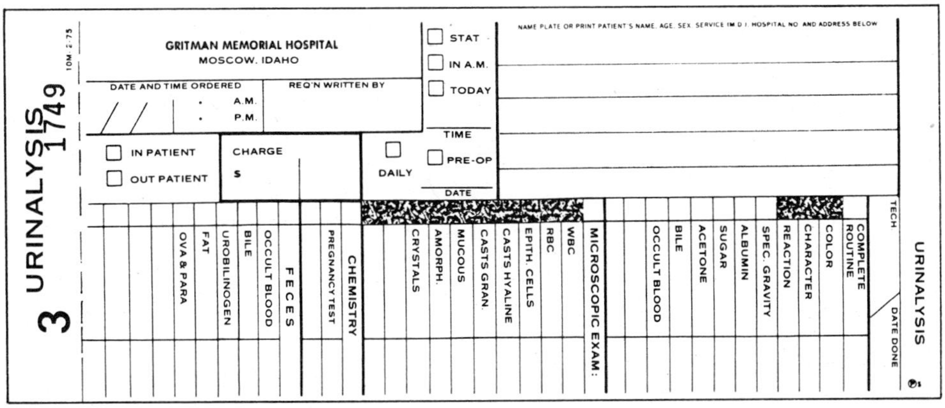

3 URINALYSIS 1749
10M 2-75

GRITMAN MEMORIAL HOSPITAL
MOSCOW, IDAHO

DATE AND TIME ORDERED / / · A.M. · P.M.
REQ'N WRITTEN BY

☐ STAT
☐ IN A.M.
☐ TODAY
TIME
☐ PRE-OP
DATE
☐ DAILY

☐ IN PATIENT
☐ OUT PATIENT
CHARGE $

NAME PLATE OR PRINT PATIENT'S NAME, AGE, SEX, SERVICE (M.D.), HOSPITAL NO. AND ADDRESS BELOW

OVA & PARA
FAT
UROBILINOGEN
BILE
OCCULT BLOOD
FECES
PREGNANCY TEST
CHEMISTRY
CRYSTALS
AMORPH.
MUCOUS
CASTS GRAN.
CASTS HYALINE
EPITH. CELLS
RBC
WBC
MICROSCOPIC EXAM:
OCCULT BLOOD
BILE
ACETONE
SUGAR
ALBUMIN
SPEC. GRAVITY
REACTION
CHARACTER
COLOR
COMPLETE ROUTINE
TECH
URINALYSIS
DATE DONE
Ⓟ

Specimen Laboratory Reports

ENERGY SERVICES
INCORPORATED
A Forsgren-Perkins Engineering Subsidiary
Energy Engineering and Development
Airport Plaza #2 1084 North Skyline Drive
Idaho Falls, Idaho 83401
(208) 529-3064

JFK-158-80
October 6, 1980

Mr. Ralph Wright
Utah Roses, Inc.
835 E. 700 S.
Bountiful, Utah 84010

Subject: Monthly Report for September, 1980.

SEPTEMBER REPORT

I. Progress

1. Well Performance Analysis

On September 16, EG&G and Energy Services personnel reviewed the technical data on well drawdown and concluded that 200 gpm operation probably can be sustained continuously for the coldest months of the winter, with drawdown not exceeding 1,000 feet for that period. However, further corrections need to be made to the analysis before specifying a safe setting depth (1,100 to 1,400 foot range) for the submersible pump.

2. Environmental

The principal issue is the question of whether a reinjection well is needed or not. A brief report examining the downstream course and chemical content of the Galena Canal is being prepared, for a forthcoming meeting with the Utah Pollution Control Board. September flows by the Utah Roses property were:

	Flow	pH	Conductivity	Temp.
Aaranall Ditch	2 cfs	7.0	1,400 μmhos/cm	60°F
Galena Canal	7 cfs	7.0	1,600 μmhos/cm	58°F

Specimen Progress Report

Mr. Ralph Wright
October 6, 1980
Page 2

II. Plans for October

Complete the report on the discharge route of the Galena and the dilution estimates. Hope to present this case to the Utah Pollution Control Board.

Sincerely yours,
ENERGY SERVICES, INC.

Jay

Dr. Jay F. Kunze, P.E.
Vice President and
General Manager

/vl

Specimen Progress Report, continued

To: Dr. Charles Stratton
Professor, English 317-01
University of Idaho

Date: December 9, 1981

From: Ben McCarroll

END OF SEMESTER STATUS REPORT

Purpose

The purpose of this report is to inform Dr. Stratton of my final status in English 317, Technical and Engineering Report Writing.

Writing Assignments

Assignment	Title	Submitted	Grade
1	Matrix Printer Proposal	Oct. 19	B+/A-
Resubmitted	" " "	Nov. 30	A
2	Filter System Recommendation	Sep. 30	B/B+
Resubmitted	" " "	Nov. 18	A-
3	Essay on English Grammar	Nov. 6	B-
4	Advanced Calculator Recommendation	Nov. 16	B
*5	Oscilloscope Procedures Manual	Nov. 24	A
6	Employment Package	Dec. 7	

* Denotes Formal Report

Excercises

I have completed all of the writing excercises that you have assigned. Also, I have completed three style analyses on my writing and four on other people's writing.

Attendance

I have attended every class period to date.

Comments on Textbooks

Both of the textbooks were helpful. Style is very informal and, therefore, is easy to read. It approaches English usage in a functional manner—it doesn't stress knowing all of the grammatical "tags" (infinitive, appositive, etc.) but illustrates what usage is awkward and what is smooth. Technical Reporting is more formal and complete; I plan to keep it and use it for a reference in the future.

Specimen Status Report

Expected Grade

Since I have completed all of the assigned excercises (as well as an employment package) and have received A's on three papers and B's on two papers, I expect I will receive an A for a final grade; that is, unless you find my English usage so poor in this report that you lower my grade.

Summary

At first, I was skeptical about this course. I had been told that just about everyone hates English courses; especially engineers. Engineers don't need to write; all they have to do is put down an equation, draw a graph, and they're set. Right? As the semester passed, I began to realize that written communication is essential in just about any field—especially technical ones. Now, I am glad that I took this course; I feel I got a lot out of it. By the way, the person who told me that everyone hates English will be enrolled in this course next semester.

Specimen Status Report, continued

☆ U.S. GOVERNMENT PRINTING OFFICE – 1974-551-303

U.S. DEPARTMENT OF AGRICULTURE

SUPERVISOR'S REPORT OF ACCIDENT

1. REPORT STATUS: [x] 1. ORIGINAL [ ] 2. SUPPLEMENTAL [ ] 3. CORRECTION [ ] 4. DELETE

2. NO. OF EMPLOYEES INJURED IN ACCIDENT: 010

3. NAME OF EMPLOYEE (LAST, FIRST, & MIDDLE INITIAL)	4. AGE	5. AGENCY CODE	6. SOCIAL SECURITY NO.	7. EMPLOYMENT GROUP CODE	8. REPORTING OFFICE (CITY AND STATE)
	23	1 1	4 5 3 2 7 8 5 0 8	0 1	Moscow, Idaho

9. KIND OF ACCIDENT	10. DATE AND TIME OF ACCIDENT (YR. MO. DAY HR. MI.)	11. NO. HRS. ON DUTY	12. WAS OCCURRENCE OUTSIDE REGULAR WORK HOURS	13. EMPLOYEE PERSONAL FACTORS	14. TREATMENT BY PHYSICIAN	15. INJURY/ILLNESS CLASS
2 (CODE)	80 06 21 11 00	05	[0] 0-NO [ ] 1-YES	00 (CODE)	[ ] 0-NO [1] 1-YES	2 (CODE)

16. DATE WORK STOPPED (YR. MO. DAY)	17. DATE RETURNED TO WORK (YR. MO. DAY)	18. NO. FULL WORK DAYS LOST	19A. TRANSFER TO ANOTHER JOB OR TERMINATION OF EMPLOYMENT	19B. LOSS OF CONSCIOUSNESS OR RESTRICTION OF MOTION	19C. EMPLOYEE UNABLE TO PERFORM NORMAL ASSIGNMENT
80 06 21	80 06 23	0002	[0] 0-NO [ ] 1-YES	[ ] 0-NO [1] 1-YES	[ ] 0-NO [1] 1-YES

20. NO. PRIVATE PERSONS INJURED OR KILLED	21. MOTOR VEHICLE DAMAGE A. GOVERNMENT	21. B. PRIVATE ON OFFICIAL USE	22. VEHICLE INFORMATION A. OWNERSHIP (1. GOVT. 2. PVT. ON OFF. BUS.)	22. B. TYPE	23. OTHER PROPERTY DAMAGE A. GOVERNMENT	23. B. PRIVATE
0 0	$ 0	$ 0	n/a (CODE)	n/a (CODE)	$ n o n e	$ n o n e

24. REPORTING UNIT ORGANIZATION IDENTIFICATION (WRITE IN): Intermountain Forest and Range Experiment Station, Moscow

25. UNSAFE ACT: P

26. FUNCTIONAL ASSIGNMENT: Forest inventory

27. SPECIFIC SITE OF ACCIDENT: Nezperce National Forest, Salmon River Ranger District, T 25N, R3E, Sec. 31

28. HAZARDOUS CONDITION: Steep hillsides

29. ACCIDENT CAUSE: Loose branches on forest floor

30. COMPLETE DESCRIPTION OF BODILY INJURY AND/OR OCCUPATIONAL ILLNESS (INCLUDE EXACT PART INJURED): Right ankle sprain

31. COMPLETE DESCRIPTION OF ACCIDENT (WHAT HAPPENED, HOW IT HAPPENED, WHAT WAS INVOLVED, WHAT LED UP TO ACCIDENT)

The job required field crew members to remove live branches from the crowns of 30 to 50 ft tall trees by cutting them with a pole pruner. While doing this Consuelo lost her footing on the steep slope and twisted her right ankle. The field crew took Consuelo to Riggins, Idaho where they were able to talk with a registered nurse who recommended a doctor look at the injury. The nearest doctor and x-ray equipment is at McCall, Idaho. Consuelo was treated there. The doctor told her to stay off the foot completely for the next 2 days and then to resume work as the ankle began to feel better. The use of crutches for at least 2 days was prescribed.

32. ACTION TAKEN TO PREVENT SIMILAR ACCIDENTS: When using the pole pruner, clear the area of branches and other debris. Establish footing before looking up and do not adjust footing without looking down again to see where your feet are going.

33. WHAT OTHER ACTION SHOULD BE TAKEN: Inform other crew members of safer procedures to use when cutting branches

34. MANAGEMENTS COMMENTS ON CORRECTIVE ACTION:

35. MANAGEMENT ORGANIZATIONAL FACTORS (CODE FROM INSTRUCTIONS):

SIGNATURE OF WORK SUPERVISOR	DATE	SIGNATURE OF REVIEWING OFFICER	DATE
Dennis E. Ferguson	June 30, 1980		

Specimen Accident Report

Writing Exercises

1. Obtain copies of a half-dozen or so monthly activity report forms from nearby firms or organizations. Comment briefly on similarities and differences.
2. Develop a standard fill-in-the-blanks form based on a recent laboratory experiment or exercise you have performed.
3. Collect a couple examples each of (a) excellent activities reports, progress reports, data reports, and status reports and (b) unacceptable or minimally adequate activities reports, progress reports, data reports, and status reports.

Writing about Writing

1. Obtain two quarterly progress reports or annual status reports prepared by a research group on your campus. Write critical reviews of each, pointing out strengths and weaknesses and making suggestions for improvement, as appropriate.
2. Interview the manager of a technical section in a nearby firm or organization to find out the nature and extent of the archival data reports they prepare and the use to which these reports are put. Prepare a written report to be distributed to your classmates.

Writing Assignments

1. Prepare an activities report on your school-related activities for last week.
2. Prepare a status report on your educational program to be presented to your department chairman or dean. Cover such items as catalog curriculum you will graduate under, courses transferred from other schools, required courses taken and remaining to be taken, electives taken and remaining to be taken, grades—both overall and in major field, withdrawals and incompletes, and anything else that might be of interest or value.
3. Consider the most recent summer or part-time job you have held. Prepare a status report that you might have written for the person who replaced you.

13

Abstracts, Summaries, and the Like

A common variety of routine writing is one in which you prepare a brief summary of something you have written or something another person has written. The scope, content, and tone of such summaries vary considerably, depending on who is to read them and why. The labels attached to such summaries vary even more. Don't worry too much about whether what you're writing is "really" an abstract, a summary, a precis, a brief, or a synopsis (there's precious little agreement, anyway); rather, get a firm hold on your audience and purpose (see Chapter 3). If you know who is going to read the summary you write—and what they're going to do with the information you present—you'll have little difficulty coming up with an appropriate direction and level of detail.

When you get right down to it, there are only two reasons people read abstracts, summaries, and the like: (1) as an aid for deciding whether or not they want to read the entire document, or (2) as a substitute for reading the entire document. All other categories of purpose are variations on these themes.

Abstracts

The term *abstract* is most commonly used to refer to the summary that is written in conjunction with a scholarly journal article or a scientific or technical report. Abstracts, then, are read mostly by people doing library research on scholarly, scientific, or technical topics. Let's look in some detail at the three most common forms of abstract.

Descriptive Abstract

The descriptive abstract is the one you prepare to accompany an article or report you have written yourself; it is a short (100- or 200-word) summary, listing, or description of the information contained in your article or report. It's the thing most people mean when they say "abstract." The primary purpose of a descriptive abstract is to enable readers of the abstract to decide whether or not they want to read the entire article or report. Since the descriptive abstract often appears in various compiled lists of abstracts and serves as the basis for computerized information retrieval systems, it must be brief, clear, and self-contained. In fact, descriptive abstracts are read most often by people who do not have immediate access to your article or report.

Here are some tips for writing descriptive abstracts:

- □ Don't attempt your abstract until after you have completed your article or report. Your abstract needs to be descriptive rather than *pre*scriptive.
- □ Make sure your abstract is self-contained and can be understood fully without reference to your complete work. Most people who read your abstract won't have your article or report to refer to.
- □ Avoid jargon and concepts that are peculiar to your field. Your abstract will be read as often as not by experts in other fields, so use language that is common to all fields of science and technology.
- □ Start with a one-sentence summary of each of the major sections in your article or report. If necessary, add an introductory or topic sentence and a concluding sentence. If your basic organizational scheme is sound, sentences like these should make a good descriptive abstract.
- □ Read a couple of dozen abstracts of articles and reports in your field. Styles, lengths, and approaches for abstracts differ somewhat from one field to another, and there's no better way to see what is current and accepted in your field.
- □ Revise your abstract as often as it takes to make it excellent. Make sure it represents the clearest and most effective written communication you are capable of producing. Remember that for every person who reads your article or report, dozens will read your abstract.

Informative Abstract

An informative abstract is typically somewhat longer than a descriptive abstract, ranging from 2 to 10 percent of the length of the article or report it summarizes. It is read most often by people who don't have the time or the inclination to read the entire work. Usually, you will prepare informative abstracts of articles or reports written by other people, and often you will do so in preparation for a literature review or state-of-the-art paper (see below). Typically, the readers of your informative abstract seek scientific or technical information that pertains to a specific and limited topic that they are interested in; your goal with the informative abstract is therefore not just to summarize the target article or report but also to summarize it with respect to your readers' topic of interest. To write a good informative abstract:

- □ Start with a brief descriptive abstract of the entire work. (The one written by the author of the article is probably as good as any.)
- □ Expand on those aspects of the article or report that are likely to be of particular interest to your readers.
- □ Finish by noting the significance of the information you have abstracted to the overall topic of interest. (Does it support or contradict the findings of other researchers? Is the information central or peripheral to the main topic? Has the article in question been cited often by other researchers?)

Evaluative Abstract

An evaluative abstract is basically an informative abstract plus your own evaluation of the quality of the work done, the significance of the findings, and the pertinence of the information to the overall topic of interest. Evaluative abstracts are generally prepared by a junior member of a research team and are read by other members of the team so that they can decide whether or not they want to read the complete work. Thus, evaluative abstracts are internal working papers rather than documents prepared for publication. A consistent outline is in order, the most common being:

I. Summary of the purpose of the research reported in the original article
II. Summary of the methods and procedures used
III. Summary of the research findings or results
IV. Summary of the author's conclusions and discussions
V. The abstractor's evaluation of the work being reported

Many research team leaders, however, prefer to see the abstractor's evaluation first, as this is the part of the evaluative abstract they are most interested in.

Summaries

Nowadays, a summary differs little from an informative abstract. The chief distinction (to the extent there is a distinction at all) lies in the fact that the term *abstract* is used with research-related activities, while the term *summary* is used more with documents that are administrative in nature. Thus, where a research report has an "abstract," a feasibility report or a proposal would be more likely to have the corresponding component labeled "summary." Traditionally, a summary appeared at the end of a document and recapitulated its main points. It served mainly to remind the readers of what had been said. Although this type of summary is still common in textbooks and reference works, it no longer has much of a place in the technical reports of business, industry, and government. What we find instead is a condensed version of the report that appears at the beginning, usually under the heading of "Executive Summary" or "Introductory Summary."

Executive Summary

An executive summary or introductory summary is a component of a formal or semiformal report (see Chapter 4) that is designed to give busy administrators, executives, or decision-makers the gist of a report so that they don't have to read the whole report in order to decide whether or not they *want* to read the whole report. Typically, an administrator will read the executive summary and then turn the body of the report over to technical experts or administrative assistants for detailed study. Alternatively, the administrator will read the executive summary to determine which portions of the report he or she wants to read in detail.

The executive summary should be the last part of your report you write—not the first. It should represent your clearest and most effective writing; remember, it is not only the first thing most people will read but also the only thing that some people will read. Concentrate on the what, who, and how much of your report. Leave the why and the how for the main body. Limit yourself to a single typewritten page for reports of a dozen pages or fewer, and limit yourself to 5 percent or so of total report length (exclusive of appendixes) for longer reports. Avoid choppy paragraphs in your executive summary, but at the same time avoid loose-jointed and rambling paragraphs. Strive for cohesiveness and continuity. (See Chapter 19, "Style at the Paragraph and Section Level.")

Concluding Summary

In a few of the technical reports you write—particularly those of a more persuasive nature—you may want to conclude by recapitulating the main points you have made in the body of your report. Be careful, though, not to introduce new information or new arguments—just summarize. You may wax a bit more eloquent in your concluding summary, and a certain amount of exuberance is allowable; just don't get corny or slushy. And don't go on and on.

□ Literature Reviews

A literature search is an exhaustive survey of the published literature on a given topic, and a literature review is the document that comes out of such an activity (although many use the term *literature search* to refer to both the activity and the document). A literature review can be a stand-alone document, or it can appear as part of a thesis or dissertation, part of a journal article, or part of a research grant request. With the stand-alone literature review, your primary purpose is to bring another scholar or researcher up to date on a particular topic; hence, you want to be both exhaustive and current. In form, the stand-alone literature review is much like the undergraduate term paper. You must be careful to quote accurately, cite thoroughly, and label your own opinions and observations as your own. There's a key difference, however; with a literature review you cannot get by with just a sample of the available sources. You must be thorough. And you should stick pretty much to primary sources. Don't load a literature review with references to *Time* magazine, *Popular Mechanics*, and the *Reader's Digest*.

The literature review component in a thesis, article, or grant request serves to orient the readers to the research being reported or proposed by providing a background against which the research can be considered. Here, you want to be selective rather than exhaustive, summarizing only the major, pertinent work that has been done before. Your purpose is not just to demonstrate that you have read all the material but also to show how your research relates to research that has been done by others. For this reason, it is better to organize your literature review according to topics and subtopics than reference by reference. Your readers want to know what has been said about certain topics—not just what certain people have had to say. On the other hand, avoid statements like

> Wolf predation on ungulates has been studied widely (Smith, 1957; Jones, 1962; Johnson, 1965; Black and White, 1974; Grey, 1981).

or

> Oil shale can be developed economically [2], [5], [19], [26], [33].

At best, these seem like cheap ways to work a long list of references into your review. Instead, include a sentence or two about each reference, pointing out the particular aspect of the problem discussed and the major conclusions presented by each author or set of authors. If a given reference isn't worth several sentences of discussion, it isn't worth including in your review at all.

Begin your literature review and each major section of it with a general overview of what has been done and what has been said. A general statement like this serves to orient your readers and gives them an introduction to what follows. Develop the main body of each section by summarizing various published sources that confirm or support the generalizations you have made. Be sure to document each of your sources properly and consistently (see Chapter 9), and regardless of the bibliographic format you use mention authors' names and publication dates—particularly for the key works you cite. Conclude each section with an evaluative summary of your own, stating your views of what is germane to the topic of interest or how it relates to your own research. Be careful, though, to label your opinions and conclusions for what they are: your own opinions and conclusions.

□ State-of-the-Art-Papers

Just as the summary is the business/industry/government counterpart of the academic abstract, so the state-of-the-art paper (or position paper, or white paper) is the business/industry/government counterpart of the literature review. It is a paper designed to bring someone up to date on a particular topic so that he or she may discuss it intelligently. A bank loan officer, for instance, might read a state-of-the-art paper on residential solar heating systems to help him decide whether (or how much) to lend for construction of such systems. Or a legislator may ask a staff assistant to prepare a position paper on child care facilities in order to be better prepared to debate an upcoming bill to regulate such facilities.

Depending on the background knowledge of its readers, the state-of-the-art paper may begin with a historical survey of work done in a particular area or it may limit itself to the current work being done. Often, the state-of-the-art paper goes beyond formally published research results and includes data from preliminary and interim reports and personal interviews. Like the literature review, the state-of-the-art paper takes the form of an undergraduate term paper; but unlike the typical term paper, the state-of-the-art paper must be thorough in its coverage of sources and—most particularly—must be as currrent and up-to-date as possible. In preparing a state-of-the-art paper, you want to be objective and unbiased both in your selection of references to summarize and in your summary comments themselves. Depending on your readers, however, and the uses to which they will put your information, you may wish to include observations, opinions, and conclusions of your own—either on the topic as a whole or on certain references you summarize. Once again, be careful to distinguish between fact and opinion and between someone else's conclusion and your own.

□ Annotated Bibliographies

A bibliography is nothing more than a list of books; or, more specifically, a list of titles, authors' names, and publication data for a set of books, articles, and other publications. An annotated bibliography is a bibliography that contains some commentary on each reference as well. Such commentary takes one (or occasionally both) of two forms: a descriptive abstract of the work, or an evaluative comment prepared by the writer of the bibliography. An annotated bibliography consisting of descriptive abstracts is easier to prepare—particularly if most of the works already have descriptive abstracts accompanying them. One consisting of evaluative comments is more difficult to prepare—you have to *read* all of the sources and make judgments about them—but it generally proves a more useful document for readers.

As you set out to prepare an annotated bibliography, decide first whether you want to be exhaustive or selective. This, of course, depends on your purpose and your readers' purposes. Are you saying, in effect, "Here's a list of everything that has been published on topic X," or "Here's a list of (in my opinion at any rate) the key sources you should consult to learn something about topic X"? Either way, begin your annotated bibliography with an explanatory note telling your readers what to expect. As with everything else you write, keep in mind that you're not performing an empty exercise—you are preparing a working document someone else is going to use in conjunction with his or her own work.

□ Meeting Minutes

A final type of written summary that just about everyone is called upon to prepare at some time or another is the minutes of a meeting. Meeting minutes serve three basic purposes:

- □ To remind people who attended the meeting what went on
- □ To allow people who didn't attend the meeting to get some idea of what went on
- □ To document for archival purposes the key actions that took place at the meeting

To accomplish these purposes, a good set of minutes must include the following items of information:

- □ **Data on the meeting itself:** Time, place, who was in attendance (and perhaps who was absent), who presided over the meeting, the nature of the meeting (regular, special), and if it isn't obvious, the purpose of the meeting.

- □ **A record of formal motions made:** The precise wording of motions made, the maker and seconder of each motion, whether the motion passed or failed, and (if customary) the number of votes for and against.
- □ **A record of other formal actions taken:** Reports heard and accepted, resolutions passed, items tabled or referred to committees, points of order raised, questions of information raised, and the disposition of any agenda item not covered elsewhere.
- □ **A summary of discussions:** Discussion on motions, resolutions, and amendments; questions on and discussion of reports; comments for the good of the organization; anything else that someone might want to recall at a later date.

Use ancillary documents to help you in writing the minutes of the meeting. The meeting announcement and the agenda will provide data on the meeting itself. If people are present whom you don't know, circulate a sign-up sheet to get names (and correct spelling) and affiliations. Get copies of all written reports submitted and all formal statements read.

Ask makers of motions to give you their motions in writing, or if this is inconvenient, get their approval for what you have written *before* the vote is taken. If no one objects, use a tape recorder to aid you in summarizing discussions. Do not, however, let the recording become a substitute for written minutes. If you're not sure that your summary is accurate, review a draft of your minutes with the people involved in the discussion. Be generous with your note-taking. It is better to boil rambling notes down for concise minutes than to attempt to expand cursory notes from memory to generate complete minutes.

Keep in mind that no one (save possibly a typist) ever complained that minutes were too complete, or too detailed, or too easy to understand. Avoid abbreviations. Avoid vague references to issues. Something like:

> Re: impact stmt, ref. DAC for further consid.

will have little meaning six weeks or six months after the meeting (if it had any at all when you wrote it). Something like this will have a much better chance of being understood in the future:

> The impact statement on the proposed parking ramp was referred to the Director's Advisory Committee for further consideration.

□ Specimen Abstracts, Summaries, etc.

7.20. Endocrine glands

349. Quality requirements of basal S-TSH assays in predicting an S-TSH response to TRH - Wide L. and Dahlberg P.A. - Dept. Clin Chem., Univ. Hosp., Uppsala SWE-*SCAND. J. CLIN. LAB. INVEST.* 1980 40/Supp. 155 (101-110).

Different variants of a solid phase TSH radioimmunoassay method (having optimal precision at different S-TSH levels) were investigated for their capability in predicting an S-TSH response to TRH from the basal S-TSH levels. Sera from 283 patients with, or suspected of having a thyroid disorder, were tested. There was a highly significant correlation between the basal S-TSH level and the increase of S-TSH after TRH for a group of individuals with normal S-TSH levels. The best discrimination of responders to TRH from non-responders was obtained with a TSH assay method with prolonged preincubation (144 h) of serum specimens and antibody. With this method assay of the basal S-TSH level was found to be useful in predicting an S-TSH response to TRH in untreated patients suspected of having hyperthyroidism. The assay offers a less expensive diagnostic alternative to the TRH test in this clinical situation. We demonstrated the effect of varying the weight of the ratio of false positive to false negative results on the optimal decision level. The basal S-TSH level should be measured by different variants of the assay method whether this is used in the diagnosis of hyper-or hypothyroidism.

350. Effect of experimental hyperthyroidism on skeletal-muscle proteolysis - Carter W.J., Benjamin W.S.V. and Faas F.H. - VA Med. Cent., Little Rock, Ark. 72206 USA - *BIOCHEM. J.* 1981 194/3 (685-690)

It is not clear whether the muscle wasting commonly observed in hyperthyroidism is due to alteration in the rate of protein synthesis of degradation. The effect of experimental hyperthyroidism on skeletal-muscle proteolysis in the rat was studied by measuring alanine and tyrosine release from isolated skeletal muscles in vitro and 3-methyl-histidine excretion in vivo. Alanine release from the isolated epitrochlaris-muscle preparation was increased as soon as 24 h after a 25 μg dose of L-tri-iodothyronine in vivo. Conversely, alanine release from muscles of hypothyroid rats was decreased, but restored by L-tri-iodothyronine supplementation before death. Furthermore, 3-methylhistidine excretion was increased in hyperthyroid rats throughout an 18-day treatment period. The increased amino acid release from isolated muscles and the increased 3-methylhistidine excretion

Specimen Descriptive Abstracts

THE CRYSTALLIZATION OF GLASSES IN THE Mg-Si-O-N SYSTEM Order No. 8200269

SHAW, THOMAS MCCARROLL, PH.D. *University of California, Berkeley*, 1981. 134pp.

The general crystallization behavior of glasses in the Mg-Si-O-N system has been investigated.

In preparation of the glasses, it was found that Si_3N_4 was a necessary constitutent of melts for glass formation, but that decomposition reactions limited the range of compositions which produced useful glasses. Melts of SiO_2 rich compositions separated into SiO_2 rich and MgO rich glassy phases on cooling. Compositional analysis of the phases established that phase separation results from extension of the miscibility gap in the MgO-SiO_2 system into the Mg-Si-O-N system.

Crystallization in a glass of composition 59 wt% SiO_2, 32 wt% MgO and 9 wt% Si_3N_4 was investigated in detail using the techniques of differential thermal analysis, x-ray diffraction, optical microscopy and transmission electron microscopy. Crystallization occurred mainly by internal nucleation of crystals, although some surface crystallization was also observed. Intermediate temperature heat treatments in the temperature range 1000°C-1300°C crystallized the MgO rich phase of the glass to spherulites of enstatite, but had no effect on the SiO_2 rich phase. The spherulites developed at intermediate temperatures consisted of fan like arrangements of plate shaped enstatite grains separated by thin films of uncrystallized glass. After heating to temperatures above 1300°C, faceted grains of enstatite were formed adjacent to the SiO_2 rich pockets by recrystallization of the spherulites. Heating to above 1300°C also precipitated crystals of Si_2N_2O residual glass between enstatite grains in the spherulites. Volume changes on crystallization of the glass caused porosity to form throughout the microstructure.

The effect of two stage nucleation and growth heat treatments on microstructure were investigated. Nucleation treatments at temperatures of 800°C-1000°C brought about a considerable refinement of the microstructure and optimum nucleation occurred at about 850°C. The effect of nucleation treatments were limited, treatments longer than 2 hours producing no further refinement of the microstructure. A brief study of crystallization in glasses of other compositions showed that nucleation was also highly sensitive to the composition of the glass. Circumstantial evidence suggests that nucleation was caused by silicon impurity particles precipitated from the glasses on cooling from the melt.

The results indicate that Mg-Si-O-N glasses may be useful as starting materials for the production of glass ceramics of new compositions. Additional experiments for exploration of this possibility are suggested.

Specimen Informative Abstract

Abstract

Kummer, Hans
1968 Reproduction. in <u>Social organizations of hamadryas baboons</u>. Chicago: The University of Chicago Press, pp. 175-179.

Kummer's book describes his study of free-living hamadryas baboons in Ethiopia. While the chapter on reproduction includes a short section describing maternal behavior at birth, the vast majority of it discusses the synchronization of reproductive cycles.

Kummer collected data on seven troops between January and October of 1961. Among these hamadryas, whose cycles average 30 days in length, three reproductive stages were identified: females with young infants; females with perineal swellings; and anoestrous females, including those who were pregnant or who had nursing juveniles. In June, when it was noticed that females of a one-male unit were often in the same phase of oestrus, detailed recordings of the reproductive states of females in one-male detached units began.

This study focused on 41 one-male units, with two adult females each. Most often, it was found that the two adult females plus the two and three year old juveniles were in the same reproductive phase. The observed frequencies of reproductive phase combinations were compared with the frequencies expected if these phases resulted from random combination. The differences were marked, indicating clearly that reproductive phases were not randomly distributed. After the May - June mating peak, the oestral synchronism of the units declined somewhat; an increasing number of females lacked signs of sexual swelling when they were expected to be in oestrus. The possibility exists that many or all of these females were, in fact, pregnant. However, with the births from this mating period not expected unit November or December, Kummer did not have a chance to record them.

Kummer concluded that social factors do affect female oestrus cycles within single male units. It was also apparent that sychronization of reproductive cycles was an intragroup rather than an intergroup phenomenon.

This chapter of Kummer's has not been cited in the other few articles discussing synchronization, probably because book chapters are not listed in <u>Index Medicus</u> or <u>Biological Abstracts</u>. It is, however, a valuable addition to the meager literautre on this subject. Kummer's work, since it focused on other primates, is certainly much more relevant to human research than are the rodent studies. The fact that Kummer's observations were conducted in the wild adds to their significance.

Specimen Evaluative Abstract

analyzing technical writing, however, we find that many of the "faults" are justifiable if not carried to extremes, and if not used inappropriately when writing for nontechnical audiences.

Scholnick, E. K. and Adams, M. J., Relationships Between Language and Cognitive Skills: Passive-Voice Comprehension, Backward Repetition, and Matrix Permutation, *Child Development, 44:*4, pp. 741-746, 1973.

*Semantic and cognitive factors governing passive-voice comprehensions were explored. Ninety-six children, divided among three grade levels (kindergarten through second), identified the object referred to by nonsense-syllable actors and recipients in passive sentences and reversed active sentences and a classification matrix. Comprehension of passives did not require reversal to an active form since children who understood passive sentences did not necessarily pass the language and matrix reversal tasks. Level of comprehension of passives, but not backward repetition, was affected by actor content. But, there were significant intercorrelations among the tasks, and, therefore, some overlaps in linguistic and cognitive performance.

Siegel, D., Nonsources of Unpassive, in J. P. Kimball, *Syntax and Semantics, 2,* Seminar Press, New York, pp. 301-317, 1973.

Siegel identifies an unpassive as a sentence "characterized by the presence of an auxiliary verb (either "be" or "go"), by the presence of a past participle prefixed by "un-," and by the optional presence of an agent phrase." Nonpassive sentences rarely derive from well formed active sentences ("The president's blunder was unreported" has no acceptable active equivalent: "The press unreported the President's blunder"). She concludes with three features any explanation of unpassives must include.

Stanley, J. P., Passive Motivation, *Foundations of Language, 13:*1, pp. 25-39, May 1975.

Stanley summarizes the current inadequate explanations of the meaning and function of the passive, believing that the focus on word-order only limits explanations and understanding. Rather, she will analyze how writers use the passive, its contexts, the situations where the writer needs it, and those uses judged acceptable. She will further attempt to answer questions about why it exists at all. She reviews representative discussions of the passive, offers contemporary uses of it, and presents explanations of these uses. A theoretical analysis follows this material. She suggests two reasons why the speaker/writer would choose passive: (1.) don't know logical subject; and (2.) do know it but don't want to provide it. She concludes by asserting that the use of the passive voice reflects a philosophical orientation, allowing the writer to appeal to some universal source for validation that the reader may or may not know.

Stanley, J. P. and Robbins, S. W., Forced Inferences: Uses and Abuses of the Passive, *Papers in Linguistics, 10,* Edmonton, Canada, pp. 299-311, 1977.

The authors are concerned about why the passive "has become one of the most widely used sentence patterns in the language." They believe that the passive is a device the writer can select to persuade the reader/hearer as

Specimen Annotated Bibliography

UNIVERSITY CURRICULUM COMMITTEE MINUTES

(1976-77 Meeting #4, 4 October 1976)

The meeting was called to order by the chairman at 4:10 p.m. in the Education Conference Room. Members attending: Knight (chmn.), Byers, Carson (w/o vote), Falter, Gunning, Iriarte, Lawrence, Siems, Stauber, Tovey for Jacobsen, Members absent: Jacobsen, Jenkins, Telin, Wohler. Observer: Professor D. O. Everson. The minutes of the preceding meeting were amended to show that Professor Byers, not Professor Lawrence, seconded the motion to table (p. 2) and were approved as amended.

Proposed Program in Computer Science Ready (UCC-76-047). Professor Siems reported that the Information Science Subcommittee has reviewed the revised course outline for the proposed computer science course, CS 101, Introduction to Computer Algorithms; the subcommittee recommends that UCC approve the entire program for forwarding to Faculty Council.

MOTION: That item #1 of UCC-76-047, the addition of CS 101, be approved, and that the entire proposal be forwarded. Motion passed unanimously.

Proposal Program in Applied Statistics Approved (UCC-77-006). The report from the College of Agriculture entitled "Proposed Program in Applied Statistics," 15 September 1976, was taken up for action.

MOTION (moved and seconded by the Information Science Subcommittee): That the Program in Applied Science be approved as presented in UCC-77-006. (See below.)

Members of the committee, particularly those who have sat on the Information Science Subcommittee during the years that the proposal was being formulated, were enthusiastically in favor of the proposal. The discussion focused on details of implementation of the program, especially the proposal to jointly number the proposed course ApSt 301, Principles of Statistics, in four or five departments. Questions were raised concerning (1) keeping the basic statistics course in the course listing for the College of Business and Economics so as to facilitate accreditation of the college; (2) providing for multiple sections to be listed in the time schedule of one of the "sections"--renumbering the sections 301A, 301B, etc., was considered, but that would defeat the idea of jointly numbering them; (3) teaching the course at the sophomore level in the business college and at the junior level elsewhere--everyone seemed to agree that reducing the level of the course to 200 is a good idea because, among other reasons, if a graduate student takes the course, he shouldn't be able to get credit for it; (4) correclating the descriptions of the various sections of a jointly-numbered course so that they are in fact the "same as"; (5) coordinating the assignment of instructors for applied statistics courses.

MOTION TO AMEND (Siems, Byers): That the section of the proposal entitled "Proposed courses" (on pp. 7-8) and section 4-d (on pp. 8-9) be deleted, and that the remainder of section 4 be relettered. Motion to amend passed unanimously. When the vote was taken on the main motion as amended, it too passed unanimously.

The meeting was adjourned at 5:10 p.m.--Franklin D. Carson, Secretary

Specimen Meeting Minutes

Writing Exercises

1. Consult a dozen or so scholarly or trade journals in your field or discipline. Prepare a list or table indicating which ones use abstracts, what kind of abstracts (descriptive or informative), average lengths of abstracts, and any other data that strike you as interesting or valuable.
2. Collect three examples of (a) executive or introductory summaries that you judge to be excellent and (b) executive or introductory summaries that you judge to be unacceptable or minimally adequate. Document your sources.

Writing about Writing

1. Prepare an annotated bibliography on the topic "Writing abstracts and summaries." Aim the bibliography at college students majoring in some field of science or technology.
2. Write critical reviews of several literature reviews or state-of-the-art papers. Check some of the references for accuracy. As best you can, evaluate the completeness of the documents. Point out strong points and weak points and make suggestions for improvement as appropriate.
3. Write critical reviews of several sets of meeting minutes for student or faculty organizations at your school. Point out strong points and weak points and make suggestions for improvements as appropriate.
4. Write an essay on "The difference between college term papers and state-of-the-art papers." Aim the essay at faculty members in your major field.

Writing Assignments

1. Prepare a 200- to 250-word summary of a 2000+-word article in your field. Attach a copy of the article on which you have underlined the key sentences and noted the organizational structure of the article.
2. Write a state-of-the-art paper on some field of study in which you have recently done extensive library research. (Don't start from scratch; the research will take too long.) In your bibliography for this paper, include the library call numbers for each of your references.
3. Prepare an annotated bibliography on some topic in your field. Include at least a couple dozen references. In your bibliographic entries, include a brief descriptive abstract, a subjective evaluation of the work, and the library call number.
4. Make arrangements with the chairman or secretary of some faculty or administrative committee on your campus to attend two successive meetings for the purpose of taking minutes. Review the minutes you write with the regular secretary and revise as necessary. Submit your original minutes, the secretary's comments, and your revised minutes as your assignment.

5. Assume you are on the staff of a state legislator. Prepare a position paper on some upcoming piece of state legislation or on some issue that is likely to come before the legislature this year.

14

Instructions and Manuals

As you advance through the management ranks in a firm or organization, you will be faced with more and more occasions for lateral and downward communication (see Chapter 3). One of the more common forms of lateral—and especially—downward communication is instructional writing. You will be called upon to provide specific and detailed instructions for procedures and operations that others will be expected to perform. You may have to outline office or laboratory or shop procedures; you may have to provide instructions for using or maintaining simple pieces of equipment or apparatus; you may well have to provide simple written directions just to get things done the way *you* want them done. In the first part of this chapter we will focus on writing basic instructions, saving the full-fledged instruction manual for later in the chapter.

Writing Style for Instructions

Good instructions are complete, clear, and well-illustrated. The success of a set of instructions hangs not on elegance of diction and smoothness of phrasing but on completeness of information and clarity of expression. People rarely read instructions for pleasure; they read them because they have a need to know. And as often as not they begin a procedure on their own, referring to the written instructions only when they get confused or go wrong. People typically do not read instructions in sequence as a continuous narrative. For these reasons, the writing style you use for instructions differs markedly from that of all other forms of technical writing. Consider the following simple instructions:

Locating the Class Party

These are instructions for finding the party that is to be held for members of the tech writing class. The party will be held at 122 Mary Street in Springfield. The instructions assume that you will be driving into Springfield from the west on Highway 8 (see map).

1. Go under the railroad bridge and on past the Lumberjack Café to Front Street.
2. Turn left on Front Street.

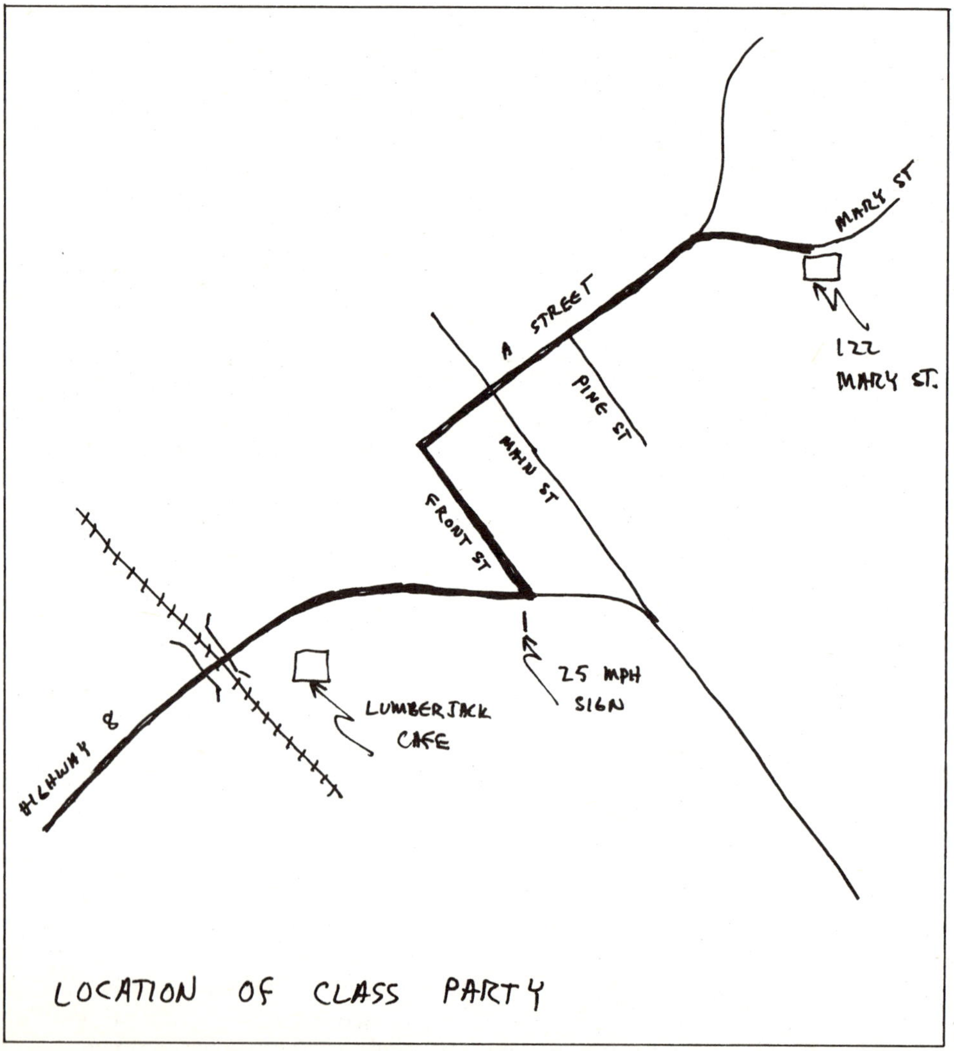

LOCATION OF CLASS PARTY

Note

Front Street is just opposite the first 25-mph speed limit sign you encounter coming into Springfield on Highway 8.

3. Proceed northwest on Front Street for two blocks until you come to "A" Street.
4. Turn right on "A" Street.
5. Proceed northeast on "A" Street for approximately four blocks until "A" Street runs into a wishbone-shaped intersection, with Monica Street curving off to the left and Mary Street curving off to the right.
6. Swing right onto Mary Street.

Caution

The speed limit drops from 25 mph on "A" Street to 15 mph on Mary Street. This is strictly enforced.

7. Proceed about a half block on Mary Street, until the pavement turns to gravel.
8. Find a place to park; 122 Mary Street is on your immediate right.

Note

The house you're looking for is dark brown with white trim and a cedar shake roof. Also, look for the blue horse trailer parked in front.

Notice that these instructions are simple, direct, clear, and complete. This is the way you should write all instructions. Don't worry about insulting your readers' intelligence. Don't worry about mentioning something they already know. In more than twenty years in the technical communications field, I have *never* heard anyone complain that a set of instructions was too clear. Or too complete. Or too easy to understand. Here are some additional suggestions for instruction writing.

Use Groups of Numbered Steps

As you organize your instructions for any sizable project or job, break the job down into individual tasks and provide a separate heading or subhead for each of these tasks. As a rule, you should have no more than eight or nine individual steps in each task; if you have a dozen or more individual steps, try to break the operation into two tasks. Look for phases that different people might perform, or look for convenient stopping points in the total sequence. You might begin the discussion of a task with a sentence or two in the indicative mood and paragraph format explaining the importance of the task or providing background information. When you get into the actual instructions, however, number each step. And limit yourself to one step per number. This accomplishes two things: it makes it easy for your readers to keep their place as they follow through your instructions, and it tends to keep you as writer from leaving steps out.

Use Active Voice and Imperative Mood

Use the active voice and imperative mood for your individual instructions. "First, do this. Then, do that," rather than "First this is done. Then that is done." A sentence like "Check the oil before you start the engine" leaves no doubt in the reader's mind that *he or she* is supposed to check the oil. On the other hand, a sentence like "The oil should be checked before the engine is started" suggests a number of other possibilities: "The oil should be checked, but doesn't really have to be." "Somebody else should have checked the oil, so I don't have to."

Use Normal English

Avoid the kind of broken English that shows up in soup-can-label recipes and on some instruction plates (see Figure 14–1), the kind of broken English where definite articles and other short function words are omitted. Admittedly, one sees a lot of this broken style in instruction writing, but there is no reason for it, really. The space saved isn't worth the possible confusion introduced. And reading time is actually increased

Remove suction-hot gas and liquid line adapters from outdoor and indoor unit packages. Sweat solder to respective lines. Remove seal nuts from indoor unit and make suction-hot gas and liquid line connections, useing Teflon gaskets provided. Repeat procedure at outdoor unit. Remove seal nuts from suction-hot gas and liquid line connections of indoor unit. Make charged tubing connections, using Teflon gaskets provided.

NOTE: If sight glass is to be installed in liquid line, connect male adapter on sight glass assembly to breakaway valve on liquid line. Tighten connection. Loosely connect sight glass assembly to liquid connection on indoor unit. Momentarily open valve on liquid line tubing to purge connection. Tighten connection.

Figure 14–1 Example of English with Short Function Words Missing

for the choppy version, even though it contains fewer words than the full version (see Figure 14–2).

```
   Remove the suction-hot gas and the liquid line
adapters from the outdoor and the indoor unit pack-
ages.  Sweat solder them to their respective lines.
Remove the seal nuts from the indoor unit and make the
suction-hot gas and liquid line connections, using the
Teflon gaskets that are provided.  Repeat this procedure
at the outdoor unit.  Remove the seal nuts from the suc-
tion-hot gas line and liquid line connections of the in-
door unit.  Make the charged tubing connections, using
the Teflon gaskets provided.
   NOTE:  If a sight glass is to be installed in the
liquid line, connect the male adapter on the sight glass
assembly to the breakaway valve on the liquid line.
Tighten the connection.  Loosely connect the sight glass
assembly to the liquid connection on the indoor unit.
Momentarily open the valve on the liquid line tubing to
purge the connection.  Tighten the connection.
```

Figure 14–2 Example from Figure 14–1 with a Full Complement of Function Words

Be Religiously Consistent with Terminology

Each time you refer to a particular part, call it by the same name. Don't refer to something as a shutoff valve here, a stopcock there, and a cutoff somewhere else; your readers will wonder whether you have one, two, or three such devices. Include all modifiers each time you refer to a particular component. If it's a retaining spring here, it should be a retaining spring there, and not just a spring. Be particularly careful about inconsistencies in terminology between your body copy and your figure captions and callouts; it is very easy to slip up here because you process them at different times.

Illustrate Profusely

Instructions are like show and tell. Don't overlook the showing as you get involved in the telling. Don't force your verbal descriptions to carry your communication when a map, drawing, or schematic will do just as well (if not better). Don't use just one drawing when several would be better. Don't limit yourself to line drawings when photographs would show things more clearly (see Chapter 8). Once again, I have never heard anyone complain that instructions were too well illustrated.

Make Ample Use of Notes, Cautions, and Warnings

Adequate instructions tell people what to do. Excellent instructions go beyond this to tell people why things should be done (or not done) and offer suggestions for easy ways (or alternate ways) to do things. One way to do this is to make use of notes, cautions, and warnings in your instructions. In general, a note (see Figure 14–3) offers the reader useful

o. When the cut-in tool is installed correctly and the underspeed override valve is closed, reduce the engine speed (rpm) to 2325 ± 10.

NOTE: This is the engine speed used to adjust the cut-in point of the underspeed valve.

p. Straighten the lock tab and loosen the locknut on the underspeed cut-in adjustment screw.

NOTE: Adjustment is made with a screwdriver and the adjustment screw is located on the right side of the machine.

q. Turn the adjustment screw out (counterclockwise) two full turns.

NOTE: This step is necessary because the adjustment must always be made by turning the screw in (clockwise).

r. Pull the FORWARD-PARK-REVERSE lever into RESET (forward or reverse). The continuity light must not come on. If it does, turn the underspeed cut-in adjustment screw out until it goes off.

Figure 14–3 Representative Notes

If the ambient temperature is 50°F (10°C) or less, it is necessary to increase the temperature of the R−12 refrigerant in the can. This is necessary to insure that the refrigerant enters the system as a vapor (gas) and not as a liquid.

CAUTION

Do not allow refrigerant as a liquid to enter the system through the suction side of the compressor or compressor damage and failure will occur.

To increase the temperature of the refrigerant in the can, it is necessary to place the can in a pan of heated water for 5 to 10 seconds during the charging procedure (until the can is empty of refrigerant).

Figure 14–4 Representative Caution

1. Install tool (D) to hold the lift arms up if it is necessary to remove only one lift cylinder. Install tooling (B) to hold the lift arms up if it is necessary to remove both lift cylinders.

⚠ WARNING

Oil that is hot and under high pressure can cause injury. To prevent injury all pressure in the hydraulic system must be released before any hydraulic lines are disconnected. To release the pressure in the hydraulic system stop the engine and move the implement control levers to all positions. Let the hydraulic oil become cool before work is done on the hydraulic system.

2. Remove bolts (1) from the bracket that holds the oil tubes to the lift cylinder.

Figure 14–5 Representative Warning

information—a reason why, an alternate method, a potential problem to avoid. A caution (see Figure 14–4) gives the reader a suggestion to follow in order to avoid possible damage to the equipment. The warning (see Figure 14–5) is reserved by careful writers for those situations where error might lead to personal injury. Safety, of course, is a most important aspect of giving instructions, so don't bury your safety information in some obscure place—put it right in the instructions themselves.

As you write the instructions for a particular task, do not assume that your readers have indeed read the instructions for all earlier tasks. In fact, assume that they have not. If information presented elsewhere in your instructions is crucial to the understanding and performance of the task at hand, refer your readers specifically to the section in question. Or better yet, if the information needed is fairly brief, repeat it. Put it before the reader in exactly the place he or she needs it; don't send your reader chasing for something you could include in a few lines. Above all, don't be reluctant to repeat cautions and warnings every time they are applicable.

□ Success Strategies for Student Writers □

Pick a procedure that you are thoroughly familiar with as a topic for your instructions. Aim your instructions at readers who are totally unfamiliar with the procedure. Avoid procedures such as Gram staining or titration or setting up a transit that are described in standard textbooks; it's too tempting to just follow someone else's model rather than think through the procedure on your own. Look to your hobbies and your summer or part-time work experience for topics.

□ Checking the Completeness and Accuracy of Your Instructions

As you write instructions for a procedure or task that you are intimately familiar with, it is very easy to leave out crucial steps—steps that seem obvious to you because of your detailed knowledge of the process. Similarly, it is very easy to mislabel components, say "left" instead of "right," or get things out of order. One of the best ways to check for problems like this is to give your written instructions to someone and ask him or her actually to perform the procedure following your instructions. In large firms and organizations, instruction manuals are field-tested in this manner; for smaller operations you as a writer should conduct this field testing. Don't just ask someone else who is quite familiar with the procedure to read over your instructions and comment on them; get someone who is *not* familiar with the procedure and ask that person to actually go through each step according to your instructions. You'll be surprised what will turn up.

The Manual as Reference Piece

"When all else fails, get out the instruction manual!" This is often said and more often believed, for it is a fact of human nature that most people regard written instructions as a last resort, a sort of insurance policy to fall back on just in case they can't get the procedure or operation figured out on their own. It is certainly the case that people are even less likely to curl up by the fire with a glass of warm milk and read an instruction manual than they are to do so with a technical report. Readers of manuals are at best a reluctant audience.

By and large, people use instruction manuals the same way they use dictionaries and encyclopedias: as references. When a person has a particular task to perform and doesn't know for sure how to perform it, he or she will consult that portion of the manual that covers the particular task in question. Rarely if ever will one of your primary readers go through an instruction manual from cover to cover—any more than a person reads a dictionary from cover to cover. Rather, your primary readers—the people who need to perform the tasks you discuss—will refer to your manual on a need-to-know basis, consulting various sections of it at different times as they are faced with different tasks.

The key to writing a successful instruction manual, then, is compiling it in such a way that it is easy to use as a reference piece. Package your information in small easily managed chunks. Use extensive headings and subheads to identify the packages of information. Repeat the same information in different places in the manual, if that will help your readers learn what they need to know. Cross reference extensively. And include a section that tells your readers how to use the manual—how to find information in it.

General Outline for Instruction Manuals

I. Introduction
- A. Purpose of the Manual
- B. Scope of the Manual
- C. Additional References

II. Physical Description of the Mechanism

III. Functional Description of the Mechanism

IV. Detailed Instructions (sections as needed)
- A. Assembly
- B. Installation
- C. Initial Startup

D. Operation
E. Maintenance and Trouble-Shooting
F. Repair and Servicing

V. Parts List

□ Introduction

The introduction to your instruction manual is the one section you can assume everyone will read. In the introduction, you want to outline just what the manual is intended to do and how you expect people to use it. Remember, this is the introduction to the *manual*, not the introduction to the apparatus or hardware the manual deals with. If your manual is self-contained, say so. If the total package of instruction is broken up into several separate manuals, cite each companion piece completely and let your readers know what information is included in each. List in some detail the packages of information contained in your manual and the order in which they are presented. (If your manual is longish—fifteen pages or more—include a table of contents.) If you know of some general outside references that provide useful background information or discuss theory of operation, share this knowledge with your readers. In general, provide in your introduction everything your readers should know in order to use your manual most profitably.

□ Physical Description of the Mechanism

In this section and the next, you give your readers the big picture of what your mechanism looks like, what it does, and how it works. (For a procedure manual you would instead give your readers the big picture of the overall operation and how the various steps of the process fit in.) The reason for this is simple: the more people know about how a mechanism works, the better they will be able to operate it. The more they know about how the mechanism is supposed to work, the better they will be able to repair it when it malfunctions. As you put your physical description together, keep two things in mind: (1) your readers probably have the mechanism in front of them to look at, and (2) they just want to know enough about it to operate it; they aren't going to build one of their own.

Often, an illustration or two with components labeled will be sufficient for physical description, as in Figure 14–6. If the mechanism you are describing has internal components, provide an exploded view, cross section, or cutaway view so that your readers can see inside the mechanism (see Chapter 8). Complement your illustrations with verbal descriptions but stop short of describing what any fool can see by looking at the mechanism or your pictures of it. You can organize your verbal descriptions by following outlines suggested by Steven E. Pauley:

II. Physical Description of the Mechanism
  A. Purpose of the Mechanism
  B. Overall Appearance
  C. Major Components
  D. Description of First Component
    1. Purpose of the Component
    2. Size, Shape, and Material
    3. Location and Relation to Other Components
  E., F., etc. Descriptions of Second, Third, etc. Components

As you follow this generalized outline, use specific rather than the generalized headings and subheads. For instance, write "Purpose of the Oscilloscope" rather than "Purpose of the Mechanism," and write "Amplifier" rather than "Description of First Component." You probably won't need the third-level heads "Purpose of the Component," "Size, Shape, and Material," and so on at all; they're just in the outline to remind you to include such information in your description. The same principles apply to functional descriptions as well. Write "Function of the Two-Cycle Engine" rather than "Statement of the Mechanism's Overall Function," and write "Compression-Power Stroke" rather than "Description of the Second Phase of Operation." Avoid the third-level heads but include information about purpose, results, and relationships to other phases in your descriptions of each phase in the operation.

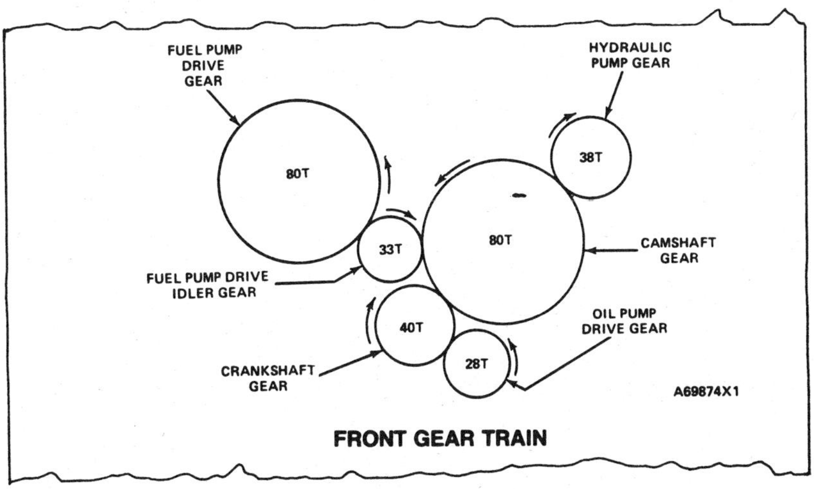

Figure 14–6 Mechanism Illustrated with Components Listed

□ Functional Description of the Mechanism

For some mechanisms (such as a light bulb, a can opener, a pencil sharpener) a physical description is sufficient; if you know what the components look like, you know pretty well how the device works. For other mechanisms, however (a fluorescent light, a transistor, a plate of photographic film), you need more; a detailed physical description of the mechanism's components still leaves you perplexed as to just how the device accomplishes what it does. Some kind of functional description or explanation is therefore necessary. And of course with any mechanism exhibiting a fair degree of complexity (a microwave oven, an oscilloscope, a single-lens reflex camera), most readers will need both a physical and a functional description. The organization of your functional description should be similar to that of your physical description, except that you focus on phases of operation rather than physical components:

III. Functional Description of the Mechanism
- A. Statement of the Mechanism's Overall Function
- B. Explanation of Operating Principles
- C. Major Phases of Operation
- D. Description of the First Phase of Operation
  1. Purpose of the Operational Phase
  2. Description and Results of the Action
  3. Relationship to Other Phases
- E., F., etc. Descriptions of Second, Third, etc., Phases
- G. Summary of Entire Cycle of Operation

The decision to present a physical description, a functional description, or both depends, of course, on the nature of the mechanism you are describing and on the knowledge and needs of your audience. Sometimes you may find it convenient to combine physical and functional aspects in a single description. This is fine so long as you stick with a consistent organizational structure—either component-by-component or phase-by-phase. Don't mix them together. Always remember that the only purpose of the mechanism description is to aid your readers in assembling, operating, or servicing the apparatus or hardware covered in your manual. Don't get carried away with the description and let it become an end in itself.

□ Detailed Instructions

The heart of an instruction manual is the instructions—the detailed, step-by-step accounts of what the reader should do in order to accomplish the various tasks outlined in the manual. Since the manual in general and the detailed instructions in particular are not intended to be read in sequence as a continuous narrative, the writing style you use for instructions differs markedly from that of all other forms of technical writing. At this point, you should reread the section on "Writing Style for Instructions" earlier in this chapter. Everything said there is applicable to the detailed instructions portion of your manual. In particular, you should:

- □ Use numbered steps for all but the simplest of instructions
- □ Write instructions in active voice and imperative mood
- □ Use normal English with a full complement of function words
- □ Be religiously consistent with your nomenclature and terminology
- □ Illustrate your instruction manual profusely
- □ Make ample use of notes, cautions, and warnings.

The ordering of the subsections within your detailed instructions warrants a little thought. If your firm or organization has published other similar manuals, consult one of these earlier manuals for organizational guidelines. If you're plowing new ground, consult several similar manuals from other firms or organizations whose publications you respect. Get your ideas where you can find them; don't waste energy reinventing the wheel. Here are some guidelines to keep in mind, however:

1. Keep introductory and read-once information separate from step-by-step instructions.
2. Explain oft-repeated operations first in a section or subsection, saving those procedures that are performed only occasionally until later.
3. Put emergency information in a position of prominence—at the beginning or end of a section or on a page all by itself. When a technician's world is coming apart, he doesn't want to waste precious minutes thumbing through a manual. Consider using colored paper or colored ink for emergency information.
4. When explaining generalized procedures—say, performing an operation with a hand calculator or computing the volume of a load of wood—offer plenty of examples. People learn inductively as well as deductively.
5. Don't be overly concerned with chronological order among the various sections of your detailed instructions (but be strictly chronological *within* any set of numbered instructions). Order of importance and order of frequency are more significant than chronology in most cases.
6. Use clear headings and subheads and ample cross references to help your readers find the information they need.

□ Success Strategies for Student Writers □

Don't try to write an instruction manual for a mechanism unless you are intimately familiar with it. And don't try to work from memory; have the mechanism right in front of you. If you want to write instructions for a particular tool or instrument used in your field, see if you can borrow one from your department.

Write your detailed instructions as if you had a photographer available. Set aside space for photos and provide figure numbers, captions, and callouts. Then in the space for the photo, write a brief description of what the photograph would show.

A full-blown instruction manual can get rather lengthy; check with your instructor to see if you can submit the introduction and the description portions and the detailed instructions portions as separate assignments. Write each portion as if your readers had the other portions available.

□ Parts List

Although you have identified individual components of your mechanism in the description sections of your instruction manual, you should gather together in one place a list of all the serviceable, or replaceable, components. Start with an illustration (or several illustrations) in which each serviceable part is called out. Then set up a table (see Figure 14–7) in which you give the illustration callout number for the part, the quantity needed, the official part number, and a brief description of the part.

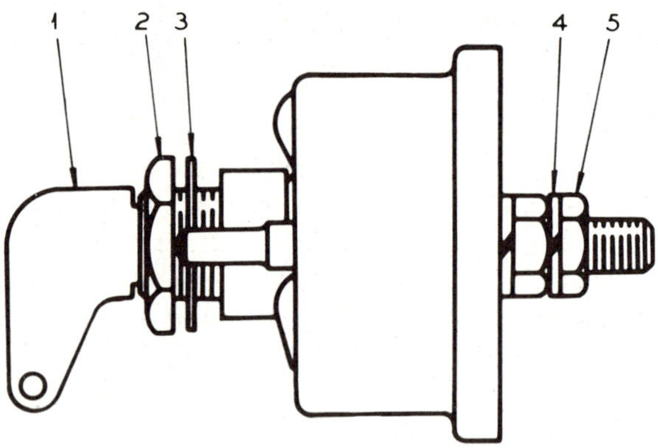

REF. NO.	PART NUMBER	DESCRIPTION	QTY
1	8H5306	KEY----------------------------	1
2	8H5303	NUT----------------------------	1
3	2H3922	LOCKWASHER------------------	1
4	7F7259	LOCKWASHER------------------	4
5	6F7045	NUT----------------------------	4

211499

Figure 14–7 Parts Illustration and Table

DISASSEMBLY AND ASSEMBLY **POWER TRAIN**

FINAL DRIVES AND SPROCKET ASSEMBLIES

12. Remove plate (11) and the shim(s) from the spindle.

13. Use pry bars to loosen ring gear (12) and the ring gear hub on the spindle. Use two men to remove ring gear (12) and the hub as a unit. Weight is 23 kg (50 lb.).

14. Remove ring (13) from ring gear (12). Remove hub (14) from the ring gear.

15. Install tooling (C) on the sprocket hub and fasten a hoist. Remove sprocket hub (16). Weight is 84 kg (185 lb.). Remove bearing cone (15) from the sprocket hub.

16. Carefully remove Duo-Cone seals (17) from the sprocket hub and spindle. Be careful not to damage the contact surfaces of the seals.

17. Inspect the rubber seals of the Duo-Cone seals for any cuts or damage.

18. Remove bearing cups (18) from the sprocket hub. Remove the O-ring seal for the sprocket hub.

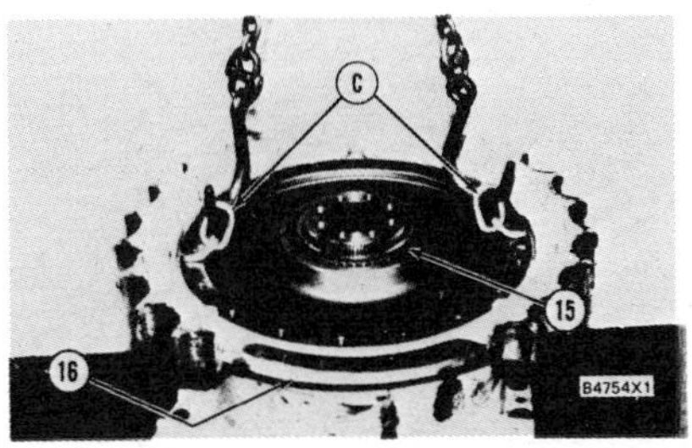

2-63

Specimen Detailed Instructions (Caterpillar Tractor Co.)

FINAL DRIVES AND SPROCKET ASSEMBLIES

19. Use tooling (D) to remove bearing cone (19) from the spindle.

20. Remove bolts (20) that hold the spindle to the final drive case.

21. Install two ½''-13 NC forged eyebolts (21) in spindle (23) and fasten a hoist. Install two 5/8''-11 NC forcing screws (22) in the spindle and remove the spindle.

22. Remove bearing race (24) from the spindle. Remove the O-ring seal from the spindle.

23. Remove gear (25) from the final drive case.

CAUTION

Bearing (26) can be removed as shown in Steps 24 or 25. Remove the bearing only if a replacement is necessary because the bearing will be damaged when it is removed.

24. Use tooling (E) to remove bearing (26).

Specimen Detailed Instructions, continued

FINAL DRIVES AND SPROCKET ASSEMBLIES

25. Install two 3/8"-16 NC forcing screws (27) in the back side of gear (25) and use the forcing screws to push the bearing off the gear.

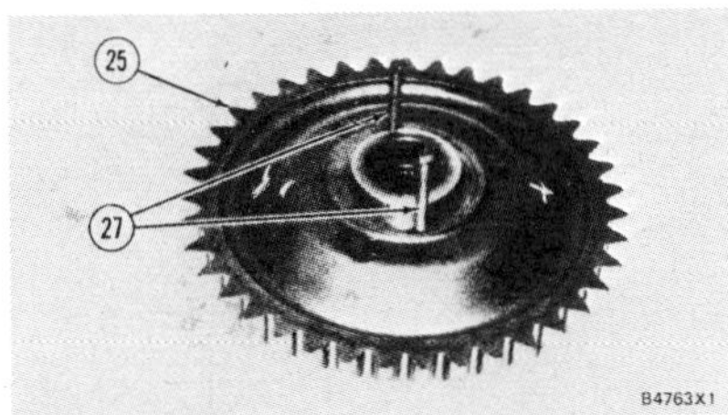

CAUTION

Do not let pinion (29) fall out of the case when it is removed.

26. Install a 5/8"-11 NC forcing screw (28) in the final drive case and use the forcing screw to push pinion (29) out of the case.

27. Install a 3/8"-16 NC forcing screw in thrust plate (30) and remove the thrust plate.

NOTE: Thrust plate (30) can be turned over and installed to get a new wear surface.

28. Remove bolt (32) and the washer from shaft (31). Remove shaft (31) from the pinion. Remove the O-ring seal from the shaft.

29. Use tooling (F) and a press to remove the bearing cones from the pinion.

30. Remove final drive case (33) from tooling (A) and turn up side down. Weight is 72 kg (160 lb.).

31. Use tooling (G) to remove the bearing cap for the pinion bearing from final drive case (33).

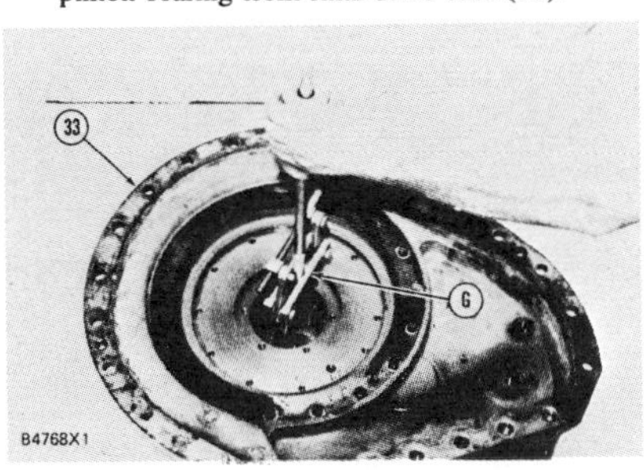

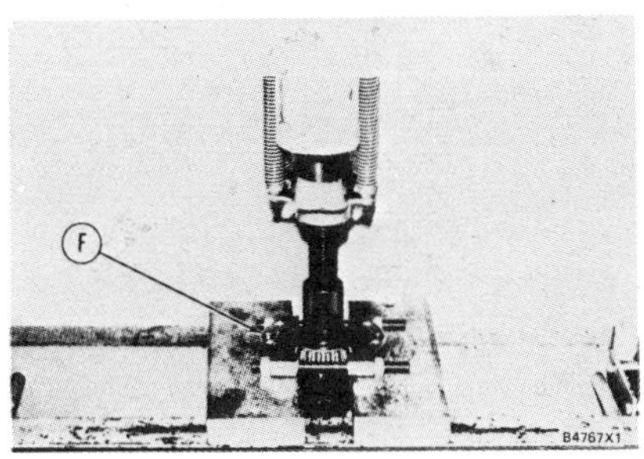

Specimen Detailed Instructions, continued

DISASSEMBLY AND ASSEMBLY POWER TRAIN

FINAL DRIVES AND SPROCKET ASSEMBLIES

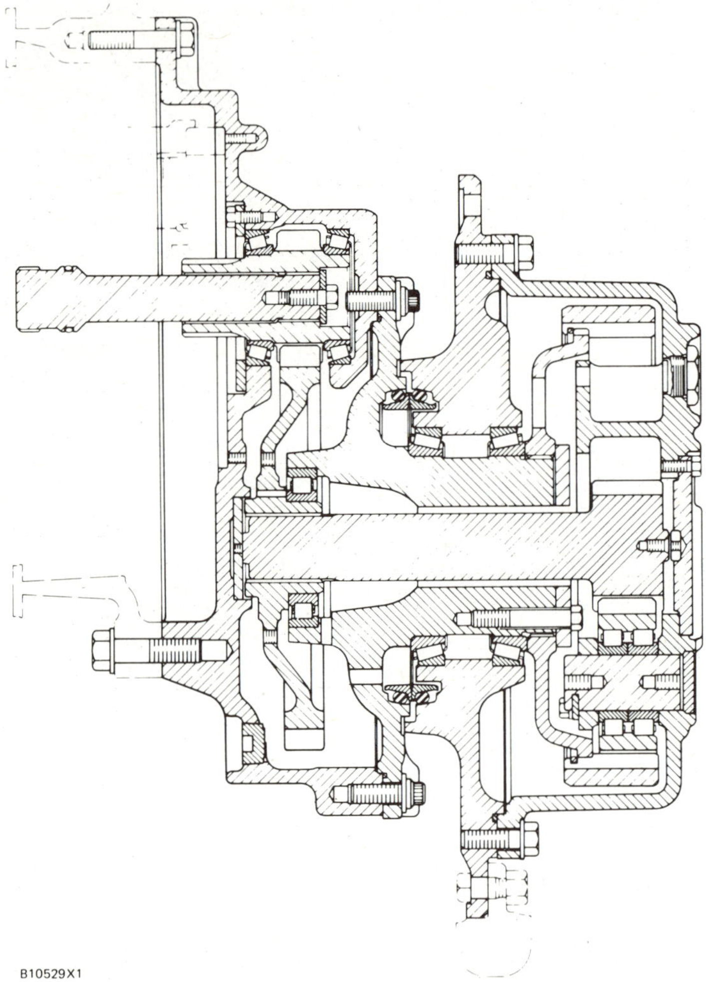

B10529X1

2-66

Specimen Detailed Instructions, continued

DISASSEMBLY AND ASSEMBLY POWER TRAIN

FINAL DRIVES AND SPROCKET ASSEMBLIES

ASSEMBLE FINAL DRIVES AND SPROCKET ASSEMBLIES 4050-16

	Tools Needed	A	B	C	D
1P2420	Repair Stand	1			
FT996	Positioning Group	1			
FT957	Adapter Group	1			
1P520	Driver Group		1		
8M9395	Installer			1	
5P9736	Bracket-Link				2

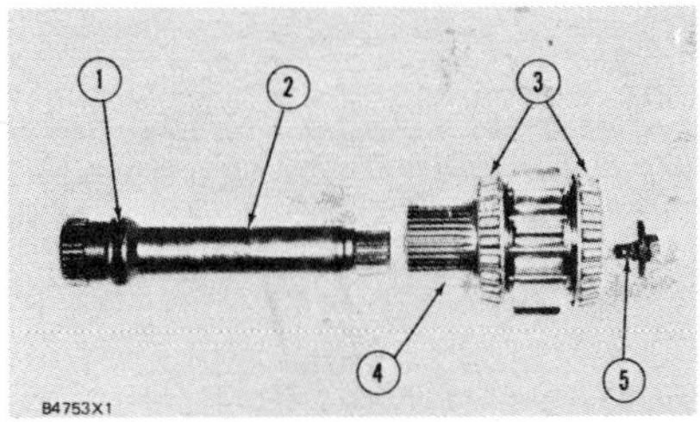

1. Heat bearing cones (3) to a maximum temperature of 138° C (280° F) for a maximum time of 8 hours. Install bearing cones (3) on pinion (4). Use a press to push bearing cones (3) tightly against their seats after the bearing cones have become cool.

2. Install O-ring seal (1) on shaft (2).

3. Put shaft (2) in position in pinion (4) and install bolt and washer (5) to hold the shaft in the pinion.

4. Use tooling (B) to install the bearing cup for the pinion in the final drive case.

5. Put clean oil on the bearing cones on the pinion. Install pinion (4) in the final drive case. Put the bearing cup and retainer (6) in position and install two 3/8"-16 NC bolts 1 ¼ in. long in the retainer to push the bearing cup far enough into the final drive case to install the correct bolts. Remove the 1 ¼ in. long bolts and install the correct bolts in the retainer. Tighten the bolts and be sure retainer (6) is tight against the final drive case. Check the torque needed to turn the pinion. Go to Step 6 if the torque is 1.1 N·m (10 lb.in.) or less. If the torque is greater than 1.1 N·m (10 lb.in.) add one shim at a time under retainer (6) until the torque is correct. Add a maximum of three shims. If the torque is still too high check for foreign material under the bearing cups and cones.

6. Turn the final drive case over and put in position on tooling (A). Use four bolts and nuts to hold the final drive case in position on tooling (A). Weight is 72 kg (160 lb.).

NOTE: Thrust plate (7) can be turned over and installed to get a new wear surface.

7. Use a brass bar and hammer to install thrust plate (7) in the final drive case.

Specimen Detailed Instructions, continued

FINAL DRIVES AND SPROCKET ASSEMBLIES

8. Heat the bearing for gear (8) to a maximum temperature of 138°C (280°F) for a maximum time of 8 hours and install the bearing on gear (8). The side of the bearing with the larger radius on the inside diameter must be toward the shoulder on the gear. After the bearing has become cool use a press to push the bearing tightly against its seat. Put clean oil on the bearing.

9. Install gear (8) in the final drive case.

10. Install O-ring seal (10) on the spindle.

11. Lower the temperature of race (9) and install the race in the spindle.

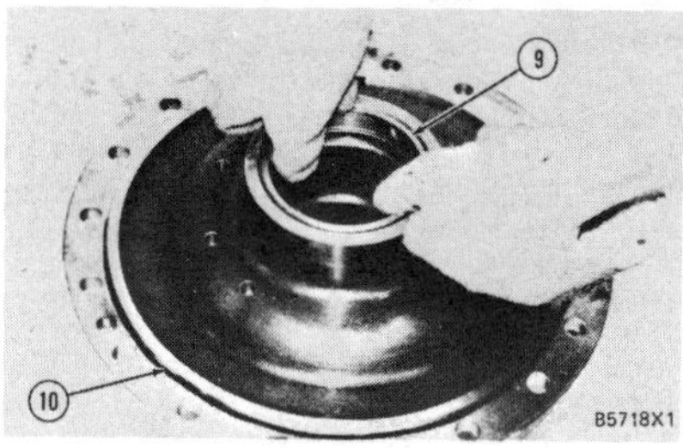

12. Install two 5/8"-11 NC guide bolts (13), 6 in. long, in the final drive case. Install two ½"-13 NC forged eyebolts in spindle (11) and fasten a hoist. Install the spindle on the final drive case with the hole between bosses (12) in alignment with hole (14) over the pinion. Weight of the spindle is 36 kg (80 lb.).

Specimen Detailed Instructions, continued

CHAPTER 24

SERVICE OPERATIONS

SYSTEM PUMPDOWN

System pumpdown removes almost the entire refrigerant charge from the low pressure side of the system and isolates it in the condenser. This permits service or repair work to be performed on the compressor, evaporator, liquid line and suction line with no significant refrigerant loss.

The system pumpdown procedure is as follows: (See Figure 24-1)

1. If the system is not equipped with permanent gauges, install a pressure gauge in the backseat port of the compressor suction valve and crack the valve clear of its backseat.
2. Close the liquid line shutoff valve.
3. Install a jumper wire across the terminals of the low pressure control. Since the suction pressure is to be pumped below the setting of the low pressure control, the jumper is used to short the control out of the circuit.
4. Lower the setting of the conditioned air or chilled water temperature controller to be assured that the system will continue to operate through the entire pumpdown cycle.
5. Start the system.
6. Watching the suction pressure gauge, allow the system to function until the pressure reading is reduced to 2 psig. At this point, stop the compressor and front seat the compressor discharge and suction valves. The refrigerant charge is now trapped in the condenser between the front seat of the compressor discharge valve and the liquid line shutoff valve.
7. *Remove the jumper wire from the low pressure control.*

Always allow a slight positive pressure to remain in the system. If the pressure were reduced below atmospheric, large amounts of moisture and air would be drawn into the system when it is opened.

Caution: If the system is to be opened to the atmosphere, allow all piping and refrigeration components to warm to the ambient air temperature before the opening is made. This will prevent moisture from condensing on the inside surfaces of the system.

PARTS REPLACEMENT

Once the system is pumped down and the equipment has warmed to ambient temperature, the defective part may be removed from the pumped down portion of the system. As soon as the part is removed, the open refrigerant piping should be plugged to prevent the needless entry of air and moisture.

Before the new or repaired part is installed, the surfaces to be soldered must be cleaned thoroughly, using fitting brushes or crocus cloth.

When the part is being installed, it is advisable to sweep air from the piping to prevent the formation of oxides on the interior pipe surfaces. The air is displaced by passing a low volume flow of nitrogen through the line. Using the example system in Figure 24-1, if the expansion valve were being replaced, the nitrogen would be admitted to the line through the charging valve, passed through the opened solenoid valve and allowed to exit through the front seat port of the compressor suction valve.

The repaired area is leak tested by cracking the liquid line shutoff valve for a moment, causing the refrigerant to pressurize the piping. The newly made joints are then tested with a refrigerant leak detector. After the joints are found to be tight, the test pressure is relieved.

Note: It is always good practice to replace the core of the filter-drier any time the system is opened to the atmosphere.

Connect a vacuum pump to the liquid line charging valve and evacuate the pumped down portion of the system, break the vacuum with oil-pumped, dry nitrogen and re-evacuate to 500 microns. Close the charging valve and stop the pump.

START-UP AFTER REPAIR

After the repaired areas are found to be tight and the pumped down portion of the system is evacuated, the system may be prepared for start-up.

The compressor suction and discharge valves are backseated and the liquid line shutoff valve is opened, allowing refrigerant to re-enter the system. If the solenoid valve(s) was opened manually, return it to automatic control.

Start the system and check the compressor oil and refrigerant sight glasses to make certain that the system contains an ample charge of each. Reset the conditioned air or chilled water temperature controller, as necessary.

ADDING REFRIGERANT

Refrigerant may be added to the system, in the gaseous state, through the backseat port of the compressor suction valve. Because this is not a particularly fast method of charging, it is usually reserved for the addition of small amounts of refrigerant needed to "trim" the charge.

The compressor suction valve is back seated and a charging line is loosely connected to the back seat port. The valve on the refrigerant cylinder is cracked, causing refrigerant to leak past the loose connection, purging air from the charging line. The connection is then tightened.

132

Specimen Detailed Instructions (The Trane Company)

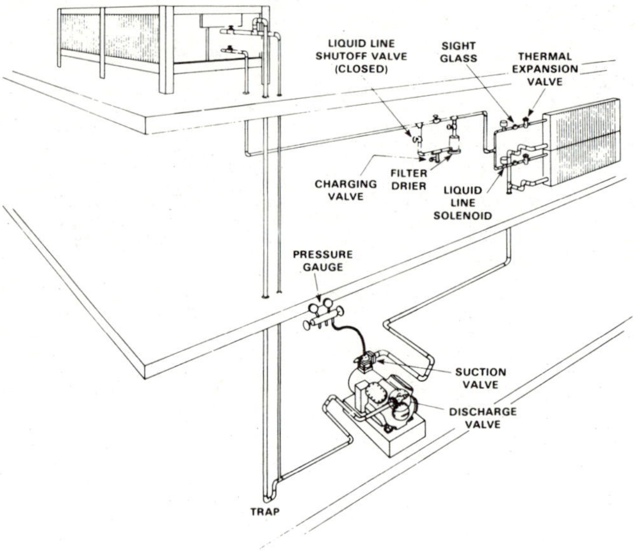

FIGURE 24-1

The compressor is started and the suction valve is cracked off its backseat. With the refrigerant cylinder in the *vertical* position, its shutoff valve is opened. Since the suction pressure is less than the pressure within the cylinder, refrigerant gas is drawn from the cylinder into the system.

When the liquid line sight glass no longer shows evidence of flash gas, the compressor suction valve is backseated, the cylinder valve closed and the charging line removed. The plug is then replaced in the backseat port of the suction valve.

REMOVING REFRIGERANT

Occasionally it is necessary to remove refrigerant from the system to reduce the high condensing pressures and temperatures resulting from an overcharge.

Refrigerant is removed by loosely connecting an empty or partly full refrigerant cylinder to the liquid line charging valve. The line is purged before the connection is tightened.

With the system in operation, the charging and cylinder valves are opened, allowing the refrigerant to enter the cylinder. To speed this operation the refrigerant cylinder may be packed in cracked ice and salt, to reduce its vapor pressure.

Once the condensing pressure returns to normal, the valves are closed and the refrigerant connection removed.

ADDING COMPRESSOR OIL

Since a certain amount of oil remains in the refrigerant piping, condenser, evaporator, etc., at all times, additional amounts are generally required to return the crankcase oil level to normal.

A general rule states that approximately 1 additional pint of oil is required for each 10 pounds of refrigerant charged into the system.

One method for adding oil is to pump it into the compressor crankcase while the system is in operation. The advantage of this method is the actual operating oil level can be established.

Loosely connect a suitable hand operated, piston type oil pump to the compressor crankcase oil charging valve. Operate the pump until oil appears at the loose connection and then tighten it. Open the crankcase charging valve and pump sufficient oil to raise the level on the sight glass to that specified by the manufacturer. Close the oil charging valve and remove the oil charging equipment.

Specimen Detailed Instructions, continued

TRANE REFRIGERATION MANUAL

Note: Always use the oil specified by the compressor manufacturer. Take the oil from the original, sealed container. If oil is to be saved, seal the unused portion from the atmosphere to prevent moisture contamination.

REMOVING EXCESS OIL

Since the oil charging valve of Trane Compressors is located below the crankcase oil level, it may be used to remove oil.

To remove oil, pump down the system in the manner described at the beginning of this chapter. Connect a flexible tube to the oil charging valve. Open the valve slowly, allowing the oil to flow into a suitable container.

Note: Oil removed from the compressor crankcase generally contains some refrigerant. The release of pressure causes the refrigerant to boil from the oil, resulting in the production of a considerable amount of foam. Therefore, the container used to collect the oil must be of sufficient size to contain the foam as well as the oil.

When sufficient oil is drained, close the charging valve tightly and start the compressor. The compressor oil level is then checked periodically for several hours to make certain that the correct operating level is being maintained.

COMPRESSOR VALVE REPLACEMENT

The compressor heads may be removed for the purpose of valve inspection or replacement in the following manner. Notice that it is not necessary to pump down the entire system. It is only necessary to pump down the compressor.

1. Install a pressure gauge in the backseat port of the compressor suction valve and then close the valve tightly against its frontseat.
2. Install a jumper wire across the terminals of the low pressure control. Start the compressor.
3. Operate the compressor until the suction pressure is reduced to 2 psig. Stop the compressor and frontseat the discharge valve immediately.
   Note: It may be necessary to repeat this procedure several times before the suction pressure stabilizes at 2 psig. At this reduced pressure, refrigerant boils from the oil to again raise the pressure within the machine.
4. Remove the cylinder heads.
   Note: The cylinder heads of the Trane Models E, F and R Compressors are under spring tension. To prevent the head from being released suddenly, remove all but two bolts, located on opposite sides of the head. Alternately, loosen these bolts a few turns at a time to relieve the spring tension. If the head does not follow the bolts, tap the head with a rubber or rawhide mallet to loosen it.
5. Once the head is removed, the valve cages may be removed for inspection or repair. The valve cage assembly of the Trane Models E and F Compressors is illustrated in Figure 24-2.

FIGURE 24-2

Note: Refrigerant will continue to boil from the oil for a period of time, maintaining a slight positive pressure within the lower portion of the machine. Consequently, if the work is completed promptly, excess air and moisture can be prevented from entering the compressor.

6. When repairs are complete, reinstall the heads, using new gaskets, and torque the head bolts to specification.
7. Connect a vacuum pump to the backseat port of the discharge valve and a vacuum gauge to the similar port of the suction valve.
8. Operate the vacuum pump until the vacuum gauge reading is reduced to 500 microns. This may take some time because refrigerant will continue to boil from the oil as the pressure is reduced.
9. Once the 500 micron vacuum is achieved, close the suction valve on the vacuum pump and stop the pump.
10. Crack the compressor suction valve off the frontseat and pressurize the machine to 2 psig.
11. Quickly remove the vacuum pump and vacuum gauge connections. Replace and tighten the plugs in the backseat ports of the valves before the positive pressure is lost.
12. Backseat the compressor suction and discharge valves and start the compressor.
13. Check the head gasket areas with a refrigerant leak detector.

REMOVING NON-CONDENSABLE GASES

If the heat transfer surfaces of a condenser are clean and the condensing air or water temperatures and flow rates are normal, however the condensing pressure is high, the system probably contains non-condensable gases, such as air.

Non-condensable gases are swept into the condenser along with the refrigerant vapor. Here, a separation takes place when the refrigerant is condensed. The non-condensable gases then remain in the condenser, rendering part of the condensing surface ineffective. This loss of condens-

Specimen Detailed Instructions, continued

CHAPTER 24 — SERVICE OPERATIONS

ing surface is the cause for the high condensing pressure and temperature.

To verify the presence of non-condensables, pump down the system in the manner described at the beginning of this chapter and allow the condenser to cool to the ambient air temperature. If a water cooled condenser is involved, drain the water from the tubes to speed the temperature leveling.

When the condenser has cooled, record the condenser pressure. If the condenser pressure is more than 10 psig above the refrigerant saturation pressure that corresponds to the ambient temperature, the system contains non-condensable gases.

Non-condensables are eliminated by removing the refrigerant charge, evacuating and recharging the system.

The refrigerant charge is removed by packing a refrigerant cylinder in chipped ice and salt and connecting it to the liquid line charging valve. The system is then started and the cylinder and liquid line valves are opened. Since the temperature at the cylinder is low, the pressure within the cylinder is considerably lower than that within the warm liquid line, causing the liquid to flow into the cylinder.

To cause the compressor to operate at its minimum stage of loading, raise the setting of the system temperature controller. This will help stabilize the suction pressure. (Note: If the compressor capacity is controlled by suction pressure, it will unload automatically as the suction pressure falls.)

Because of the draw-off of refrigerant, the compressor will be stopped eventually by the low pressure control. When this occurs, close the liquid line charging valve. In a short period of time, the suction pressure will probably increase sufficiently to restart the compressor. At this point, reopen the liquid line charging valve. When the suction pressure is reduced to the point where the compressor is short cycling, stop the compressor and close the liquid line charging and cylinder valves. The recoverable portion of the refrigerant charge has now been removed from the system.

The non-condensable gases are removed by evacuating the system, breaking the vacuum with oil-pumped, dry nitrogen and then re-evacuating to 500 microns. Finally, the system is recharged with refrigerant.

COMPRESSOR SEAL INSPECTION (OPEN COMPRESSOR)

The shaft seal of an open compressor consists of a spring loaded carbon ring which runs on an opposing steel ring. The faces of the rings are finely machined to form perfectly flat mating surfaces. The seal is lubricated by the compressor oil pump.

Occasionally, the shaft seal of an open compressor may show signs of refrigerant leakage immediately after start-up. This may be due to the pressure of the seal spring having forced most of the oil from between the sealing surfaces during shipment or during a long storage period prior to installation. In most cases, oil pumped from the crankcase will relubricate the seal, stopping the leak.

If the leak persists after a short running period, inspect the seal in the following manner.

1. Pump down the compressor in the manner described under "Compressor Valve Replacement" in this chapter.
2. Remove the compressor-motor coupling.

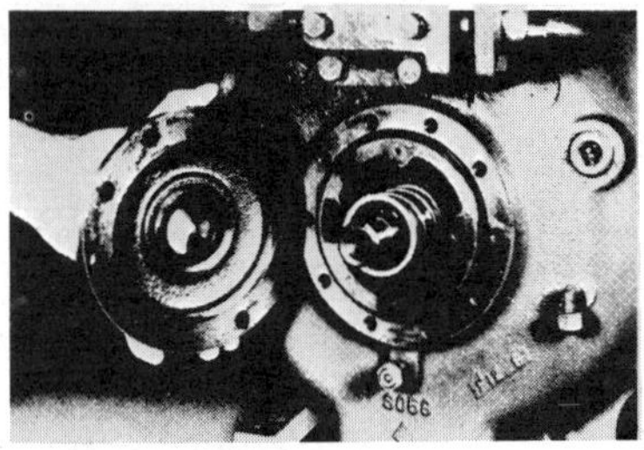

FIGURE 24-3

3. Remove the seal cover, Figure 24-3. Note that the cover contains the steel ring.

FIGURE 24-4

4. Remove the carbon ring, Figure 24-4. Clean the sealing surfaces of the rings with a clean, soft chamois.
5. Inspect the surfaces of both rings. If either surface is scratched or grooved, replace the entire seal. Never attempt to replace individual parts of this assembly. If, however, the surfaces appear smooth, lubricate the seal faces of the rings with compressor oil and reassemble.
6. Evacuate the compressor in the manner described under "Compressor Valve Replacement" in this chapter.
7. After a short running time, stop the compressor and leak test the seal. If the leak persists, replace the seal assembly.

Specimen Detailed Instructions, continued

ADJUSTMENT OF UNDERSPEED VALVE

TOOLS NEEDED: 1-6V2070 Underspeed Cut-in Tool.
1-1P5500 Portable Phototach Group.
1-8N5945 Adapter Assembly.
1-8S4627 Continuity Tester.

1. Preparation of the Machine

   a. Remove the cap on the oil filler tube and leave it off for the rest of the test. Cap must be replaced when the test is completed.

   b. Change the location of the U-TUBE on the brake instruction plate from RUN to TEST.

   c. Install a S1617 (3/4 X 5/16, 18) bolt in the speed-brake lever to hold it in the full speed position.

   d. Install an accurate tachometer in the tach drive location on the front of the engine. Use an 8N5945 Adapter Assembly and 1P5500 Portable Phototach Group.

NOTE: Use the following procedure to get the temperature of the oil in the HPCU up to 80 ± 5°C (180 ± 10°F).

   e. Start the engine. Pull the governor control lever into the HI IDLE detent. Check the engine speed. It must be 2540 ± 50 rpm. If the engine speed is low and the engine is cold, let it get warm before making any adjustments.

   f. When the engine speed is 2540 ± 50 rpm, pull the transmission control lever into the full FORWARD position and hold it for 1 minute. Move the lever back to PARK and leave it for 30 seconds. Pull the lever into full REVERSE position and hold it for 1 minute. Move the lever back to PARK and leave it for 30 seconds. Repeat these operations as often as necessary to get the desired oil temperature of 80 ± 5°C (180 ± 10°F) after the 30 seconds in PARK.

   g. After the oil temperature gets to 80 ± 5°C (180 ± 10°F), shut the engine off and raise the cab to the 24° position. See topic PROCEDURE TO RAISE CAB.

2. Underspeed Valve Check and Adjustment

   a. Adjust the depth of the underspeed valve cut-in tool to approximately 127 mm (5 in.) long.

NOTE: Depth is measured from the bottom of the large locknut to the tip of the tool.

   b. Disconnect the linkage to the underspeed-override valve.

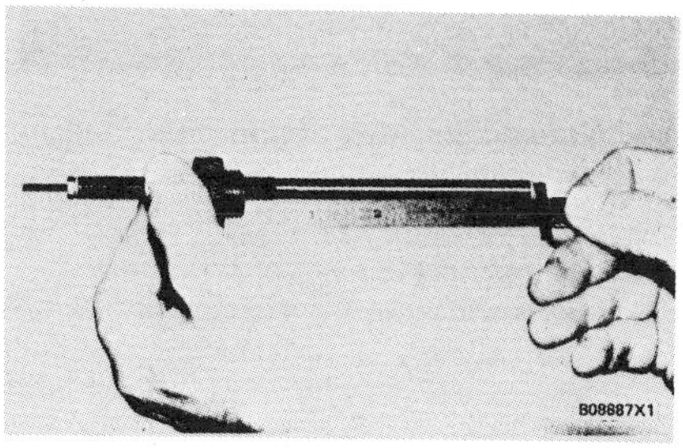

CHECK OF UNDERSPEED VALVE CUT-IN TOOL

NOTE: This action will let the lever drop and open the override valve.

   c. Remove the plug from the top of the HPCU cover.

   d. Install the cut-in tool in the top cover. Be sure the large locknut makes contact with the cover and the rod is toward the left side of the HPCU.

   e. Put the machined flats at the top of the large threads parallel with the sides of the cover.

NOTE: The threaded hole that holds the rod is drilled "off center".

   f. Turn the entire tool counterclockwise approximately 30 degrees and tighten the large locknut. Do not tighten the small locknut at this time.

   g. Check the 8S4627 Continuity Tester by touching the two ends together. The light must come on.

   h. Connect the continuity tester between the top of the cut-in tool and any good machine ground. Put the tester where it can be easily seen.

   i. With the continuity tester correctly installed, start the engine and pull the governor control lever into the HI IDLE detent. HI IDLE is 2540 ± 50 rpm.

   j. Push the FORWARD-PARK-REVERSE lever into the FORWARD RESET position.

NOTE: In the reset position there must be a small load on the engine. This can be heard easier than it can be seen. However, if it is done correctly there will be a small decrease in engine rpm, but it will still be within specifications. An indication of brake pressure does not insure that the transmission is in full reset.

Specimen Detailed Instructions (Caterpillar Tractor Co.)

⚠ WARNING

Use extreme care while doing these adjustments. It is possible under conditions of wear, for the track brakes to slip and let the machine move. See Procedure for Testing Brakes before doing any tests or adjustments.

k. With the transmission in RESET, turn the threaded rod in the cut-in tool in (clockwise) until the light in the continuity tester comes on bright.

l. Hold the cut-in tool stationary and let the FORWARD-PARK-REVERSE lever return to PARK position. Tighten the locknut on the rod in the cut-in tool.

NOTE: The underspeed cut-in tool is correctly installed and adjusted and should give an indication each time the underspeed valve is in the raised position and the transmission is in RESET.

m. Check the adjustment of the cut-in tool again. Slowly pull the transmission control lever into FORWARD or REVERSE until the light in the continuity tester is bright. If the light does not come on in either position, the cut-in tool is not adjusted correctly.

n. Raise the lever on the underspeed override valve and tie it in this position. This closes the valve.

o. When the cut-in tool is installed correctly and the underspeed override valve is closed, reduce the engine speed (rpm) to 2325 ± 10.

NOTE: This is the engine speed used to adjust the cut-in point of the underspeed valve.

p. Straighten the lock tab and loosen the locknut on the underspeed cut-in adjustment screw.

NOTE: Adjustment is made with a screwdriver and the adjustment screw is located on the right side of the machine.

q. Turn the adjustment screw out (counterclockwise) two full turns.

NOTE: This step is necessary because the adjustment must always be made by turning the screw in (clockwise).

r. Pull the FORWARD-PARK-REVERSE lever into RESET (forward or reverse). The continuity light must not come on. If it does, turn the underspeed cut-in adjustment screw out until it goes off.

s. With the HPCU in RESET, and the engine speed set at 2325 ± 10 rpm, turn the underspeed cut-in adjustment screw in (clockwise) until the continuity light comes on. Tighten the locknut on the adjustment screw.

t. Check the adjustment of the cut-in screw. Pull the FORWARD-PARK-REVERSE lever into RESET. Increase the engine rpm. Slowly decrease the engine rpm until the continuity light goes out. Check the engine rpm. It must be 2325 ± 25. If it is not, do the procedure again. If it is, be sure the locknuts are tight and bend the lock tab over the locknut.

u. Stop the engine and remove the tools and test equipment. Check the adjustment for the underspeed-override valve linkage and reconnect it.

v. Move the U-TUBE on the brake instruction plate back to the RUN position. Lower the cab. Remove the S1617 bolt from the speed-brake lever and connect the linkage.

PROCEDURE TO REMOVE AIR FROM THE SYSTEM (Power Bleed Method)

TOOLS NEEDED: 5P5224 Pressure Gauge Kit.

There are two ways to get the air out of the brake system; the manual method and the power method. Of the two, the power method is the fastest and easiest.

To power bleed the brakes, the U-TUBE on the brake instruction panel must be in the RUN position.

⚠ WARNING

Use extreme care while doing these adjustments. It is possible under conditions of wear, for the track brakes to slip and let the machine move. See Procedure for Testing Brakes before doing any tests or adjustments.

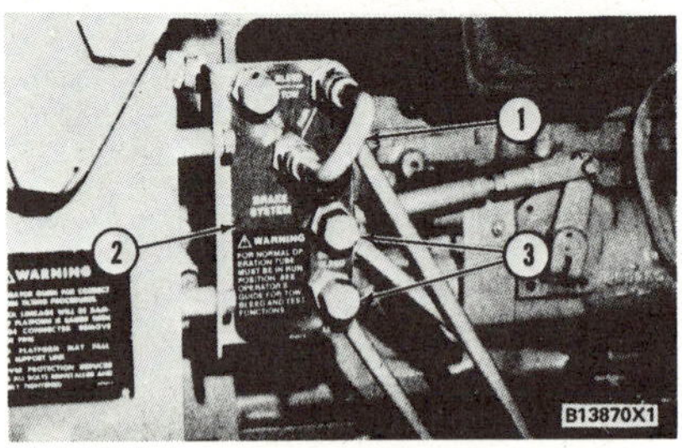

BRAKE INSTRUCTION PANEL

1. U-TUBE. 2. Brake instruction panel. 3. Caps.

Specimen Detailed Instructions, continued

1. Remove the cap on the transmission filler tube.
2. Remove one of the caps (3) on the brake instruction panel.
3. Connect the 5P4775 Hose to the outlet where the cap in Step 2 was removed.
4. Connect the 1S8937 Valve to the hose. Check to be sure the valve is fully closed (turn hand fully clockwise).
5. Take the valve and container and climb up to the operator's seat.

⚠ WARNING

Be sure container does not cover Speed-Brake pedal. Check to be sure that there are no unauthorized personnel around the machine.

6. Start the engine and pull the governor control lever into the HI IDLE detent.
7. Very carefully pull the transmission control lever into the RESET position.

NOTE: Watch the hose as the lever is pulled from PARK to RESET. When the hose makes a sudden movement and becomes stiff, the transmission is in RESET. Be careful not to pull the lever through RESET because the machine will move.

8. With the drive system in RESET, slowly open the valve and let the air-oil mixture run out into the container. Let it run until there is no more air present in the oil and then close the valve.
9. Shut the engine off. Disconnect the hose and replace the cap.
10. Connect the hose to the other bleed port and repeat the procedure for the other brake.
11. When both brake lines are free of air, disconnect the hose and replace the cap.
12. Replace the cap on the transmission filler tube.

PROCEDURE TO REMOVE AIR FROM THE SYSTEM (Manual Bleed Method)

TOOLS NEEDED: 5P5224 Pressure Gauge Kit.

To manually bleed the brakes, the U-TUBE on the brake instruction panel must be in the BLEED-TOW position.

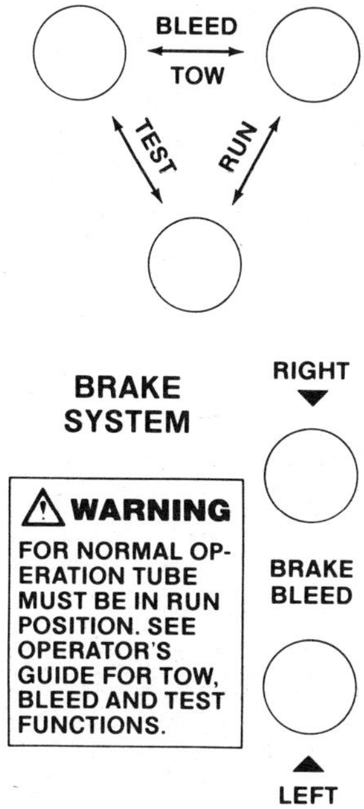

B8751X1

BRAKE INSTRUCTION PANEL

1. Remove the cap on the transmission filler tube.
2. Remove one of the caps (3) on the brake instruction panel.

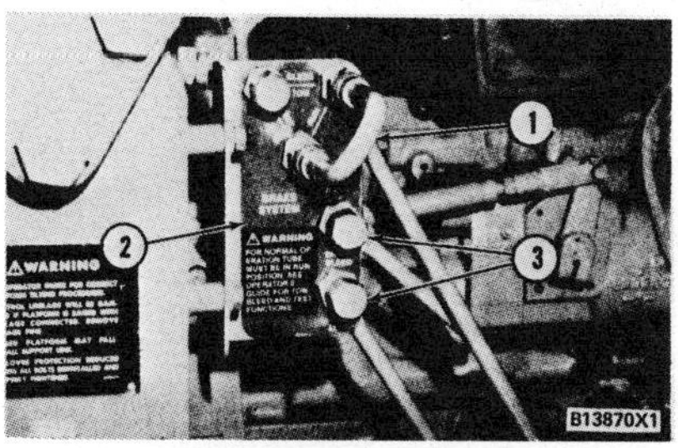

BRAKE INSTRUCTION PANEL
1. U-TUBE. 2. Brake instruction panel. 3. Caps.

3. Connect the 5P4775 Hose to the outlet where the cap in Step 2 was removed.

Specimen Detailed Instructions, continued

TESTING AND ADJUSTING

POWER TRAIN

4. Connect the 1S8937 Valve to the hose. Check to be sure the valve is fully closed.
5. Operate the hand pump after slowly opening the valve.

NOTE: Hold the valve over a container to catch the oil.

6. Continue to operate the hand pump until the oil comes out free of air.
7. Close the valve.
8. Move the hose to the other outlet and do Steps 5 and 6 until the oil comes out free of air.
9. When both brake lines are free of air, disconnect the hose and replace the cap.
10. Replace the cap on the transmission filler tube.

PROCEDURE TO TILT THE CAB 24°

Some machine adjustments and service procedures require the cab to be tilted forward to the 24° angle position. For example, any adjustments to the top linkage in the HPCU when it is in chassis; to change or inspect the oil filter for the HPCU. The following procedure gives the necessary information needed to raise (tilt) the cab.

1. There are five linkages that must be disconnected before the cab can be raised. They are; the linkage that connects the transmission control lever to the FORWARD-PARK-REVERSE lever on the transmission; the two steer linkages; and the two hydraulic implement control linkages. The linkage for the governor control and the speed brake control valve linkage do not have to be disconnected.

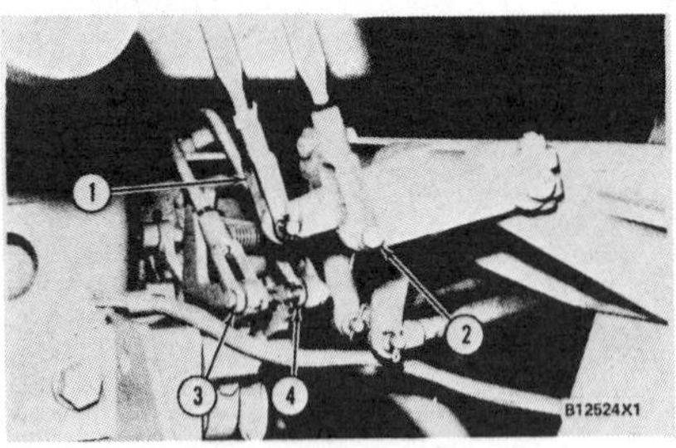

CONTROL LINKAGES
(Located on right side of machine)

1. Linkage for bucket lift control. 2. Linkage for bucket tilt control. 3. Linkage for control of left steer. 4. Linkage for control of right steer.

CONTROL LINKAGE
(Located on left side of machine)

5. Control linkage to FORWARD-PARK-REVERSE lever on transmission.

2. The tilt support bracket (10) is needed when the cab is tilted and is located inside the access door on the right side.
3. Remove three bolts (6) that hold the platform to the support structure.

NOTE: Bolts (6), shown in the illustration, are on the left side of the machine. There are three more on the right side that must also be removed.

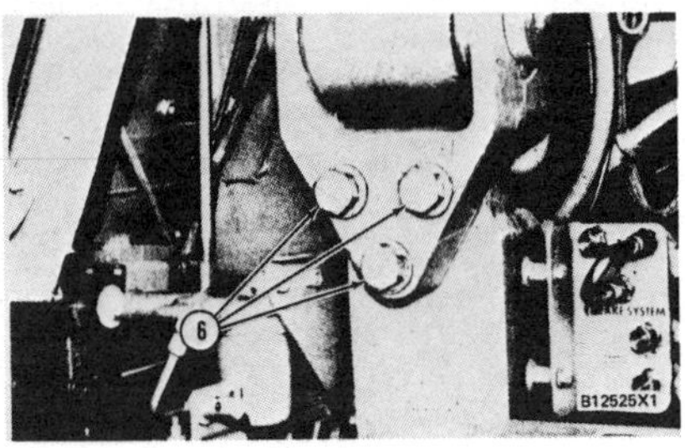

LOCATION OF BOLTS (Left side)

6. Bolts.

4. Connect the top of jack (7) to flange (8).
5. Use the handle provided with the jack and check valve (9) to be sure it is closed. (Turn fully clockwise).
6. With the six support bolts (three on each side) removed and the linkages disconnected, the cab is ready to be raised without component damage.
7. Insert the jack handle into the jack and start raising the cab.
8. Continue to raise the cab until the jack is fully extended.

2-56

Specimen Detailed Instructions, continued

LOCATION OF JACK (Right side)
7. Jack. 8. Flange. 9. Valve.

NOTE: With the jack fully extended, the cab will be tilted slightly beyond 24°.

9. Install the support bracket (10) that was stored on the right fender. Use the bolts at the bottom of the bracket.

SUPPORT BRACKET INSTALLED
10. Support bracket.

10. Use the jack handle and slowly open valve (9) until the holes in the top of the bracket are aligned with the holes in the platform support. Close the valve and install the bolts in the bracket.

11. Tighten all three bolts on both sides for safety purposes, while the machine is being serviced.

12. To lower the cab, remove support (10) and slowly open the valve on the jack. Care must be taken when lowering the cab to insure that components and linkages are not broken or bent.

13. When the cab is in the down position, tighten the platform bolts to specifications. Connect all the linkages.

HPCU EXTERNAL LINKAGE ADJUSTMENT

Transmission Control and FORWARD-PARK-REVERSE Levers

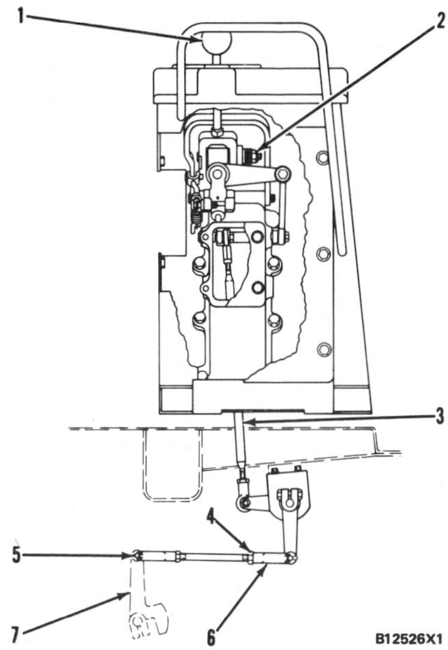

TRANSMISSION CONTROL LINKAGE
1. Knob. 2. Nut. 3. Rod. 4. Nut. 5. Pin. 6. Yoke. 7. FORWARD-PARK-REVERSE lever.

1. Remove cover (10).
2. Remove pin (5).

TRANSMISSION CONTROL LINKAGE
4. Nut. 5. Pin. 6. Yoke. 7. FORWARD-PARK-REVERSE lever.

Specimen Detailed Instructions, continued

3. Tighten nut (2) until the force needed to move knob (1) is 21 ± 4 N (4.7 ± .9 lb.).
4. Adjust rod (8) until the force needed to move safety lever (9) over center is 40 ± 5 N (9.0 ± 1.1 lb.).
5. With lever (9) in the position shown, loosen nut (4) and adjust yoke (6) until pin (5) fits freely.

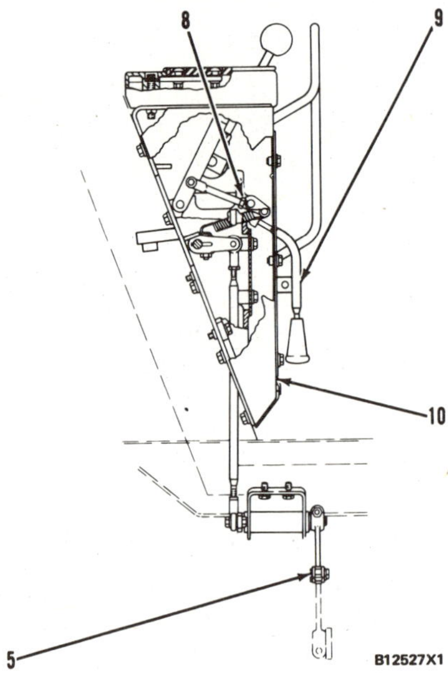

B12527X1

TRANSMISSION CONTROL LINKAGE
5. Pin. 8. Rod. 9. Safety lever. 10. Cover.

Governor Control and Underspeed Override Levers

1. Remove the cover on the left hand control console.
2. Tighten nut (3) until the force needed to move knob (1) is 20 N (4.5 lb.).
3. Remove pin (4).
4. Pull knob (2) to the rear until lever (3) is in the HI IDLE detent.

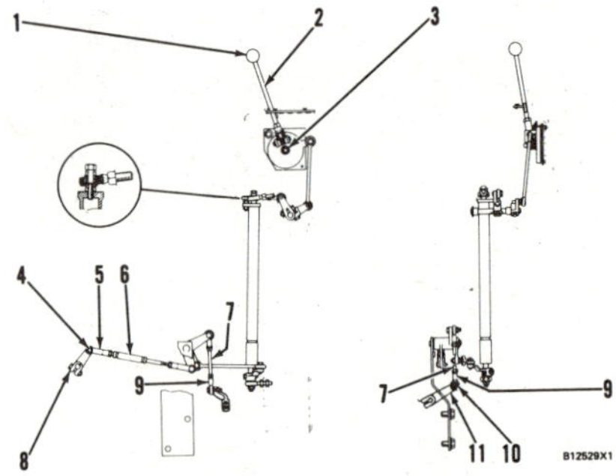

GOVERNOR CONTROL LINKAGE
1. Knob. 2. Lever. 3. Nut. 4. Pin. 5. Yoke. 6. Link assembly. 7. Rod. 8. Lever. 9. Rod end. 10. Bolt and nut. 11. Lever.

5. Loosen the nut and adjust yoke (5) so that link assembly (6) is extended 13 to 15 mm (0.5 to .62 in.) when yoke (5) is attached to lever (8). Lever (8) must be in the HI IDLE position.

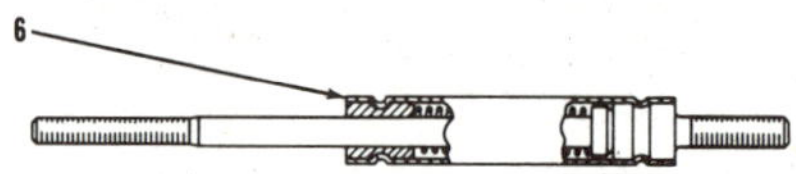

B14080X1

DETAIL OF LINK ASSEMBLY
6. Link assembly.

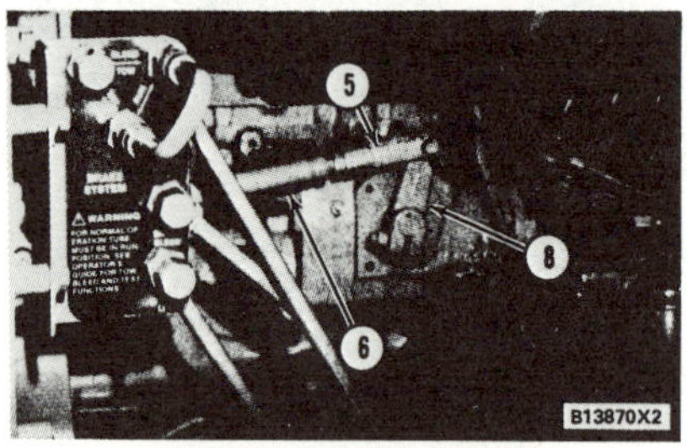

GOVERNOR CONTROL LINKAGE
5. Yoke. 6. Link assembly. 8. Lever.

6. Remove bolt and nut (10).
7. Use rod end (9) and adjust the length of rod (7) until bolt (10) can be installed through rod end (9) and lever (11) freely. Hold lever (11) up against the spring on the inside of the override valve.
8. When correctly adjusted, tighten bolt and nut (10).

Specimen Detailed Instructions, continued

Introductory Exercise for Learning to Use

CMS, EDGAR, and SCRIPT

Morgan Herrington

Specimen Procedure Manual

TABLE OF CONTENTS

Specimen Procedure Manual, continued

INTRODUCTION

Before you can just sit down to a terminal and type in a report, you will probably have to spend an hour or two learning to use the system. By "the system" I mean the IBM terminals, the editor program (EDGAR), and the SCRIPT program. Each of these three topics is introduced and briefly explained in the three following sections; if you already understand some of the system, skip the appropriate material.

TERMINALS

Introduction

First, you'll need to know a little about the physical machine that you will be working with. We have two new types of terminals: the 3277's, which show only uppercase, and the 3278's, which are usually used for SCRIPT. (The 3278's have an orange rocker arm on the left side of the screen for turning the terminal on and off; the 3277's have a round push-pull knob instead.) Since you will usually want to use the 3278's, I won't go into detail about the 3277's; however, you could use any of the terminals once you learn where the different keys are located.

Log-on Procedure

To begin, you must first find a terminal that displays "U of I" in large letters. If your terminal doesn't show this, press the ENTER key (beside the space bar) and "U of

Specimen Procedure Manual, continued

PAGE 2

I" should appear. If it doesn't, go ahead and try to log-on, but be aware that you may need to get help if you have trouble. The following steps should start the machine running and prepare it for your use.

1) Press the ENTER key to clear the "U of I" screen. All that should appear on the screen is the cursor in the lower left corner, and the words "CP READ" in the lower right corner.

2) Type LOGON, then a space, then your assigned user ID. Next press ENTER.

3) The display should ask for your password, so type it in and press ENTER. (The password is not shown on the screen.) The display could have said "INVALID USERID," in which case, you should start over with step two.

4) The display will show a short message that you don't need to worry about unless it says "INCORRECT PASSWORD." If it does, start over with step two.

5) Type STARTUP and press the ENTER key.

6) The display will respond with a cryptic message about which disks are connected to your terminal. Wait until it prints the line, "TEMP 191 WORK DISK HAS BEEN ALLOCATED AND FORMATTED." This means that you now have a temporary A disk.

You are now logged on.

Machine Status

At this point, the screen is mostly blank; however, you are only allowed to type on the two bottom lines. Typing anywhere else will lock up the keyboard and cause a small figure to appear in the lower left hand corner of the screen. If this ever happens, press RESET and move the cursor back within bounds. (The RESET key is just to the left of the space bar.)

Specimen Procedure Manual, continued

Along with this error signal on the left side, the status of the machine is usually displayed in the lower right hand corner of the screen. Before log-on, it will say "CP READ" and afterwards, it will say "RUNNING." When the screen is full, the status will change to "MORE," which means that more information needs to be displayed. In this case you can wait 60 seconds and the screen will automatically show the next page. If you don't want it to change, press ENTER and the status will change to "HOLDING." You can show the next page by holding down the ALT key and pressing CLEAR.

Using the Keypad

Besides the standard typewriter keys, these terminals provide several special keys for changing the display -- including the cursor keys, the editing keys, and the clear keys. The first of these just move the cursor around the screen without affecting the display. Especially note the key with the bent arrow; this key always moves the cursor to the beginning of the next line. The keys with the arrows that point into vertical bars move the cursor to the beginning and the end of the line.

The next keys to consider are the delete key (marked by an "a" with a curly line through it), and the insert key (marked by an "a" with a "hat" over it.) Type some letters onto the screen and move the cursor back to the middle of this line. Push the delete key and notice that it deletes

Specimen Procedure Manual, continued

PAGE 4

the character above the cursor and moves the rest of the line to the left to fill in the space. The insert key provides just the opposite function by inserting characters above the cursor.

The last two keys are the ERASE EOF key , which erases all characters from the cursor to the end of the line, and the CLEAR key which clears the entire screen. (Note that the ALT key must be held down to use the CLEAR key.)

Log-off Procedure

As a last note on using the terminal, always be sure to log-off when you are finished. To log-off:

1) Type LOGOFF

2) Press the ENTER key -- The display will show some statistics about this session.

3) Press the ENTER key to get the "U of I" ID shown.

Specimen Procedure Manual, continued

PAGE 5

EDGAR

Introduction

CMS stores program and data on disks using a "file" system to identify the separate pieces of information. These files are identified by a name, a type, and a mode. The name may be any eight characters of your choosing, the type should be 'SCRIPT', and the mode will be 'A' or 'Z'.

As a slight digression, the mode refers to the disk where the file is to be stored. Each of us will have a temporary A disk to use as a scratch pad, and we will all share a Z disk for permanent storage.

As one more point of confusion, beware that the insert key works slightly differently when in EDGAR. To add some characters, follow these next steps:

1) Move the cursor to the position where the characters should be added.

2) Press PF9 (hold the ALT key down and press the number 9)

3) Press the insert key

4) Type in the characters -- note that the right side of the line is shifted to the right.

5) Press RESET to get out of insert mode. (The RESET key is just to the left of the space bar.)

Specimen Procedure Manual, continued

PAGE 6

Creating a File

Edgar is a program to create or modify a file. To run this program -- and edit a file -- follow this example.

1) Type EDGAR name SCRIPT A (where 'name' is of your choosing) Then press ENTER.

2) EDGAR should now show an arrow near the top of the screen with 'top-of-file' and 'end-of-file' indicators below this. The area beside the arrow will be the 'command line' and it should now say 'CREATING NEW FILE.'

3) Type ADD 10 and press ENTER. This should add 10 blank lines between the top and end markers. (DELETE 'n' would have deleted n lines.)

4) Move the cursor down between the top and end markers and type in several lines of prose. (This would be a good time to practice using the delete and insert keys.)

5) If you would like to add some lines to the text, move the cursor under the first star at the end of a line. Next, type in the number of lines to be added, followed by the letter 'a'. Then press the ENTER key and the lines will be added (with the text moved down to accomodate the new space.) A 'd' will similarly delete lines.

6) Since you just pressed the ENTER key, the cursor will be up in the command line. Type BOTTOM and press ENTER and notice that the file moves up so that only the last line is displayed. Type TOP and press ENTER and the file will move back.

7) Use the commands UP and DOWN to move the file by one line at a time, and use FORWARD and BACKWARD to move it by one page at a time. (These and other commands are fully explained in the CMS USER'S GUIDE.

8) Type FILE into the command line and press ENTER. This stores your file on the A disk and terminates the editor program. (QUIT would have ended the editing session without storing the file, and SAVE would have stored the file without stopping the editor.)

Specimen Procedure Manual, continued

PAGE 7

Storing a File

Once a file is saved on the temporary A disk, you can store it on the permanent Z disk by typing

STOREZ name SCRIPT

and pressing the ENTER key. It will respond with a query for the disk password, so type it in and press ENTER.

To retrieve the file from the Z disk, type

GETZ name SCRIPT

and press ENTER.

To erase the file from the Z disk, type

PURGEZ name SCRIPT

and press ENTER. This will also require a password.

Specimen Procedure Manual, continued

PAGE 8

SCRIPT

Introduction

SCRIPT is a program that reads a file from your A disk, adjusts its format according to your specifications, and prints out the final copy on a terminal or on paper. As added functions, it can check spelling and hyphenation, add page numbers, and set up footnotes and headings.

Creating the SCRIPT File

To begin this exercise, use EDGAR to start a file. Then just type your paper into this file, inserting SCRIPT commands as you go. These commands consist of a period in the left hand margin followed by a short command abbreviation. I've given a small example below.

INPUT FILE	EXPLANATION
.ds	Double space
.ce How To Use SCRIPT	Center the title
.ti # 05	Translate tab char to #
.sp 1	Space one blank line
#This is the body of the text	Begin text -- note the irregular margins.
to be printed. Although it	
has a ragged edge margin,	
SCRIPT will reformat it to	
have straight margins.	

Try creating a file like this, but omit the headings

Specimen Procedure Manual, continued

PAGE 9

and the comments; they are just explanations for the example. When you are finished, type FILE into the EDGAR command line to store this on your A disk. Then type

SCRIPT name

and press ENTER. In a few seconds the display should show 'MORE', so press the ALT key and CLEAR. When you are done reading your example, clear the screen again so that you can type in new commands.

To revise this file type

EDGAR name SCRIPT

and press ENTER. Now use EDGAR to add a blank line before the title and insert the command '.US 1' at the left hand margin. (This means, 'underscore one line.') As a note of caution, the terminals and the small printers cannot underline correctly; however, the system printer can. Again, type FILE on the command line and press ENTER to store the edited file.

Note: I've listed some more SCRIPT commands at the end of this instruction set.

Printing a File

You will probably want to print this file out on paper now, but you'll need a short explanation first. You can use any one of five printers on this campus by specifying the correct name from the following list:

Specimen Procedure Manual, continued

SYSTEM
GAUSS
ADMIN
FORESTRY
AGSCI

The first is the main printer in the computer center and the rest are the smaller, cluster printers. To link to one, for example the cluster printer in the administration building, type:

SPOOL PRINTER ADMIN

and press ENTER. Then type:

SCRIPT name (PRINT)

and press ENTER. It will respond with 'ENTER OUTPUT CLASS' and you should normally enter T. (T specifies upper and lower case on green paper; C specifies upper and lower case on white paper.)

Special Options

The word 'print' in the last command is an example of using the option field. You can list several options if you separate them with blanks. For example:

SCRIPT name (PRINT BIND(1i))

will move the printed page one inch to the right to allow for binding. The command

SCRIPT name (SPELLCHK)

specifies that the machine should check each word for spelling errors.

If you want a list of all options, just type SCRIPT and press ENTER.

Specimen Procedure Manual, continued

PAGE 11

NOTES ... IN CASE OF TROUBLE

--- If you type a command and nothing happens -- remember that you must press ENTER to send the command to the computer.

--- If the screen status is 'MORE' or 'HOLDING' -- hold down the ALT key and press ENTER.

--- If the keyboard locks up -- press RESET. If necessary, press ENTER to move the cursor back into a valid area.

--- If you forget your Z disk file name, follow these steps:

1) Type LINK * 199 199 (press ENTER)
2) Type ACCESS 199 z (press ENTER)
3) Type LISTFILE * * z (press ENTER)

This will list out the files on the Z disk.

--- If you want to use the Z disk, use

GETZ name SCRIPT
STOREZ name SCRIPT
PURGEZ name SCRIPT

Specimen Procedure Manual, continued

PAGE 12

FREQUENTLY USED SCRIPT COMMANDS

.ce	Center the following material
.us n	Underscore 'n' lines
.sp n	Space 'n' lines
.ti	05 Translate the tab character to ' '
.uc n	Underscore and capitalize the following 'n' lines
.tm n	Set top margin to 'n' lines
.bm n	Set bottom margin to 'n' lines
.ll n	Set line length to 'n' characters
.pl n	Set page length to 'n' lines
.pa	Page eject
.cp n	Conditional page -- page only if there are fewer than 'n' lines left
.in n	Indent text 'n' spaces
.im 'n'	Imbed a file of filename 'n' into the text
.li	Literal -- allows a period in the first column
.cm	Comment line
.fo le	Format left -- allows the right margin to be ragged
.hy on	Hyphenation -- allows the internal dictionaries to be used to split words at the right margin
.fo off	Format off -- stops left and right formatting. allows text to be printed as entered.

Specimen Procedure Manual, continued

Writing Exercises

1. Prepare a list of twenty-five steps covering getting out of bed in the morning and getting ready to go to class. Use notes, cautions, and warnings as appropriate. Then arrange the steps into four or five groups, using subheadings and brief introductions, as appropriate.
2. Collect several samples of each of excellent instructions and minimally adequate or unacceptable instructions.
3. Supply the missing function words in the following instructions:

    1. Enter building from main door on south end.
    2. Go up stairs at left to second floor.
    3. Walk to window marked "Cashier."
    4. Present registration slip and student ID card.
    5. Obtain fee sheet and fill in blanks.
    6. Check fee sheet with clerk.
    7. Write check to cover fee (or pay cash).
    8. Obtain receipt.

4. Collect a half-dozen or so instruction manuals for different types of hardware or equipment. Compare the overall organizational structures. Report your findings to the class.
5. Think of some situations other than in an instruction manual where you might be asked to write a physical or functional description of a mechanism. Describe them briefly.
6. Collect two examples each of excellent instruction or procedure manuals and unacceptable or minimally adequate instruction or procedure manuals.

Writing about Writing

1. There are subtle but significant differences between explaining how a process is performed and instructing someone to perform the process. Write an essay discussing these differences.
2. Write critical reviews of two sets of instructions, pointing out strengths and weaknesses and making suggestions for improvement, as appropriate.
3. Some instruction manuals are more educational than referential. They seem to be designed to be read once or twice and then thrown away. Others seem to be designed not to be read until you have a specific need for the information in them. Collect several examples of each and compare notes with your classmates. Can you detect any patterns? Describe in detail the differences in approach that you as writer might take in preparing each of these different types of manual.
4. Obtain a copy of a fairly detailed instruction manual or procedure manual. Write a critical review of the manual, pointing out strengths and weaknesses and making suggestions for improvement, as appropriate.
5. Consider the merits of humor in instruction manuals. Write an essay on the topic.

6. Consider the merits of obscenity in instruction manuals. (Example: "The top mounting bolt is a son of a bitch to get out unless you have a special wrench.") Write an essay on the topic.

Writing Assignments

1. Write a set of directions for finding your home (or your parents' home, or your dorm room, or some other destination).
2. Write a set of instructions for some simple procedure that you are familiar with; for example:

   Changing a tire on a Datsun B 210 (or other particular automobile).
   Starting a lawnmower (or motorcycle, or outboard motor). Don't talk about lawnmowers in general—write instructions for a particular make and model.
   Saddling a horse.
   Catching a subway or city bus.
   Getting registered for fall semester.

3. Write a set of instructions for assembling a toy or a tool or a household utensil. Better yet, rewrite an existing set of instructions, expanding and improving upon them as necessary.
4. Write a set of procedural instructions for your replacement on your present (or most recent) job.
5. Prepare a section or two for a "Bachelor Living Handbook": washing dishes, sewing on a button, baking bread, doing the wash, and so on.
6. Write an instruction manual for a particular piece of hardware whose operation you are familiar with. Begin with a physical and/or functional description of the mechanism and then proceed with detailed instructions, as necessary. Include a parts list.
7. Rewrite an existing instruction manual, expanding and improving upon it as necessary. (Assume you are employed by the manufacturer and have been given the assignment of rewriting the instructions. Use whatever you can, but change whatever you feel should be changed.) Attach a copy of the original manual as an appendix.
8. Write a procedures manual for one of the following situations (or some similar situation):

   - ☐ You are supervisor of a hospital med tech laboratory. Outline the procedures for gathering specimens, running tests, and reporting results. (Don't go into detail on specific tests; instead refer your readers to standard lab procedure manuals.)
   - ☐ You are supervisor of a wastewater treatment laboratory. Proceed as above.
   - ☐ You are office manager for an insurance company. Write a procedures manual for the typing, filing, and other clerical and secretarial duties that must be performed by the office staff. (Assume a staff of five to eight.)

15

Recommendation Reports

The recommendation report (also known as the feasibility report or the analytical report) is the key tool organizational managers use to help them select among alternatives. Faced with the need to choose from a number of competing products or programs, a manager will instruct one of his subordinates to conduct a feasibility study and write a recommendation report. A feasibility study is a staff activity in which an employee or a small group of employees gathers some data on two or more alternatives, analyzes the data, and draws some conclusions. The recommendation report is the end product of such a study: a document that conveys a recommendation, together with supporting data and interpretation, to management for its consideration and decision. The recommendation report tells which alternative the employees recommends, and why.

For example, suppose you are a law enforcement officer in a small city and have some background with radio equipment. The chief of police has decided to ask the city council to authorize purchase of a dozen hand-held two-way radios that officers can use when they are away from their patrol cars. The chief knows that there are several makes and models of hand-held radios on the market, but he doesn't have the time to write for literature, talk to sales representatives, and make up charts and tables comparing costs and benefits of the different types of radios. Moreover, he lacks your specialized and up-to-date knowledge of radio equipment. So he asks you to prepare a recommendation report for him. What he wants from you is a document that will help him make up his mind; a document that states which make and model radio you believe the department should purchase and, more important, *why* you believe the department should purchase that particular make and model.

A recommendation report most often discusses several alternatives, in terms of specified criteria or standards; although it is possible for a recommendation report to discuss a single alternative to the status quo, and it is possible for a recommendation report to discuss alternatives generally in terms of advantages and disadvantages. Usually the report concludes that one of the alternatives is superior and recommends that management select that alternative. Occasionally, however, a recommendation report may conclude that the alternatives are essentially equivalent or that none of the alternatives is acceptable. In such cases, the report may recommend that the selection be made by tossing a coin, or that a new and more detailed feasibility study be conducted. For the radio report, for instance, you would probably identify three or four alternatives; say, the Motorola Model B-1, the General Electric Model 422-A, the Regency Portable, and the Datacomm Mark IV. You would identify several criteria: perhaps ease of use, durability, cost, and availability of parts and service. After gathering and analyzing data on each of the four radios, you might conclude that the Datacomm Mark IV is best because it has the most convenient controls and there is a local dealer who can service them. You would recommend purchase of the Datacomm.

□ The Study Versus the Report

It is important (see Figure 15–1) to keep the data-gathering and analysis phase separate from the reporting phase, and equally important to keep these two separate from the decision-making phase. The decision—the selection itself—is a management function that is based on information in your report. The report itself does not decide. Furthermore, you cannot begin writing your report until after you have gathered and analyzed your data and drawn your conclusions. The study phase must be completed before the reporting phase begins; similarly, the report must be completed

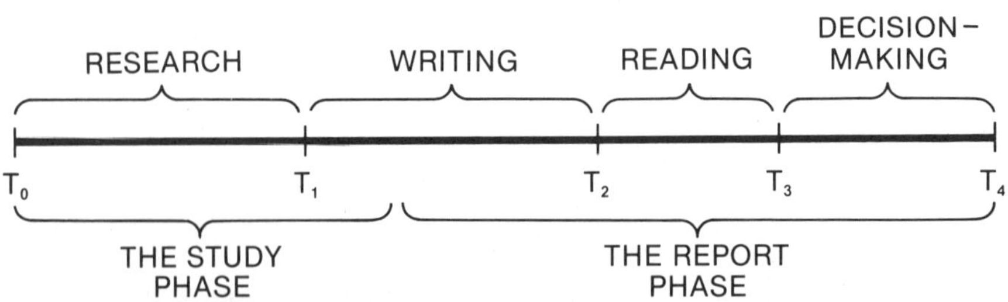

Figure 15–1 The Phases of Studying and Reporting

and submitted before the management decision can be made. Now, this is not to say that you cannot make up your *own* mind as to which alternative is superior; in fact, as you analyze your data and draw your conclusions, you will have to make up your mind! You begin your report, then, knowing what your conclusions are and what your recommendation will be. Your job is to state your recommendation and then justify it.

□ Alternatives and Criteria

Begin your study by identifying a set of reasonable alternatives and a number of evaluation criteria that fit the problem as you perceive it. Don't waste time with alternatives that are manifestly unsuitable as solutions to the problem. Similarly, don't waste time with marginal criteria that will not affect the outcome of your recommendation. As you gather data, however, you may find yourself adding or deleting alternatives or criteria. In fact, quite often the alternatives and criteria you finally discuss in your report are different from those you start with at the beginning of your study. What seems reasonable at first often proves unworkable in the final analysis. Do not fall into the trap of feeling obliged to include in your report every jot and tittle of data you gathered.

The way you select and weight your criteria will totally influence the outcome of your study and hence your report. The way you justify your selection and weighting will strongly influence the acceptance of your recommendation. Do not treat this aspect of your study lightly. Tie your criteria to the problem as presented; make sure each criterion is pertinent to the problem that faces the organization. Do not assume that the criteria should be given equal weight. In most real-world situations they are not. Assigning numerical values to each criterion may help, but be careful not to let your numbers run counter to your judgment.

□ Success Strategy for Student Writers □

In the on-the-job writing situations you must deal with the problem as given, but with classroom situations you can manipulate alternatives and criteria to your advantage. First of all, provide yourself with an optimal number of alternatives and criteria—enough to make for a reasonably rich paper but not so many that you will need fifteen or twenty pages to do your topic justice. Three alternatives and three to five criteria are about right for a three-page paper. Let the results of your data-gathering influence the final set of alternatives and criteria you include in your report. If certain of your preliminary alternatives prove unworkable, discard them. If your paper looks like it will get too long, discard some criteria. If weighting your criteria differently will make for a more interesting paper, by all means do so.

General Outline for Recommendation Reports

The following general outline (suggested by Steven E. Pauley) is suitable for most recommendation reports, although it may be modified somewhat if the report is a brief one in the form of a letter or memo or a lengthy formal report (see Chapter 4).

I. Introduction
- A. Purpose of the Report
- B. Definition of the Problem
- C. Statement of Scope
  - 1. Alternatives
  - 2. Criteria
  - 3. Limitations

II. Description of Alternatives (only if necessary)

III. Presentation & Interpretation of Data for 1st Criterion
- A. Explanation of Criterion
- B. Presentation of 1st Alternative
  - 1. Presentation of Data
  - 2. Interpretation of Data
- C. Presentation for 2nd Alternative
  - 1. Presentation of Data
  - 2. Interpretation of Data
- D., E., F. Presentation for 3rd, 4th, 5th Alternatives, etc.

IV. Presentation & Interpretation of Data for 2nd Criterion
- A. Explanation of Criterion
- B. Presentation for lst Alternative (as above)
- C. Presentation for 2nd Alternative (as above)
- D., E., F., etc.

V., VI., VII. Presentation & Interpretation for 3rd, 4th, 5th Criteria, etc.

VIII. Conclusions

IX. Recommendations

Introduction

In your introduction, you state the report's purpose, describe the problem that led to the study and the report, and discuss the alternative solutions to the problem and the criteria you used to compare the alternatives. These need not be given separate headings unless the entire introduction comes to more than one typewritten page.

Purpose

Remember your overall purpose with a recommendation report: to state your recommendation and your reasons for making it. Begin by stating your recommendation:

- □ This report recommends that color-slide film be used for taking pictures of the homecoming parade.
- □ In this report I recommend that keg beer be purchased for the fraternity picnic, with six cases of canned beer kept in reserve.
- □ After considering the alternatives, I recommend that you purchase a Homelite XL Automatic chain saw for making firewood.

Many novice writers are reluctant to be so bold at the beginning of their reports, preferring instead to keep their recommendation secret until the end of the document. They usually cite objectivity as the reason for this. Now, objectivity is necessary for valid scientific and technological research; if you have your conclusion already drawn before you begin your investigation or study, you will most decidedly bias your results. But keep in mind the distinction between your study and your report. Begin your *study* with an open mind. But remember that when you begin your report, you have completed your study, you have drawn your conclusions, and you are ready to make a recommendation. So don't be afraid to make it. Avoid weasel-word statements like:

> The purpose of this report is to determine which kind of film to use for taking pictures of the homecoming parade.
>
> This report analyzes canned, bottled, and keg beer and decides which should be purchased for the fraternity picnic.
>
> The purpose of this report is to recommend which kind of a chain saw you should purchase for making firewood.

Problem

Your description of the problem serves two major purposes: (1) it demonstrates your understanding of the situation that led to your feasibility study and report, and (2) it provides background for anyone in your primary audience not familiar with the situation. It also has the minor purpose of reminding your boss of the circumstances behind the report. (The boss has many things on his mind and may not remember in vivid detail an assignment made several weeks before—in spite of the fact that the assignment has occupied most of your waking moments since you got it.)

□ Success Strategy for Student Writers □

Exercise your imagination a bit to generate a realistic and well-developed problem for your class writing assignment. Create a role for yourself to play in a hypothetical but realistic organization. The student recommendation report with a shallow and ill-defined problem statement is at best an answer in search of a question. Such papers, even if well-written, are rarely judged to be excellent.

Scope

In this portion of your introduction you must at least list the alternatives you considered in your study and the criteria or standards you used as a basis for comparison. An excellent scope, however, goes beyond a simple inventory. Mention other possible alternatives and indicate why you eliminated them. Do the same for criteria. If your criteria are weighted equally, state this fact explicitly; if they are rank-ordered, state the rank and the weighting factors. If any external limitations were placed on your study or developed in the course of your investigations, here is the place to describe them. What you are doing in the scope, basically, is anticipating questions or objections readers might have and dealing with them in advance.

□ Success Strategy for Student Writers □

For your class writing assignments, you have control over the alternatives you identify and over the weighting of criteria. Avoid selecting alternatives that have very little real difference—say, different brands of pain reliever, or different makes of light utility pickup trucks. Instead, select different models within a given product line. Similarly, avoid selecting your alternatives in such a way that you have a runaway winner; this makes for a shallow paper. Also, consider adjusting the weighting of your criteria to make the outcome a little closer. Remember, the richer your hypothetical communication situation, the easier it is to generate an excellent paper.

□ Description of Alternatives

Whether or not your alternatives need detailed description is determined primarily by how you assess your readers and their knowledge of the situation. Suppose you are recommending a timber harvesting method. If your readers are all professional foresters, you hardly need to offer textbook descriptions of seed-tree, shelterwood, and clear-cut techniques. On the other hand, if your readers are, say, the city council members of a town that owns some forest land, you may well have to describe the alternative harvesting methods in some detail for them to understand your presentation and interpretation of the data. If you are in doubt or if you have a group of readers with mixed backgrounds, go ahead and include a detailed description of the alternatives. The reader who doesn't need the information can skip over it to the next section. Don't overlook the possibility of illustrating your alternatives here as in Figure 15–2. Photographs or drawings of the alternatives can save you a good bit of writing and make for more effective communication as well.

DESCRIPTION OF ALTERNATIVES

Here the units are illustrated and briefly described; see Appendix C for specification sheets, sales literature, and operating instructions.

The 5P2150 Engine Horsepower Meter

The 5P2150 Engine Horsepower Meter (see Figure 1) can measure engine speed from the tachometer drive on each of the engines in question.

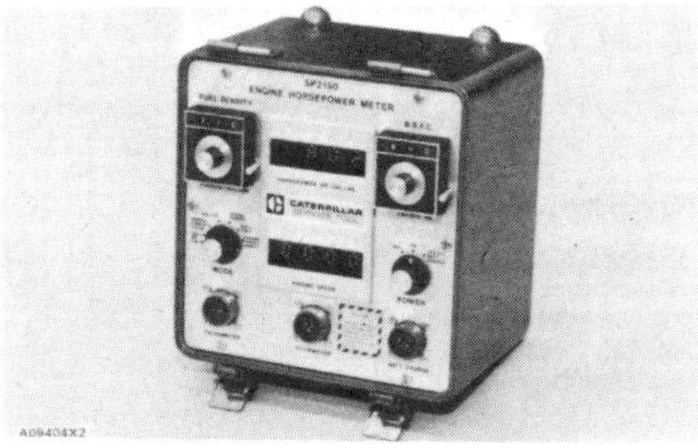

5P2150 ENGINE HORSEPOWER METER

1P5500 PORTABLE PHOTOTACH GROUP

The 1P5500 Portable Phototach

The 1P5500 Portable Phototach group (see Figure 2) measures engine speed from visual engine parts in rotation.

The 6V3121 Multitach Group

The 6V3121 Multitach Group (see Figure 3) measures engine speed mechanically, electrically, or visually.

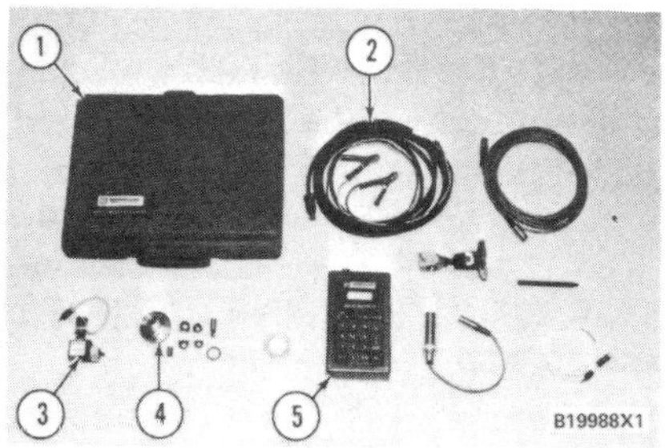

6V3121 MULTITACH GROUP

Figure 15–2 Example of Illustrating Alternatives

□ Presentation and Interpretation of Data

The presentation and interpretation of your data are the body of your recommendation report. Notice that this section is organized by *criteria*, with the alternatives falling as subsections under each criterion. If your criteria are weighted, begin with the most important. Present data on the alternatives in the same order under each criterion. If you set up patterns of expectation and stick to these patterns, it will be easier for your readers to comprehend your report. If your data run long to statistical or dimensional information, present them in tabular form (see Chapter 8).

As you follow the generalized outline for recommendation reports, use specific headings in the main body rather than the generalized headings in the outline. For example, write AVAILABILITY OF PARTS AND SERVICE rather than PRESENTATION AND INTERPRETATION OF DATA FOR 1ST CRITERION, put the explanation of the criterion in without a subhead, and write "Homelite XL-12 Automatic" rather than "Presentation of 1st Alternative." The generalized headings and subheads are there to remind you what you should cover and where—not to serve as the actual headings and subheads for your report.

Interpreting your data is necessary; contrary to popular belief, the facts do not speak for themselves. Or if they do, they speak with a mixed voice. All of us have witnessed several people drawing different conclusions from the same set of facts. The difference, of course, lies in interpretation. By including your explicit interpretation of the data presented in your recommendation report, you are accomplishing two goals: (1) you are doing your readers a service by summarizing the salient aspects of your data, and (2) you are maintaining control of your report by focusing on what *you* feel to be important. In your interpretation, point out to your readers which differences are significant and which are not. Draw their attention to the implications of the data. Focus their attention on those portions of the data that will have the most impact on the decision they will have to make. Help your readers anticipate your conclusions so that no readers are surprised by the time they finish reading your report. You are writing a report, not a mystery novel. Your goal is not to keep your readers in suspense but rather to guide them inexorably to the same conclusion that you reached during your feasibility study.

□ Conclusions and Recommendations

The general outline lists conclusions and recommendations separately, but they are often combined—particularly when the recommendations are obvious from the conclusions. Your conclusions should be a summary of the data you presented in the main body of your report; avoid introducing new information in the conclusion. If you have several recommendations, you might want to number them and add a sentence or two of justification to each. It is permissible to include personal opinion in this section, but be sure to label it as such.

□ Specimen Recommendation Reports

THE PROCTER & GAMBLE COMPANY

RECOMMENDATION REPORT FOR
TYPE OF HEAT EXCHANGER
TO HEAT TOOTHPASTE

Gayle L. Schaeffer
Senior Staff Engineer

April 16, 1980

Central Engineering Division
Ivorydale Technical Center
Cincinnati, Ohio

Specimen Recommendation Report

- 1 -

INTRODUCTION

Purpose

After considering three alternatives, I recommend that a scraped-wall heat exchanger be purchased for the proposed process area located at the Iowa City plant.

Problem

The new process area proposed at the Iowa City plant requires a heat exchanger to heat toothpaste from 93 to 115 degrees Fahrenheit. Presently, various types of exchangers exist within the Iowa City and Cincinnati plants. Since the toothpaste is a new product recently released by Process Development, the Central Engineering Division felt that reevaluating different types of heat exchangers was necessary to ensure product quality at the lowest cost.

Scope

I considered three types of heat exchangers in this study:

1) shell-and-tube
2) scraped-wall
3) falling-film

Criteria used in evaluating the alternatives were:

1) amount of shear thinning of the toothpaste
2) temperature effects such as scorching and loss of flavor
3) capitalized cost

Procter & Gamble tradition is that product quality receives greater emphasis than cost. Therefore, the shear thinning and product scorching criteria each receive a forty percent weighting, while cost has the remaining twenty percent. Because previous experience in both Process Development and Central Engineering is that double pipe and plate-type exchangers are not suitable for toothpaste, I did not consider those options. A report issued in November, 1979 by Process Development (WHTC 43-101A) recommended that the three types of exchangers mentioned on page 1 be considered for heating the new toothpaste. Appendix A describes and illustrates these three types of heat exchangers.

Specimen Recommendation Report, continued

- 2 -

SHEAR THINNING

Definition of Criterion

Toothpaste is a non-Newtonian fluid which has a viscosity that is quite sensitive to temperature and shear stresses. Quality standards for the product specify a narrow limit for toothpaste viscosity. Toothpaste viscosity may decrease as a result of mechanical shearing or excessive velocities. Because thinned toothpaste is not acceptable to a consumer, the product must be wasted or reprocessed. Hence, from both quality and economic standpoints, the heat exchanger must not excessively shear the toothpaste.

Shell-and-Tube

Pressure drop on the tube side varies according to the number of tube passes and the velocity of the fluid. To provide the necessary area for heat transfer, at least three tube passes are required. In the past, product shearing has not occurred to any significant degree when a properly sized, shell-and-tube exchanger with three tube passes was used.

Scraped-Wall

A scraped-wall exchanger is a one pass machine which uses mechanical mixing to enhance heat transfer. The fluid experiences relatively high shear stress as it is forced into the exchanger. Scraper blades, which rotate to keep the heat transfer surface clean, produce an additional amount of shear thinning. However, since the fluid passes through the exchanger only once, these shear stresses do not cause excessive thinning of the product.

Falling-Film

A falling-film exchanger operates in a single-pass mode with rapid flow through the exchanger. Rotating blades pick up the toothpaste as it enters the exchanger and throw it against the walls causing product shearing, as does the increased fluid velocity. Toothpaste viscosity remains within the quality limits after passing through a falling-film exchanger, although it tends to be at the lower limit of the range.

Specimen Recommendation Report, continued

- 3 -

In terms of product shearing, a shell-and-tube exchanger is better than the other two alternatives. A scraped-wall exchanger is acceptable whereas the falling film exchanger borders on being unacceptable according to this criterion.

TEMPERATURE EFFECTS

Definition of Criterion

Toothpaste is a heat-sensitive material that scorches if held in contact with a high-temperature metal for a certain length of time. Another temperature-related concern is that flavoring compounds, which are quite volatile, will evaporate in the exchanger and be lost. Quality standards for toothpaste require that the flavor be at an acceptable level. Because flavoring compounds are very expensive, their inefficient use drives raw material costs sharply upward.

Shell-and-Tube

The likelihood of scorching the toothpaste increases with increasing time of exposure and temperature. In a shell-and-tube exchanger, toothpaste contacts the hot metal wall of the tube directly. No mechanical mixing inside the tube occurs to facilitate heat transfer away from the walls. Increasing the fluid velocity through the tubes would decrease the contact time, thus reducing the scorching potential; however, an increased velocity produces a larger pressure drop, which decreases the toothpaste viscosity. In other words, a tradeoff exists between toothpaste scorching and high viscosity in a shell-and-tube exchanger. No flavoring compounds evaporate because the tubes are completely full of toothpaste with no free space for the vapors to exist.

Falling-Film

Falling-film exchangers are used often with heat sensitive materials in the process industry. The free space between the product flowing down the wall and the rotating blades allows vapors to evolve from the toothpaste. Even under positive pressure, a significant amount of the flavoring compounds likely would be lost.

Specimen Recommendation Report, continued

- 4 -

Scraped-Wall

Scraping the walls with rotating blades provides a clean surface for heat transfer; as a result, the toothpaste is in direct contact with the hot wall for only a minimal time. The action of the rotating blades mixes the toothpaste well, keeping the fluid at a homogeneous temperature. If the fluid is not well mixed, hot spots that lead to localized scorching will form. A scraped-wall exchanger has no free space between the floating scraper blades and the toothpaste. Lack of free space and positive pressure effectively prevent the loss of flavoring compounds.

In summary, shell-and-tube exchangers can scorch the toothpaste whereas falling-film and scraped-wall are ideal for heating sensitive fluids. Flavoring compounds tend to evaporate in the falling-film exchanger while they remain in the product when shell-and-tube and scraped-wall exchangers are used.

COST

Procter & Gamble requires that capitalized cost be used to economically evaluate alternatives. Appendix B gives the capitalized cost formula and applies it to each of the exchangers. The basis for the cost analysis is 1000 square feet of heat exchange surface area. In Table 1 is a summary of the cost analysis, with the length of service life, salvage value, replacement cost, and total capitalized cost listed. The time value of money specified by the company is 30 percent per year.

TABLE 1.

SUMMARY OF CAPITALIZED COST FOR EACH HEAT EXCHANGER

Type of Exchanger	n[1] (years)	Vs[2] ($)	Cv[3] ($)	K[4] ($)
Shell-and-Tube	10	2000	17,000	18,200
Scraped-Wall	8	2500	22,000	24,800
Falling-Film	9	1500	14,000	15,300

1 n - length of service life
2 Vs - salvage values
3 Cv - original cost
4 K - capitalized cost

Specimen Recommendation Report, continued

The falling-film exchanger is least expensive while the scraped-wall exchanger has the highest capitalized cost -- 62 percent greater than the falling-film and 36 percent higher than the shell-and-tube. Operating costs for steam are approximately equal for all three exchangers. I have not considered any differences that may exist in maintenance costs for the exchangers.

CONCLUSIONS AND RECOMMENDATIONS

Table 2 summarizes the overall comparison of the three types of heat exchangers on the basis of shear thinning, product scorching and loss of flavoring compounds, and capitalized cost. The first two criteria have equal weightings of 40 percent, while cost has a 20 percent weighting. For each criterion, the least desirable exchanger received a 1.0 while the most desirable exchanger received a 3.0. From the overall scores, I conclude that the scraped-wall exchanger is the best type to use for raising the temperature of toothpaste and recommend that one be purchased for the process area at the Iowa City plant.

TABLE 2.

OVERALL COMPARISON OF THE THREE TYPES OF HEAT EXCHANGERS

Type of Exchanger	Shear Thinning	W.F.[1]	Temperature Effects		W.F.	Cost	W.F.	Overall Score
			Scorching	Flavor				
Shell-and Tube	3	0.4	1	3	0.4	2	0.2	3.2
Scraped-Wall	2	0.4	3	3	0.4	1	0.2	3.4
Falling-Film	1	0.4	3	1	0.4	3	0.2	2.6

1 W.F. - weighting factor

Specimen Recommendation Report, continued

- 6 -

APPENDIX A

DESCRIPTION OF HEAT EXCHANGER ALTERNATIVES

Shell-and-Tube

The shell-and-tube heat exchanger is the most widely used exchanger in industry. One fluid flows inside a number of tubes contained in a shell while another fluid passes through the shell. Tubes may extend in a bundle straight through the exchanger or they may double back and forth from one end to the other several times (See Figure 1). Efficiency of the exchanger increases with this multiple pass pattern. Sufficient velocity in the tubes must be maintained in order to disturb and renew the clinging films of fluid that hinder uniform temperature distribution and reduce the total heat transferred. Baffles used on the shell side increase the flow velocity and turbulence of the shell side fluid. When the flow velocity and turbulence are increased, the rate of heat transfer and exchanger efficiency increases. The number and spacing of baffles contribute to the pressure drop on the shell side, an important consideration for operating costs.

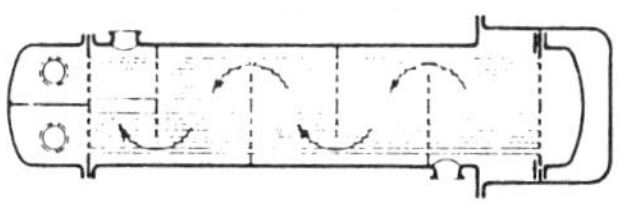

FIGURE 1.

A Shell-and-Tube Heat Exchanger with Tubes in the Form of U Bends.

Scraped-Wall

The feed enters at the top of a scraped-wall exchanger and is rapidly pushed through a small space between a revolving rotor and the exchanger walls. External pumping causes the fluid to be in a highly turbulent state upon entering the exchanger. Material flows down the heat transfer wall which is scraped clean by floating scraper blades twice each revolution.

Specimen Recommendation Report, continued

- 7 -

A heating jacket around the exchanger, as shown in Figure 2, contains the hot fluid, which is usually steam. No free space for vapors exists as the entire volume is filled with fluid. Retention time in the exchanger is fairly short, and the produce exists after only one pass through the exchanger. For some processes, a multiple-pass scraped-wall exchanger as shown in Figure 3 is more suitable than the single-pass type.

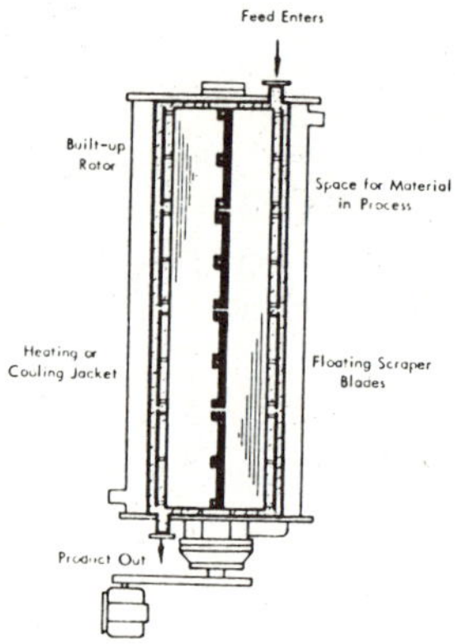

FIGURE 2.

Vertically Mounted, Scraped-Wall Type of Heat Exchanger.

Advantages of a scraped-wall heat exchanger include an extremely rapid movement of material with a high transfer of heat and a single pass which is desirable for heat-sensitive materials. Relatively viscous materials can flow through this type of exchanger without fouling it.

Figure 3.

Multiple-Pass Scraped-Wall Type of Heat Exchanger.

Specimen Recommendation Report, continued

- 8 -

Falling-Film

A falling-film exchanger is usually vertically mounted and arranged with the feed entering at the top as seen in Figure 4. Revolving blades pick up the feed and throw it against the heat transfer wall. Gravity forces the fluid downward while it still clings to the exchanger walls. On the other side of the wall, a hot gas such as steam condenses or a hot liquid circulates in the heating jacket. Rotating blades constantly agitate the fluid promoting a more uniform temperature and a clean surface for heat transfer. Like a scraped-wall exchanger, processing is very rapid with the heated product exiting at the bottom after one pass. Between the blades there is space for vapor to form and rise to the top of the unit. A centrifugal separator isolates the vapor and purges it from the system. Falling-film exchangers are especially suited for heat sensitive and viscous materials.

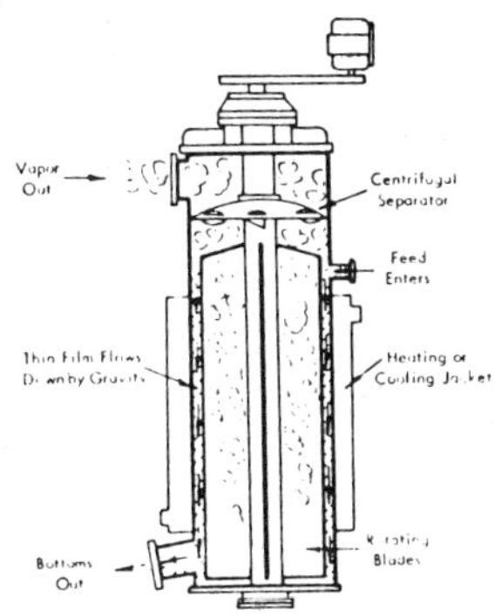

FIGURE 4.

Vertically Mounted, Falling-Film Exchanger.

Specimen Recommendation Report, continued

APPENDIX B

CALCULATION OF CAPITALIZED COST

Capitalized cost is a method to economically compare alternatives that Procter and Gamble uses for this type of project. This method gives a total cost for a piece of equipment which permits it to be perpetually replaced when its service life is over.

To calculate capitalized cost, use Equation (1):

$$K = Cv + \frac{Cr}{(1+i)^{n}-1} \qquad \text{Equation (1)}$$

where K = capitalized cost
Cv = original cost of the equipment
Cr = replacement cost of the equipment
i = interest rate
n = length of service life in years

The replacement cost is computed by Equation (2):

$$Cr = Cv - Vs \qquad \text{Equation (2)}$$

where Vs = salvage value of the equipment.

For the scraped-wall exchanger, the service life is estimated to be eight years and the salvage value \$2500. Original cost of 1000 square feet of heat exchange surface is \$22,000. Substituting these values and an annual interest rate of 0.30 into Equations (1) and (2) gives:

$$K = 22000 + \frac{(22000 - 2500)}{(1+0.3)^{8} - 1} = \$24,724 \approx \$24,800$$

Likewise, for falling-film exchanger of 1000 square feet of area and n = 9, Cv = 14,000, Vs = 1500:

$$K = 14000 + \frac{(14000 - 1500)}{(1+0.3)^{9} - 1} = \$15,301 \approx \$15,300$$

Specimen Recommendation Report, continued

Finally, for a shell-and-tube exchanger with 1000 square feet of area and n = 10, Cv = 17000, Vs = 2000:

$$K = 17000 + \frac{17000 - 2000}{(1+0.3)^{10} - 1} = \$18,173 \approx \$18,200$$

Specimen Recommendation Report, continued

SILVICULTURAL PRESCRIPTION

INTRODUCTION

Purpose: This report recommends an overstory removal-interplant harvesting technique for the timber stand 18-3-61 located on the U.S. Forest Service Red Ives Ranger District, Panhandle National Forest.

Definition of the Problem: Existing on the site is a stand which is presently producing below expected potential. It is composed of a residual overstory of larch, lodgepole pine, subalpine fir, spruce, and scattered white pine left after a shelterwood cutting in 1963. The larch and douglas-fir component is stagnant and the spruce, though showing better growth, has poor form. Overstory is fairly abundant at the top of the stand and becomes sparser toward the creek. Subalpine fir, spruce, and mountain hemlock regeneration occurred soon after harvest but rapid brush growth has inhibited additional regeneration. Although the established sapling size trees are growing vigorously, they occur in patches leaving parts of the stand unproductive. Brush is dense with an average height of four feet. Most of the overstory trees are poor quality and are not suitable for seed trees.

Scope: Our study evaluated site and stand conditions and determined which harvesting alternatives were realistic on this forested area. Predetermined constraints were: environmental, cost benefit, visual, wildlife and recreation. Given these constraints and the management objectives of the Forest Service, Kathy Roche and Cindy Lackey developed three feasible alternatives. These alternatives are 1) overstory removal-interplant 2) clearcut-burn-plant and 3) deferred action. Due to past activities in the stand, seed tree, selection, and shelterwood alternatives were not appropriate.

MANAGEMENT OBJECTIVES

The general management objectives of the Forest Service are listed in the appendix. Stand 18-3-61 has a primary management objective of producing timber.

Specimen Recommendation Report

-2-

DESCRIPTION OF ALTERNATIVES

Overstory Removal-Interplant

This alternative immediately harvests the residual trees from the prior shelterwood cut. Existing regeneration is protected as much as possible and where appropriate, fuels are jackpot burned to provide additional scarification. Planting a mix of species the spring after harvest is completed assures that all scarified areas are regenerated and a greater diversity occurs in the new rotation.

Clearcut-Burn-Plant

All merchantable material is harvested and remaining vegetation is slashed to provide additional fuels for the broadcast burn. A mixture of site-adapted commercial species is planted as soon as possible after the burn.

Deferred Action

The stand is left to continue in its present state with an entry considered in approximately 15 to 20 years. This alternative allows the subalpine fir regeneration to reach minimal merchantable sizes. At that time, remove all merchantable material and review the stand for further treatment.

ENVIRONMENTAL EFFECTS

Overstory Removal-Interplant

This harvesting operation could present a critical problem if poorly supervised. Excessive damage to existing saplings destroys their future utility. Although brush may hinder establishment of new trees, existing regeneration is already outcompeting brush species. This sapling component guarantees partial stocking on the site, and interplanting improves the species diversity and provides full stocking. Because the site remains largely vegetated, soil erosion potential is low.

Specimen Recommendation Report, continued

-3-

Clearcut-Burn-Plant

Clearcuts offer tremendous possibilities for creating a new stand with a good species diversity of highly valuable commercial species. Although planting occurs as soon as possible after the burn, there is a time lag before vegetation can get established and offer soil stability. On this site the soils are shallow and exposure for any length of time may result in excessive erosion. Due to the low volume of fuels present, a broadcast burn may not carry well. If that were the case, brush would resprout quickly and inhibit establishment of planted seedlings.

Deferred Action

Deferring action leaves the site undisturbed; thus the erosion potential is the lowest in this alternative. As the overstory continues to decline in vigor, potential for insect and disease attack increases.

COST/BENEFIT

Overstory Removal-Interplant

This alternative costs less than a clearcut because slashing is not needed thus requiring fewer manhours. Planting costs are reduced because seedlings only need to be planted where natural stocking is insufficient.

Clearcut-Burn-Plant

Although a clearcut is easier to log, it requires more time to slash and prepare the site for burning. Site preparation and planting costs are increased over that of the previous alternative.

Deferred Action

This represents an opportunity cost of capital loss. Another problem is that in 15 to 20 years reentry may not be economical.

Specimen Recommendation Report, continued

-4-

VISUAL IMPACT

Overstory Removal-Interplant

Aesthetics may be improved with the removal of poor-formed, scattered, mature timber. A vigorous stand of new growth offers a change in the texture of the landscape which is often pleasing to the eye. An overstory removal minimizes the visual impact of man's activities.

Clearcut-Burn-Plant

Small clearcuts, well-placed, can resemble natural openings and accent the overall visual texture. This stand is too large for this type of classification and is considered aesthetically undesirable.

Deferred Action

No visual change would occur in this alternative. To some people this area may look natural, but to most, past timber activity is evident. The residual stand is not of high visual quality.

WILDLIFE

Overstory Removal-Interplant

Wildlife habitat is not adversely affected with an overstory removal. Cover and forage are still available and large snags are left standing to accomodate cavity nesting birds.

Clearcut-Burn-Interplant

This harvest technique has the greatest immediate impact on wildlife habitat. Cover and forage are eliminated for the first several years after harvest. Wildlife tend to utilize only the edges of the clearcut. Snags are not available for cavity nesting birds.

Specimen Recommendation Report, continued

-5-

Deferred Action

As the stand continues to grow, browse species will eventually get out of reach or become decadent and less nutritious. As habitat deteriorates for one species, it usually improves for another. A change in animal utilization might well be noted.

RECREATION

Overstory Removal-Interplant

Overstory removal may increase hunting values by making deer and elk more visible but may turn a hiker off. Overstory removal has a moderate effect on recreational use.

Clearcut-Burn-Plant

Recreation values in general appear to decline in areas which are clearcut.

Deferred Action

Recreation values remain the same.

CONCLUSIONS AND RECOMMENDATION

Timber activities are programmed to harvest mature and overmature residual trees and return the site to full productivity by establishing a vigorous stand with good species diversity. To meet this and other objectives, our proposed activity for this stand is an overstory removal-interplant. Due to the advanced stage of the existing vigorous regeneration, a clearcut unnecessarily destroys potential volume presently in the saplings. The logging system chosen must protect the existing understory. We scheduled a few large snags to be left for cavity nesting birds. Handpiling and jackpot burning are part of the program to alleviate slash problems and give

Specimen Recommendation Report, continued

-6-

additional site prep. Planting of scarified corridors, planned for the spring after harvest, includes the following species: douglas-fir, larch, spruce and improved white pine, 2-0 bare root stock.

Specimen Recommendation Report, continued

MATERIAL SELECTION RECOMMENDATION

INTRODUCTION

Purpose

The purpose of this report is to recommend that an acrylic rod be used as the linkage member to connect the solar panel rotation gear to the activation control in the new model of the Serion Satellite.

Problem

The old Serion Satellite which is currently in orbit is malfunctioning. After considering several components of the system, I feel that the malfunction is caused by the connection rod between the solar panel and the control mechanism which activates the panel's rotation. In the original model this member was made of aluminum. According to the research group investigating the malfunction, the temperature extremes to which the rod is exposed caused it to expand. By creating undue stresses, the change in length of the rod disturbs the delicate balance of the system network. Since this aluminum rod has expanded and is not allowing complete rotation of the panel, I recommend that the next model of the satellite replace this aluminum rod with a rod made of a more dependable material.

Scope

There are many materials which could be used to replace the current aluminum rod. However, because a lighter material affords a greater fuel savings, I chose acrylic, nylon, and poly-vinyl chloride to be tested. Since the testing of these materials could not actually be carried out under the field conditions in space, they were tested under a laboratory mode which best simulated the actual loading conditions. In evaluating the

Specimen Recommendation Report

- 2 -

materials I felt the following criteria should be used:

1. Lowest maximum change of rod length between the temperature extremes of 0 F and 120 F,

2. Lowest weight per rod,

3. Lowest cost per rod.

However, because the current malfunction of the system is believed to be caused by the aluminum rod's elongation, the total maximum change in rod length was doubly weighted.

PRESENTATION OF DATA FOR ELONGATION

Because of the varying teperatures that the member will be exposed to, it is necessary to know the change in length of each material at the expected maximum temperatures of 0° F and 120° F. Actual calculations used to determine the total elogation of each material are quite lengthy but may be found in Appendix A. Several of the parameters required for these calculations include the modulus, the change in modulus due to the temperature variations, strain at the maximum expected temperatures, the design cross sectional area, and the actual rod length. Resulting total elongations, found in Table 1., show that acrylic experiences the least amount of elongation of the three materials. However, the difference between acrylic and poly-vinyl chloride is not significant.

Table 1. TOTAL ELONGATION

MATERIAL	TOTAL ELONGATION IN INCHES
Acrylic	.1128
Nylon	.2770
Poly-vinyl Chloride	.1267

Specimen Recommendation Report, continued

-3 -

PRESENTATION OF DATA FOR ROD WEIGHTS

Another important criterion is the weight of the rod. The design cross sectional area, which is directly related to the diameter of the rod, is used to calculate the weight of the rod. In order to provide sufficient operating space, physical specifications of the system require that the diameter of the rod not exceed two inches. All the materials met the maximum allowable two-inch diameter requirement. To calculate the actual weight of the rod, I multiplied the cross sectional area of the rod by the required rod length of 12 inches. Then this volume was multiplied by the specific weight of the material to obtain the actual weight. Table 2. gives values for the diameter, the specific weight, and the actual weight for each material. Notice that nylon is the lightest of the three materials by a substantial margin of 38 percent.

Table 2. TOTAL WEIGHT AND RELATED PARAMETERS

MATERIAL	DIAMETER (in.)	SPECIFIC WEIGHT	WEIGHT (lbs.)
Acrylic	1.84	1.185	1.36
Nylon	1.18	1.125	.537
Poly-Vinyl Chloride	1.35	1.400	.868

PRESENTATION OF DATA FOR ROD COST

The final criterion considered was that of cost per rod. Total cost encompasses two facets. One facet is that of the cost of production, the other is the actual cost of the materials. While both poly-vinyl chloride and acrylic are relatively inexpensive to produce, nylon must undergo a rather expensive drying process to insure that its stiffness will not change as time goes on. Therefore, there is a cost of approximately three dollars

Specimen Recommendation Report, continued

-4-

per pound which must be added to the material cost of nylon. The added production cost of poly-vinyl chloride and acrylic are 56¢ and 20¢ respectively. Both the material costs and the total costs are listed in Table 3. Even with the added production cost of Nylon, it still is less expensive than either of the other materials.

Table 3. COSTS

MATERIAL	MATERIAL COSTS ($/lb.)	TOTAL COSTS ($)
Acrylic	3.50	4.92
Nylon	4.90	4.24
Poly-Vinyl Chloride	5.20	5.07

DATA ANALYSIS

The results of all calculations for each criterion are found in Table 4.

Table 4. RESULTS

CRITERION	ACRYLIC	NYLON	POLY-VINYL CHLORIDE
Weight (lbs.)	1.36	.537	.868
Cost ($)	4.92	4.24	5.07
Elongation (in.)	.113	.277	.127

A simple fractional method of analysis allowed a comparison of the data. The material with the least value for each criterion was used as a standard for comparison and received the maximum rating of 1.0 for that particular criterion. Each of the other materials received a lesser rating according to its ratic to the best material. I double the rating for the elongation criterion in order to apply a greater weighting. Final ratings and each material's total rating are found in Table 5.

Specimen Recommendation Report, continued

-5-

Table 5. RATINGS

CRITERION	ACRYLIC	NYLON	POLY-VINYL CHLORIDE
Weight	.395	1.0	.618
Cost	.862	1.0	.836
Elongation	2.0	.814	1.78
TOTALS:	3.25	2.814	3.23

CONCLUSIONS AND RECOMMENDATION

According to the totals obtained from Table 5., acrylic exhibits the best overall preformance of the three materials for the given criteria and the suggested weighting system. However, the margin between poly vinyl chloride and acrylic is small. Based on these results, I recommend that an acrylic rod be used as the linkage member between the solar panel rotation gear and the activation control in the new Serion Satellite.

Specimen Recommendation Report, continued

Writing Exercises

1. Focus on several major problems facing your college or university, such as funding, declining enrollments, traffic and parking, faculty governance, and the like. For each, clearly identify several alternative solutions or possible courses of action. List the alternatives.
2. Do the same for several problems facing a community—either your home town or the town in which your college or university is located.
3. For any three of the problems identified above, list the possible criteria that might be used in evaluating the feasibility of possible solutions.
4. For any three of the problems identified above, identify and describe the individuals who would have primary responsibility in making the management decision regarding a solution to the problem.
5. Collect two examples each of excellent recommendation reports and minimally adequate or unacceptable recommendation reports.

Writing about Writing

1. Write an essay on the topic: the Feasibility Study versus the Feasibility Report.
2. Write an essay pointing out how the selection and weighting of evaluation criteria can affect the outcome of a recommendation report. Use some specific examples.
3. For one of the problems identified in Writing Exercises 1 or 2 above, write up in paragraph form (a) a detailed description of the problem and its background, the alternative solutions, and the evaluation criteria, and (b) a detailed analysis of the audience, including primary audience, secondary audience, and immediate audience.
4. Defend one of the following positions: (a) Primary audiences for recommendation reports are *usually* technical experts, or (b) Primary audiences for recommendation reports are *rarely* technical experts. Be specific and detailed in your defense.
5. Obtain a copy of a major recommendation report or feasibility report. Write a critical review of the report, pointing out strengths and weaknesses and making suggestions for improvement as appropriate.

Writing Assignments

1. As an employee of a firm or organization (or an as outside consultant hired by the firm or organization), you have been given the assignment of evaluating three (or more) alternate types of equipment and recommending one to be purchased or used. Present data on the alternatives, interpret the data and draw conclusions, and make a recommendation. Don't just do a *Consumer Report* or *Road and Track* sort of comparison, but tie your evaluation to a particular use of the equipment in a particular set of circumstances. Employ at least three criteria. Consider these possibilities:

- ☐ The county is developing a small roadside park. As recreation specialist, you are asked to recommend a particular type of fireplace to be installed in each of ten campsites.
- ☐ You're working for a small accounting firm. The boss has decided to purchase new hand calculators for each of the accountants and has asked you to recommend a particular brand and model.
- ☐ The outdoor program at a nearby university wishes to purchase cross-country skis (or backpacks or sleeping bags or canoes or what have you) for their rental program. They have asked for your recommendation.
- ☐ You're a lab technician for a hospital. The director has asked you to recommend a particular type of analysis equipment to be purchased.
- ☐ A large corporate farm that you work for wants to acquire a new hay-harvesting system. You have been asked to select among the several that have been proposed.

2. Instead of alternative types of equipment or hardware, you are asked to evaluate three (or more) ways of solving a technical problem and to make a recommendation. Consider these possibilities:

- ☐ You're a civil engineer for the highway department. A secondary road crosses a small creek, and you're asked to recommend either a pipe culvert, a box culvert, or a wooden bridge for the crossing.
- ☐ You are a district ranger for the USFS. Several harvesting methods (seed-tree, shelterwood, clear-cut, among them) have been proposed for a particular timber tract in your district. You must recommend one to the Forest Supervisor.
- ☐ You are a financial consultant for an investment firm. A client has a couple of hundred thousand to invest. Consider alternatives based on the client's particular investment objectives and make a recommendation.

16

Proposals

Businesses, industries, and government agencies thrive on new ideas, for without new ideas—new and better ways of doing things—organizations tend to stagnate. A proposal is a documented idea. It is possible to make an oral presentation to your immediate supervisor on some new idea you have, but the minute your idea has to jump a level in the organizational hierarchy or go outside the department (or outside the firm) documentation is necessary. Hence, the written proposal. A written proposal is a document that proposes a solution to a technical problem and provides sufficient information so that mangement decision-makers can decide whether or not the proposed solution will be accepted.

A proposal is like a three-legged stool (Figure 16–1). In order to accept a proposal, management must be convinced:

1. That the proposed solution is technically sound and operationally feasible.
2. That the person or group proposing to solve the problem is indeed capable of effecting a solution.
3. That all costs have been identified, the costs are reasonable, and the benefits of the solution are commensurate with the total cost.

These three elements of the proposal are usually called (1) the Technical Section, (2) the Managerial Section, and (3) the Financial Section. If any one is missing or weak, the proposal collapses, like a stool with one leg jerked out from under it.

Figure 16–1 The Proposal as a Three-Legged Stool

A proposal is similar to a feasibility or recommendation report (see Chapter 15) in that it deals with a solution to a technical problem, but it differs from the recommendation report in that it focuses on a single solution rather than alternative solutions and it goes beyond a discussion of *what* should be done to include information on *how* it will be done, *how long* it will take, and *how much* it will cost. Some proposals are invited by the organization that has the problem. These, called *solicited proposals,* are usually written to a set of specifications and their organizational structure is laid down by the organization doing the soliciting. Proposals initiated by the individual or group seeking to do the work of solving the problem are called, logically enough, *unsolicited proposals;* their organizational structure is freer than that of the solicited proposal. The three varieties of proposal you are most likely to be involved with are the in-firm proposal, the unsolicited firm-to-firm proposal of limited scope, and a special type of proposal called the research grant request.

□ The In-Firm Proposal

As its name suggests, the in-firm proposal is the document you use to convey ideas about your solution to a problem you see within you own organization. The emphasis here is on *your solution;* if you make a habit of calling management's attention to problems and then suggesting that someone else solve them, you won't make many points. In-firm proposals are usually informal or semiformal (see Chapter 4). The emphasis in most in-firm proposals is on the technical solution and the cost-benefit analysis. In your managerial section it is not so important to demonstrate the competence of the people who will be working to solve the problem as to demonstrate their availability for internal reassignment. Similarly, you must demonstrate the facilities and equipment needed are not only present within the firm but available for use by your project team.

In your financial section, cover both direct and indirect costs. Don't assume that just because "the maintenance department will do the work" there is no cost involved. Give a detailed breakdown of clock-hours required by various personnel categories and assign—or ask that management assign—dollar values to these figures.

□ Success Strategy for Student Writers □

Look around your campus for in-firm proposal possibilities. Consider your major department, your dorm or living group, your fraternity or sorority, or any other organization you are a member of. If you want to propose a solution to a general campus problem, put yourself in the role of a faculty or staff member.

□ The Firm-to-Firm Proposal

The one-sentence version of a firm-to-firm proposal is "X proposes to do Y for Z for W dollars." *Interwest Engineering proposes to design a new wastewater treatment facility for the City of Lewiston for $2,400,000. The Associated Students of Landscape Architecture propose to develop a 35–acre arboretum for Western University at a cost of $75,000 over a three-year period. Gyppo Loggers, Inc., will harvest 150,000 board feet of timber from Fred Johnson's woodlot for $10,585.75.* In other words, the firm-to-firm proposal is a statement of who is going to do what for whom and for how much.

In the firm-to-firm proposal—particularly the unsolicited firm-to-firm proposal—your description and analysis of the problem are vitally important. If you can't demonstrate your understanding of the problem facing another firm, your organization will get nowhere trying to outline a solution. With this type of proposal, too, the managerial section is vitally important and often overlooked. Even if your solution to the technical problem is feasible and your project cost is attractive, your firm won't get the job unless you can convince the decision-makers that, like Union Pacific, you can handle it. Your firm's track record with similar projects is important here, as are the credentials of your personnel. And your financial section must demonstrate that not only is your cost estimate reasonable but that you have identified all areas of cost. With the firm-to-firm proposal more than any other, the three legs must be sound for your proposal to stand.

□ The Research Grant Request

A request for research funding is not an act of begging. It is an offer of services rendered for a stated price, and as such it is really a proposal.

Funding agencies and organizations are, it turns out, basically eager to award research funds; their very existence depends on their ability to disperse the funds they have control over. What they need from you is something to hang their hat on: a document that can be offered to justify their decision to grant funds for your research project. They don't require a *guarantee* of success, but it's up to you to demonstrate a *likelihood* of success and a reasonable benefit from your efforts. Keep in mind that the benefits of research typically do not accrue to the funding agency itself but rather to society as a whole or to a specific segment of society.

The research grant request often includes a literature review or a state-of-the-art statement (see Chapter 13). A good job on this is part of demonstrating your competence to conduct research. A strong technical section with a detailed plan of attack is crucial; many proposals fail because they are vague on *how* the research is going to be conducted.

A strategy with grant requests that succeeds more often than not is to feed the funding agency's ideas back to them in their own words. If the agency specifies that projects "contribute to the general public's understanding of mankind's history and cultural legacy," assure them that your project contributes to the general public's understanding of mankind's history and cultural legacy. If the agency funds research aimed at "a better understanding and quantification of important processes in tropospheric chemistry," make it known that your research project is designed to provide a better understanding and quantification of important processes in tropospheric chemistry. They'll probably be flattered, and at least you'll be assured that your proposal won't be thrown out on the first round for failure to meet the general goals of the program.

One final word: An excellent guideline for the unsolicited research grant request is NSF Brochure 78–41, *Grants for Scientific Research*, available from the National Science Foundation, 1800 "G" Street N.W., Washington, DC 20550, Attn: Forms and Publications Unit.

□ Success Strategy for Student Writers □

A request for authorization to proceed with a student research project could serve as a paper for your technical writing course. Even though you wouldn't be asking for salary support (and even if your major department is picking up incidental expenses), you can draft your request as a formal proposal. Emphasize your credentials (previous courses, work experience) and identify the *areas* of cost—just as if you were seeking outside funding.

□ General Outline for Proposals

I. Introduction

A. Purpose of the Document

- B. Problem to Be Solved
- C. Scope of the Proposal

II. Technical Section
- A. Background (if necessary)
- B. Plan of Attack (Methodology)
  1. Overview of the Solution
  2. Task Breakdown by Phases
  3. Scope of Work
- C. Time and Work Schedule

III. Managerial Section
- A. Credentials and Experience of the Firm
- B. Facilities and Personnel Available
- C. Authorities and Accountabilities

IV. Financial Section
- A. Costs
- B. Method of Payment
- C. Revenues or Cost Savings Generated by the Project

V. Tangible Products of the Project

VI. Urge to Action

□ Introduction

The basic structure for the introduction to your proposal is similar to the one for recommendation reports: purpose, problem, and scope.

Purpose

Begin with a clear one- or two-sentence statement of what you propose, similar to the one-sentence version of the firm-to-firm proposal (above).

Problem

Follow with a paragraph or two describing the problem to be solved. Your problem statement serves two purposes: first, it provides the decision-makers with background information just in case they need it; second (and more important), it demonstrates your knowledge and understanding of the problem the organization faces. If you can do justice to the problem in a couple of paragraphs, do so in the introduction; if not, summarize the problem in the introduction and then develop the problem description more fully in a "Background" section immediately following the introduction.

Scope

The scope section should provide the scope of your document rather than the scope of your proposed solution. It serves as a kind of table of contents, identifying for your readers the various sections of your proposal and the order in which you present them.

□ Technical Section

Background

Use the background section, as needed, to develop your problem statement more fully. For a research grant request, use the background section for your state-of-the-art report or literature search.

Plan of Attack

Start the main body of your technical section with an overview of your proposed solution. Give your readers the big picture of where you are going with your proposal before you get down to the details. (Keep in mind that the decision-makers probably aren't going to read the details, anyway—see Chapter 3). Follow this by an identification of the major phases in the implementation of your proposed solution and a breakdown of the individual tasks to be performed. For instance, if you're proposing engineering services in connection with building a logging road, phases might be:

- □ Preliminary mapping phase
- □ Preconstruction phase
- □ Construction phase
- □ Inspection and verification phase

Individual tasks in the preconstruction phase might be: running centerline, cross sectioning, slope staking, and setting clearing stakes. Be sure to include all necessary tasks—even those to be performed by others. This will set you up for the next subsection of your proposal.

In the "Scope of Work" subsection you want to state clearly who is going to perform which tasks. For instance, if you are proposing a timber harvest, you want to state specifically who will mark the trees for cutting, who will do the skidding, who will do the loading and hauling, and who will pile and burn the slash. (Everyone assumes you'll do the felling and bucking.) If you are proposing cafeteria service for your college dormitory, state specifically who will let bids on equipment, who will supervise the installation, who will plan the menus, and who will supervise the serving staff.

For most proposals it is essential to include a time and work schedule that indicates both clock hours required for various tasks and elapsed time required for various phases. Present this information graphically (see Chapter 8). If tasks are inherently ordered, indicate the sequencing; if tasks or phases can overlap, indicate this. Often, a simple bar graph (see Figure 16–2) will suffice. For more complicated projects, consider using a PERT chart or similar management tool.

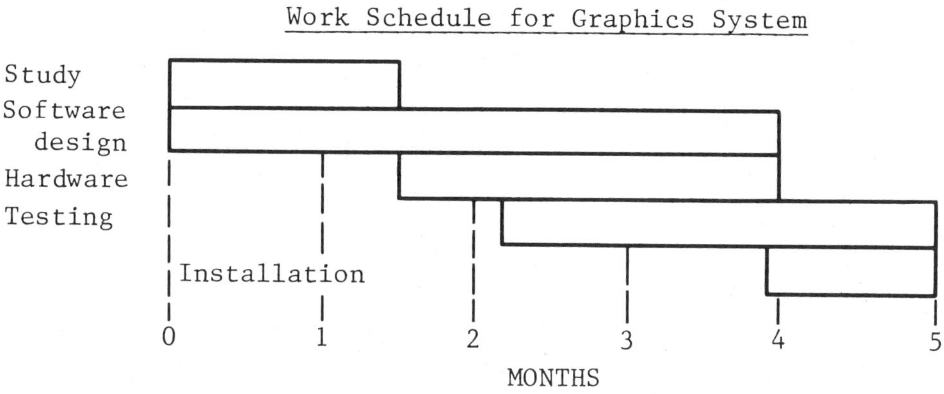

Figure 16–2 Time and Work Schedule Illustrated as a Bar Graph

□ Managerial Section

Your job with the managerial section is to convince the decision-makers that you (and the organization you represent) can get the job done. Start by describing the credentials of the firm and the people who make it up. List and briefly describe projects your organization has successfully completed, emphasizing those that are similar to the project you are proposing. Provide the names and telephone numbers of individuals who can vouch for the quality of work. Demonstrate (as best you can) a track record of getting things done on time and at cost.

If special equipment or facilities are required for the project you propose, indicate that your organization has (or has access to) them. Don't just state that the equipment and facilities exist within your group; demonstrate that they are available for assignment to the proposed project. The same for personnel. Indicate the specific people in your organization who will be involved in the project. For a firm-to-firm proposal, provide resumes.

Credentials and qualifications of personnel lead you directly into the next subsection—authorities and accountabilities. In many proposals, it is important to indicate the particular individuals who will have authority to make certain decisions and the individuals or groups who will stand accountable for various phases of the project. In some cases, this will read very much like your scope of work statement, but in other cases, the people actually doing the work might be different from those who are accountable for seeing that it gets done and gets done right.

□ Financial Section

As the general outline for proposals indicates, the financial section is more than a bottom-line, total dollar figure for the cost of your proposed project. Management decision-makers are interested, of course, in the total cost, but they are just as interested finding out whether you have identified all areas of cost and whether your itemized estimates are reasonable. *Cost over-run* is a nasty phrase these days; so is *hidden cost*.

In general, costs for a project can be broken down into four areas:

- □ Wages, salaries, and fees
- □ Operating expenses
- □ Capital expenditures for materials and equipment
- □ Overhead and other indirect costs

For personnel costs, estimate the clock hours required for each wage or salary category in each phase of the project and multiply by the average hourly rate to get subtotals (see Figure 16–3). For in-firm proposals, list specific clock-hour requirements and add (or allow room for management to add) dollar-cost figures. Don't assume that personnel within the firm or organization are "free." If you plan to rely on volunteer labor, list the value of this donated labor and subtract it from your total to arrive at the net cost of the project. Asking for $10,000 toward a $30,000 project is a lot easier than asking for $10,000.

Operating expenses include such things as travel, lodging and meals, computer time, office or lab supplies, telephone expenses, and the like. One rule of thumb for estimating operating expenses is "Make your best guess, add 20 percent, and multiply by two." For major capital expenditures, list not only prices but also the sources of the price information. If you plan to dispose of equipment on completion of the project, subtract resale value and list only net cost to the project.

Overhead and indirect costs might include such things as rent and

WAGES AND SALARIES

Category	No. of Hours	Hourly Rate	Total
Project Engineer	220	$35.00	$7700
Survey Crew Chief	150	20.00	3000
Survey Crew	150	45.00	6750
Equipment Operators	600	18.00	10800
Maintenance Men	60	12.00	720
Flaggers & Gophers	550	8.00	4400
		Grand Total:	$33,370

Figure 16–3 Cost Breakdown for Personnel Costs

utilities on office or lab space, fringe benefits for personnel, travel to a conference, costs of publishing project reports, or expenses for secretarial and support personnel. Many organizations use a flat-rate figure for overhead—usually a percentage of the total of the first three categories of cost.

For the firm-to-firm proposal, method of payment can be an important part of your financial section. Do you expect to be paid in advance? At the end of each phase? Periodically throughout the project? Do you plan to post performance bonds, or allow the client to withhold a contingency sum? Even with the in-firm proposal, you may wish to speak to budget allocations or interdepartmental transfers.

The upshot of all of this is that preparing the financial section of a proposal is quite a bit more involved than one might expect. In fact, you are probably well advised to seek the guidance of the accounting section within your firm or organization in preparing this section. Of course, past proposals for similar sorts of projects can be a good source of information as well.

□ Success Strategy for Student Writers □

Getting accurate and complete cost data could prove to be more time-consuming than writing and revising the entire remainder of your proposal. Consider identifying the various categories of cost and filling in the dollar figures with x's, as in Figure 16–4, page 334. Chances are that your instructor is more interested in the way you present your financial data than in the veracity of your figures. (But be sure to get permission from your instructor first.)

Often there is more to the financial aspects of a project than the cost. If what you propose will lead to long-term savings or will generate revenue, be sure to include this information in your proposal. Make sure your projections are realistic, however.

□ Tangible Products and Urge to Action

It is a good idea to close your proposal on a positive note; unfortunately, the financial section is often the least positive aspect of what you propose. So discuss the benefits. Summarize what you have said earlier, and restate for emphasis the most positive aspects of your proposed solution. In particular, when you are proposing professional services (including the service of conducting research), stress the tangible, visible products of your services. If you are proposing to conduct a forest inventory for a client, point out that he will receive field data, three (or one or five) copies of the cruise report, a species map (in color) of his timber holdings, and a

Table 2.	COST BREAKDOWN		
PERSONNEL EXPENSES			
Party Chiefs (2)	$X.XX/hour	Crew (4)	$XX.X/hour
	104 hours		80 hours
	$XXX.XX		$XXX.XX
2-party chiefs	X 2	4-crew members	x 4
	$XXX.XX		$XXXX.XX
Total $XXXX.XX			
OPERATING EXPENSES			
Travel			
120 miles/day for 10 days		$XXX.XX	
Computer Time		$XX.XX	
Telephone		$XX.XX	
Equipment (abney, compass, etc.)		$XX.XX	
		Total $XXX.XX	
OVERHEAD EXPENSES			
Secretarial Expenses	$X.XX/hour		
	12 hours		
	Total $XX.XX		
TOTAL EXPENSES	$XXXX.XX		

Figure 16–4 Financial Summary without Specific Figures

set of aerial photos. If you are proposing to do an architectural design, point out that tangible products include plans and elevations, perspective drawings, three (or one or five) sets of complete construction plans and specifications, and a 1:12 scale model. For a research proposal, mention the vehicles you plan to use to disseminate the research information generated: interim and final reports, technical bulletins, journal articles, and conference presentations.

Your urge to action may be enthusiastic and persuasive in tone or reserved and factual in tone, depending on your analysis of the primary audience for your proposal (see Chapters 3 and 10). But you should have one. It is a bit gross, however, to label it "Urge to Action." Call it "Conclusions," or "Summary" (if appropriate), or "Concluding Comments," or something like that.

□ Specimen Proposals

PROPOSAL: A COMPUTERIZED COSTING PROCEDURE

for
Encoder Products Company
Sandpoint, Idaho

Proposed by
Programming Success Incorporated
Moscow, Idaho

Janell Watt
English 317-01
Assignment #1

Second Submission
December 9, 1981

Specimen Proposal

2

COMPUTERIZED COSTING PROCEDURE

I. AN OVERVIEW OF THE PROPOSAL

Purpose

Programming Success Inc. proposes to develop for Encoder Products Company a customized costing program that is integrated with your firms existing computer system. The cost for this project will be $2,560.00.

PROBLEM

The problem that emerges in most rapidly expanding firms such as Encoder is a lack of control. To gain control firms need efficient methods of gathering and distributing data used in management decisions. One of the most useful forms of data is a costing analysis. Although Encoder has a manual procedure for costing, it is unable to gather, transform and distribute data quickly enough to be useful. To solve this problem, we can write a costing program to give you up-to-data reports at your fingertips.

Specimen Proposal, continued

3

CONTENTS OF PROPOSAL

II. DETAILS OF PROPOSAL

Hardware Requirements: This proposal does not require Encoder to purchase any additional hardware. It is designed to run on your OSM 3566 computer. Terminals and printers can be used in their current locations.

User: The reports generated by the program are available to any user with a correct password. For security purposes, each password can be set up to allow full access or limited access to the reports. For instance, the operators have access to programs that add, delete and change data. These programs update the costing reports.

File Layout: Existing files that we will use are Payroll Masterfile and Accounts Receivable Masterfile. Automatic transfer of data to and from these files increases the reliability of the data. This data

Specimen Proposal, continued

4

integrity feature will save the data from transfer errors. Files created by our design are Job Masterfile and Employee Job file.

Payroll File

Emp#	Name	Dept#	Rate	Deductions	Gross	Hours

Employee Job File

Emp#+Job#	Dept#	JHours	JGross	JOverhead

Job Masterfile

Job#	Co.#	THours	Labor	Overhead

A/C Receivable File

Co#	Co_Name	Address	Balance	Tax

Personnel

Title	Name	Phases
Project Director	Gene Schraut	1-A; 8-C
Systems Analyst	James McLaughlin	1-A,B; 2-B
Programmer	Janell Watt	1-A; 2-A,B; 3-6; 7-A: 8-A,B,C
Literature	Darlene Held	7-B

Specimen Proposal, continued

5

<u>Phases of Proposal</u>

Num.	Phases	Date Scheduled
1	Organizational Meetings	
	A. Managers Meeting	Oct. 20
	B. Examine Hardware	Oct. 20
2	Review of Design	
	A. Manual Procedure	Oct. 22
	B. Hardware	Oct. 24
3	Review of Specifications	Oct. 26
4	Design Documentation	Oct. 28
5	Program Coding	Oct. 30
6	Program Testing	Nov 7
7	Program Documentation	
	A. Write Users Manual	Nov. 14
	B. Print Users Manual	Dec. 1
8	Installation	
	A. Install Program	Dec. 19
	B. Train Operator	Dec. 22
	C. Brief Managers	Jan. 3

III. MANAGERIAL SECTION

<u>Credentials</u>: Programming Success Incorporated has been installing computer systems for six years. Most of our

Specimen Proposal, continued

6

experience is in the area of customized accounting and reporting programs. We have many satisfied customers. Feel free to contact any of them.

<u>Facilities</u>: Our computer facilities include an IBM 3765 computer along with many microcomputers. Programming languages available are CMS Basic, Waterlou Basic, Fortran and Cobol.

<u>Personnel</u>: The programmer assigned to your project is Jahell Watt. She has six years of experience including the programming of an accounting system much like Encoders. Project director is Gene Schraut. If any problem arise contact him with our toll free number (800-263-8541).

IV. FINANCIAL SECTION

<u>Payment</u>: Payments are to be made by monthly installments.

<u>Cost Savings of Proposal</u>: The most apparent cost savings this proposal generates is the labor saved by computerizing. Personnel that spent a week calculating costs per unit will now do the task in less than an hour. Personnel freed from the task can then be used elsewhere.

Specimen Proposal, continued

7

(40 hrs/wk) x (4.00 salary) =	160.00	weekly savings for employee
	+ 160.00	savings for employee you won't have to hire.
	$320.00	Total weekly labor savings

$$\frac{\text{(2818.00 cost of program)}}{\text{(320.00 weekly savings)}} = \text{9 weeks to pay for program with savings}$$

Savings are also generated by having a more accurate estimate of cost per unit. This statistic allows you to determine a better price for your products.

<u>Costs of Proposal</u>

Labor			
Project Director (12.00/hr.)	48.00		
Systems Analyst (10.00/hr.)	70.00		
Programmer (6.00/hr)	500.00		
Literature (4.00/hr.)	400.00		
Total Labor costs		1,018.00	
Computer time			
wordprocessing (40 hrs.)	200.00		
development (160 hrs.)	600.00		
Total computer time		800.00	
Paper		100.00	
Selling and Administrative		200.00	
Total costs			2,018.00
Profit			800.00
Total cost of Proposal			$2,818.00

Specimen Proposal, continued

8

V. PRODUCTS OF PROPOSAL

This project will produce a customized costing program on a floppy disk along with a users manual to explain how to use the program.

VI. CONCLUSION

All that is lacking in Encoders current costing procedure is speed. Our proposed costing program can be up and running in two months. I hope to you decide quickly so we can get started.

Specimen Proposal, continued

A PROPOSAL TO STUDY THE FEEDING BEHAVIOR AND POPULATION DYNAMICS OF THE BANDTAILED MONGOOSE (Trageliphus rangus)

INTRODUCTION

Studies of African carnivores have been concentrated on the larger species, largely because of their greater importance economically. Few species such as the bandtailed mongoose have ever attracted attention, and as a result basic questions such as social behavior of feeding ecology and factors that regulate populations remain to be answered. This field study of the bandtailed mongoose, a common ververid of Africa south of the Sahara, should answer these questions.

Few brief observations are found in scientific journals, many of which are summarized in James and Bond, 1963. The only field study done on the mongoose is that of Marco Polo in India, 1256. Marco Polo's research only accounted for population dynamics of one year of study, hardly long enough a time to truely obtain accurate results. Recent studies of the large social carnivores, e.g., lion (Goodall, 1972), cheetah (Cheep, 1906), hyena (Xzercosfsky, 1976), wolf (Slydoghsky, 1977), and African wild dog (Ughleeman, 1978) have emphasized the structure and importance of co-operative behavior in social hunting and aggressive behavior in regulating populations. Such behavior in small carnivores such as the bandtailed mongoose would be less obvious but almost certainly present.

The objectives of this field study are to assess the population dynamics in order to determine what factors control population size and to determine whether any social behavior accounts for co-operation in food finding.

TECHNICAL SECTION

Study area and background

We will conduct this field study in the foothills of the Wologeesee Mountain range, Liberia. Marco Polo, (1256), studied the bandtailed mongoose in India because of their numbers; however, recent findings (Sannah, 1978) show them to be more plentiful in West Africa. The study area is in foothills not more than 200 feet in height, rocky with scattered brush and grass, and covers an area of about 15 km^2 (Figure 1).

Specimen Research Grant Proposal

Page 2

Figure 1

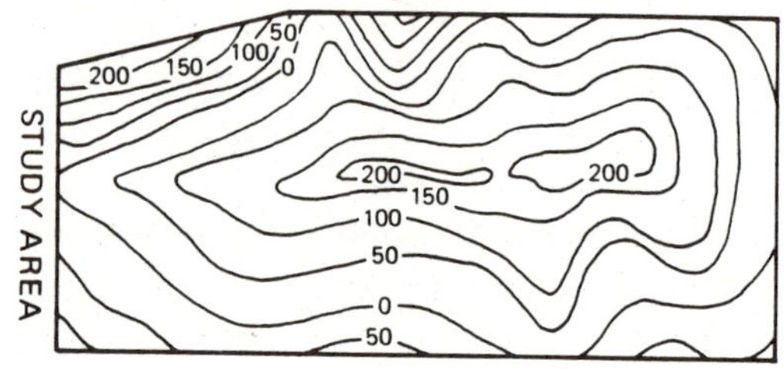

The main vegetation is medium height grassland with occasional scattered thickets of (Intelligence inumerous) growing in close association with (Dumbo insaneous), (Qualatiis effectiviitus), (Bannanus domesticus), and (Jumbo histerics). Annual precipitation is 850-1000 mm. A rainy season and dry season last each about six months; from July-December and January-June respectively. Because the equator runs through the study area, the daily temperatures remain quite constant and sunrise and sunset are respectively at 0645 and 1845 hours.

The study area is located in the center of the foothill range, which consists of a constant vegetation type and supports six mongoose packs. Results from this project should be quite representative of the total 50 km^2 area where the mongooses thrive, as well as applied to other areas in Western Africa.

Design

Over a two year period a total of a year of actual field observation will have accumulated. Three months out of each year will be spent in observation both in rainy and dry season (Figure 2). Six days out of each working week we will spend observing; each day a different one of the six packs.

Because behavior and population dynamics are the basis of this field study, hours upon hours of observation and note taking are needed. The project can be broken up into several general phases.

First and foremost the population of each of the six packs of mongooses must be tagged and their ages recorded. The next two important steps can be categorized under the second phase: field observation, including observations of co-operative

Specimen Research Grant Proposal, continued

Page 3

Figure 2.

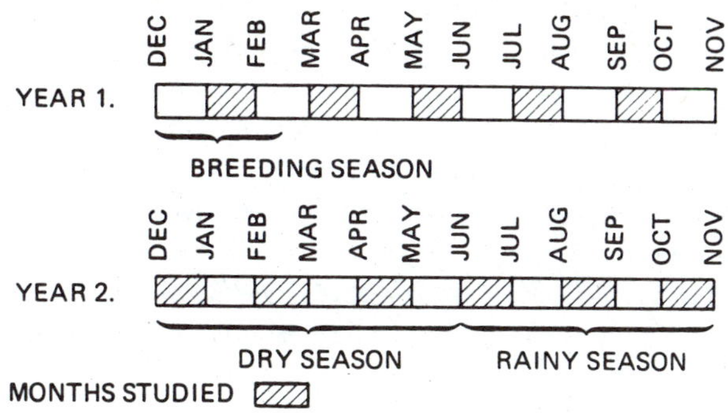

feeding behavior and of population ecology (reproduction and mortality). The last phase is the collecting of our data and its assessment.

We will do the initial tagging with color-coded Moritzon 19-A ear tags, developed for Dr. Skagg's otter study (1974). A different colored ear tag will be given to each pack. These ear tags are effective and easily seen through 10 x 35 binoculars. We will then break up the individuals of each population into the following age classes: Juveniles-one month old, Subadults-six months to one year of age, Adults-one year old. For tagging, we will live trap the mongooses using a Murphy 3.3m x 1.3m x 0.8m high wire mesh trap with a manually operated door at one end and two small doors on the sides for removing the trapped animals. We will place the trap over the main entrance of a den and block up other escape routes within a thirty meter radius. When the mongooses come out in the morning they can be caught singly or in groups by releasing a manual wire, shutting the door. This system seemed to work well for Dr. McKetta's marmot studies (1977). McKetta was able to catch 67 at one time; however, one should note that he used dynamite to drive them out. After release, we will monitor the packs' movement over the two year period.

Specimen Research Grant Proposal, continued

Page 4

Observation, during the next phase of the study, will be done with 10 x 35 binoculars from a 1975 diesel Landrover jeep. Mortality rates and reproduction rates, comprising population dynamics, will be monitored throughout the study. We will make accurate counts each morning of a pack and record the data on a check sheet (see APPENDIX 1). Shortly after the breeding season we will tag the new members and record their count. Mortality, due to predators, aggressive interactions and disease that is observed will also be recorded on the check sheet shown in APPENDIX 1, however, mortality that occurs during the months of our absence we will record separately.

We will also observe social co-operative feeding behavior from a Landrover with binoculars. Using the ear tags we will be able to distinguish between packs, and the resulting field notes we will record on a tape-recorder and on a second check sheet (APPENDIX 2). The types of behavior we will look for are as follows:

1. Two individuals co-operating in search of food.
2. Three or more co-operating in search of food.
3. Co-operation in defense of a kill or food find from another animal.
4. Co-operation in taking prey away from another predator.

In McKetta's (1977) study of the African marmot, he was able to make most of his observations from a Landrover and after several weeks the marmots seemed to become accustomed to his presence, even allowing him to drive within 100 meters of them. This phase of the study is no more than a collection of all incidents of feeding co-operation seen in a single day.

PERSONNEL

I will direct this field study and my wife, Nadia Leidenfrost, shall be a general field assistant. I have a Masters degree in Wildlife Resource Management from the University of Idaho and have been out of school for four years working for the Bureau of Land Management in Idaho, directing a population dynamics study of the Pronghorn Antelope. The study was completed last fall and successful. I was born and raised in Liberia, living there for the first eighteen years of my life, and am familiar with the Wologeesee Mountain Range. My wife, Nadia, graduated from Colorado State University two years ago with a Masters degree in statistics. Over the past two years she was working with Dr. D.K. Doe of the University of Idaho on the Spruce-bud Worm project, a federally funded study. Dr. Earl Lackison and

Specimen Research Grant Proposal, continued

Page 5

Dr. Stevenson of the Animal Behavior Department at the University of Idaho will serve as consultants for the field study. Dr. Lackison has a PhD in Ethology from the University of California, a Masters in Journalism from the University of Pennsylvania, and a second Masters in Vertibrate Biology from the University of Iowa. Dr. Stevenson has a PhD in Vertibrate Biology, a second PhD in Ethology, and a Masters degree in Statistics.

BUDGET

Gas (To and from Zorzor to get supplies every two weeks, and once a month trips to Monrovia)	XXX.XX
Binoculars (Two pair of 10 x 35)	XXX.XX
Live Trap (Murphy 2-door 3.3x1.3x0.8 meter high wire mesh)	XXX.XX
Prefabricated bungalow	XXXX.XX
Color coded ear tags (Approx. 3,000,000., the extras will be sold to the Liberian government's population study of PEACE CORPS and U.S.A.I.D. personnel)	XXX.XX
Taperecorder (Must have automatic memory and search)	XXX.XX
Six ninety minute tapes (Preferably TDK SA-90)	XX.XX
Living expenses: high	XXXXX.XX
Miscellaneous*	XXXXX.XX

*Dynamite 500 lbs.
*D-9 caterpillar--for leveling the study area

Air fare to Liberia for the two of us I did not include because my wife and I plan to go to Liberia anyway to live. I already own a Landrover (diesel) in Liberia, so that expense is taken care of also.

BENEFITS OF THE STUDY

Small carnivores such as the bandtailed mongoose in the future will become increasingly important economically as man encroaches upon the habitat requirements of the larger species such as the lion, leopard, etc. In underdeveloped countries such as Liberia most all of the protein in the diet of the native people comes from the bush, and a lot of it from small animals like the mongoose. It will not only be beneficial economically to study this animal, but also beneficial to the scientific field. Only God knows when and where the cure for cancer will be discovered; . . . maybe here in this field study.

Specimen Research Grant Proposal, continued

LITERATURE CITED

Cheep, S. 1906. Cheetah survey: a survey of the stomach contents of 760 cheetahs. The Journal of Wildlife Management. 34 (1): 56-2000.

Goodall, J. 1972. Hunting behavior of the lion (Pantherus terriblis). Oryx. 5 (4): 617-647.

James and Bond. 1963. World sightings of the mongoose. The Journal of Unheard-of Statistics. 8: 83-199.

McKetta, C. 1977. A hurried field study of the African marmot (Rangus africanus). National Simplistic. 25: 95-96.

Polo, Marco. 1256A.D. Marco Polo's Journal. Encyclopedia Britannica. 13: 13000-15780.

Sannah, D. 1978. Population Census--West Africa. Who's Who in Africa. 1178.

Skaggs, C. 1974. Otter statistics for Northwestern United States and Alaska. The Journal of Unwanted Research Results. 71: 356-377.

Slydoghsky, S. 1977. A complete analysis of three month old wolf embryos. Journal of Wolf Insidetology. 6: 581-788.

Ughleeman, Z. 1978. The African wild dog (Canis madus). The Science of Man's Best Friend. 7: 101-115.

Xzercosfsky, J. 1976. The hyena: analysis of the corpus luteus and its insertion in the uterus. The Journal of Comparative Anatomy. 61: 99-125.

Specimen Research Grant Proposal, continued

APPENDIX I

DATE:

MORTALITY PACK: TIME:

Juveniles _____ V
Subadults _____ W
Adults _____ X

Fights______________________________.
Disease______________________________.
Predators______________________________.
Natural death (old age)______________________________.
Juvenile death______________________________.
Unknown______________________________.

*Death in our absence:

POPULATION PACK: TIME:

Morning count:

REPRODUCTION PACK:

*Litter size:

*Litter size for each female in the pack.

Specimen Research Grant Proposal, continued

APPENDIX II

DATE:

CO-OPERATIVE FEEDING BEHAVIOR PACK:

2 individuals co-operating in search of food.
Notes:

3 or more individuals co-operating in search of food.
Notes:

Co-operation in defense of a kill or food supply from other animals.
Notes:

Co-operation in taking prey away from a predator.
Notes:

MISCELLANEOUS

Specimen Research Grant Proposal, continued

PARKING DEVELOPMENT PLAN

INTRODUCTION

PURPOSE

In order to expand the amount of parking available in the Boise downtown redevelopment area, Swamby Paving and Construction Company proposes to level and pave three currently vacant lots at a cost of $5,280,000. Located at the corners of Ninth and Main, Idaho and Main, and Capitol and Idaho street, these lots could provide a total of 2400 parking spaces in the immediate vicinity proposed retail shopping mall.

PROBLEM

"In accordance with the goals and objectives established by the Boise Redevelopment Agency for the development of a downtown shopping mall, it is necessary to promote high density parking within the proposed development area."* While construction of the proposed multilevel parking garage is awaiting approval, it is necessary to provide immediate parking relief for the growing number of downtown employees and shoppers. We therefore propose immediate development of the three lots for off-street parking.

SCOPE

This paper includes an overall description of the construction methods along with a task breakdown and a time schedule. Also included is a listing of the construction team management members and a financial section with a complete cost breakdown.

OVERVIEW

Because of the location of the three designated lots (see Figure 1.), three crews, one for each lot, will work to level and grade each lot. Each lot will have asphaltic surfacing, stall demarkation, lighting and two accesses on each of the bordering streets. Dimensions and the number of parking spaces provided in each lot are shown in Table 1. Major construction phases include:

clearing and grading

laying of the asphalt

asphalt sampling and testing

finishing touches of marking and lighting

*ENVIRONMENTAL ASSESSMENT, Financial Settlement of Urban Renewal Projects, IDAHO R-4 & R-5. Boise, Idaho. May 1979. pg. V-31.

Specimen Firm-to-Firm Proposal

-2-

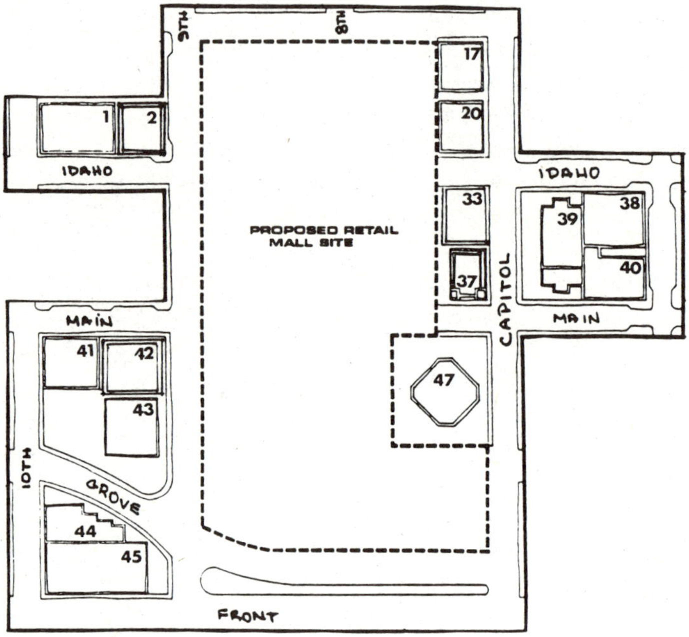

Figure 1. Project Site Plan

Table 1. Lot Dimensions and Spaces

LOT NUMBER	DIMENSIONS (ft)	SPACES
# 2	400 x 600	600
#42	600 x 600	800
#33	400 x 700	1000

Specimen Firm-to-Firm Proposal, continued

-3-

METHODOLOGY

Clearing and Grading

Crews will work to prepare the lots for the asphaltic surfacing by leveling and grading. After having conducted some preliminary testing, we feel that the subgrade needs some minor improvements. We also feel that it would be worthwhile to sterilize the subgrade to prevent weeds from germinating. Because we are using a Full-Depth asphalt concrete pavement, subgrade drainage is not required.

Asphalt Surfacing

We use a hot mixed and hot laid asphalt surfacing. Using two 12-foot pavers, crews will place a four inch layer of asphalt on the prepared subgrade. A total of 264,000 tons of asphalt will be used for the three lots. As soon as the hot mix material can be compacted without displacement, crews will begin rolling.

Sampling and Testing

In order to provide guarantees for the paving work done, we feel that it is necessary for the laid asphalt to meet the specifications listed in Table 2. Therefore lab crews will take three random samples from each of the lots for testing.

Table 2. Specifications

Core density	2.4 g/cc
Percent Permiable Voids	4%
Percent Asphalt Content	75%

Finishing

Paint crews will lay out a general angle parking pattern for each lot. Six mercury vapor lights, strategically placed, will provide the lighting for each lot.

TIME AND WORK SCHEDULE

Total completion of the project requires 30 days, with testing. Figure 2 provides a complete breakdown of the length of time required to complete each phase. Actual paving will require two days at each site, however paving requires good dry weather so a two day safety margin is included.

MANAGERIAL SECTION

Swamby Paving and Construction Company is a well established firm which has been active in the Boise Valley for 25 years. We have been involved in

Specimen Firm-to-Firm Proposal, continued

4

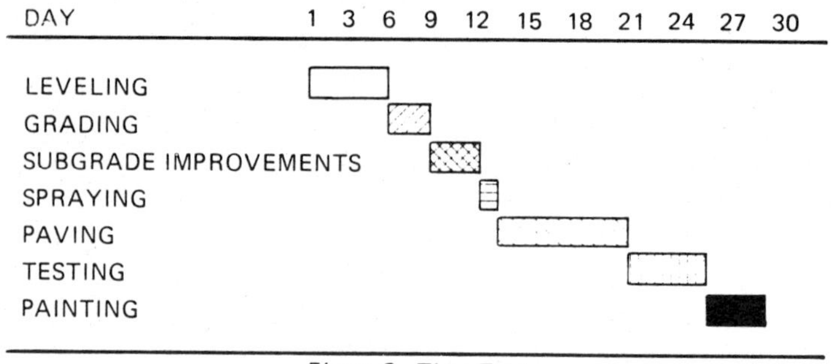

Figure 2. Time Table

several large construction projects including West Gate Mall, Boise City Center Service and Parking Garage, and the Grove street parking garage. As one of the largest paving companies in the northwest, we have developed an integrated equipment and materials system which provides a substantial depth to our company. Therefore the entire project from preliminary surveying to lighting can be directed by our management and handled by our experienced crews. Jim Jambo, parking projects engineer, will be in charge of the Parking development Plan. Steve Albert, head of the surveying and development department, will take charge of the crews responsible for grading and preparing the subgrade. Sam Steven will direct the paving crews and is also responsible for the painting crews. Resumes for these three key individuals are included in the appendix. The only facet of the project which we will not be responsible for is the sampling and testing of the asphalt. This will be done by Asphalt Testing Lab of Caldwell.

FINANCIAL SECTION

Costs

Since our company will handle all facets of development for the parking project, we can guarantee our costs. Table 3 lists a complete breakdown of costs.

Table 3. Costs

Asphalt Materials		$2,904,000
Equipment		200,000
Paint		5,000
Lighting		23,000
Labor		300,000
Overhead		1,214,000
Profit		633,600
	Total	$5,279,600

Specimen Firm-to-Firm Proposal, continued

-5-

Method of Payment

Payments will consist of an initial down payment of $1,280,000 to be made one week in advance of the beginning of the project. Following completion of the project, another payment of $1,000,000 will be made. The remaining $3,000,000 balance will be paid biannually over a six year period in sums of $500,000.

CONCLUDING STATEMENTS

We at Swamby Paving and Construction feel that the parking development plan is an essential part of the future downtown development. By providing parking in the downtown area, Boise can attrack a greater number of investors and merchants who are currently reluctant to join the downtown development team because they feel that shoppers are taking their business to the outlying areas because of availability of free parking. We hope that the city of Boise and Windmar developers will let us begin our project at once so that the concept of downtown redevelopment, through the utilization of a shopping mall, becomes a reality.

Specimen Firm-to-Firm Proposal, continued

Writing Exercises

1. Collect several examples each of in-firm proposals, firm-to-firm proposals, and research grant requests. Compare their overall organizational structures. Report your findings to the class.
2. Can you think of any types of proposals that do not fit into any of the three major categories identified in the text? If so, describe them briefly.
3. Collect two examples each of excellent proposals and unacceptable or minimally adequate proposals.

Writing about Writing

1. Consider the kind of document that says, in effect, "This is what *I* think *you* should do in order to solve a particular problem." Explain why such a document is not a proposal. Also explain why such a document has limited value (if any value at all) in the real world of business, industry, and government.
2. In the realm of management decision-making, do proposals precede recommendation reports or do recommendation reports precede proposals? Support your opinion with authoritative comments and specific examples.
3. Obtain a copy of a fairly substantial proposal. Write a critical review of the document, pointing out strengths and weaknesses and making suggestions for improvement as appropriate.

Writing Assignments

1. You are employed by a firm offering professional services, and you have been asked to prepare a firm-to-firm proposal for submission to a prospective client who has a particular problem or need. Outline your solution to the technical problem. Discuss why you or your firm should be selected to implement the solution. Provide a specific breakdown on costs, personnel requirements, and so on. Include a timetable. Here are some possible situations:
   - □ You are partner in a timber management consulting firm and have been asked by a private landowner to propose a timber-management program to suit his particular timberlands and meet his particular management objectives. (Assume your firm will be managing the timber—for a price, of course.)
   - □ A local business that has been expanding has asked your accounting firm to propose a new bookkeeping system.
   - □ You are employed by a consulting engineering firm, and the boss has given you the task of preparing a proposal for engineering services in connection with the construction of a new shopping mall.
   - □ You own a land-surveying company. A local developer has asked you to submit a proposal for doing the survey work on a fifty-lot subdivision he plans to develop.

2. You are on the staff of a sizable manufacturing or marketing firm or government agency. You suspect there might be a more efficient or more effective way to accomplish some task within the organization than the way it is currently being done, and you wish authorization to proceed with a feasibility study. Write an in-firm proposal outlining how you would go about finding out, how long it would take, who would be involved in the study, and how much it would cost. Consider these possibilities:
   - □ You are a recreation specialist and wish to develop a better method of determining the user carrying capacity for a wilderness area, a hunting or fishing area, a park, or a stretch of river.
   - □ You are an engineer with the State Highway Department and wish to develop a more effective method of inspecting highway construction projects.
   - □ You are a regional sales manager and wish to do a market survey and analysis for a new product.
3. Write a research grant request for some undergraduate (or graduate) research project that you (and perhaps one or two of your fellow students) might wish to undertake.

17

Technical Articles

The technical article needs no introduction. You are probably more familiar with technical articles than with any other variety of technical writing (save possibly letters). You have read scores of articles in the course of your education, and you have looked at hundreds more. Moreover, the term papers and research papers you have written are more like technical articles than like anything else covered in this book. You know the general outline and format for technical articles—particularly those in your field—and you are familiar with the tone and writing style exhibited by articles published in scientific and technical journals and in trade and popular magazines. (In fact, this knowledge and familiarity might be a source of problems. Have you been trying to force your reports, proposals, and manuals into the mold of the technical article?) Articles differ from the other forms of technical writing chiefly in three ways:

1. They are written to inform rather than to initiate some specific action.
2. They are published and distributed by some organization *other* than the one you work for.
3. They are directed toward an unknown and rather wide-ranging audience rather than a specific and more or less captive audience.

Each of these differences calls for some special strategies on your part to help in getting your article accepted for publication in the first place and getting it read widely once it has been published. More on these strategies below.

You may have any of a number of reasons to write a technical article. If you find yourself doing original research—at a college or university, a government laboratory, a private research foundation, or even in the R & D division of an industrial firm—you will undoubtedly be asked to publish the results of your research; in fact, your very livelihood as a researcher will depend on it. If you wind up in an applied field of science or technology, you may well find that your career can be enhanced by publishing articles on techniques that you (or your firm) have developed, new products and product concepts or new applications of existing products, advances in the technology of your particular field, or new ways to solve old problems. Or, if you just have some sound information to share and better-than-average composition skills, you may find that scientific and technical articles can both satisfy your creative urges and augment your income.

As we look through the titles of periodicals that contain articles of a technical or scientific nature (and there are something in excess of 10,000 of them) we find a spectrum ranging from the most prestigious scholarly journals and publications of professional societies through trade journals and technical magazines and on into commercial and popular magazines. We find circulations ranging from hundreds of thousands (millions for popular magazines) to scant dozens. We find annual subscription rates ranging from more then $1000 down to $2.97. Rather than attempt to classify and define technical periodicals, let me offer some examples of the range and variety. Figure 17–1 lists some periodicals in the field of automotive mechanics; Figure 17–2 does the same for clinical psychology and Figure 17–3 for wildlife biology. You can do likewise on your own for accounting, agronomy, computer science, forestry, geology, literature, physics, water pollution control, zymurgy, or any other discipline you might care to name.

International Journal of Mechanical Studies
Journal of Lubrication Technology
Automotive Engineering
Journal of Applied Mechanics
Car Mechanics
Motor Service
Automotive Age
Automotive Week
Automobile Quarterly
Motor Trend
Car and Driver

Figure 17–1 Periodicals in the Field of Automotive Mechanics

Acta Psychologica
American Journal of Psychology
British Journal of Social and Clinical Psychology
Clinical Psychologist
The American Behavioral Scientist
Human Behavior
Clinical Psychology Review
School Psychology Digest
Behavioral Science
Psychology Today
World Health

Figure 17–2 Periodicals in the Field of Clinical Psychology

Journal of Wildlife Management
Journal of Natural History
American Naturalist
American Midland Naturalist
New York Fish and Game Journal
California Fish and Game
Avicultural Magazine
Condor
North American Bird Bander
International Wildlife
National Wildlife
Audobon Magazine
Natural History

Figure 17–3 Periodicals in the Field of Wildlife Biology

□ General Strategies for Articles

Articles are written to inform. That means you've got to have something to say—something that hasn't been said before and something that at least a fair number of readers might be interested in. List some possible topics or titles. Consider the research projects you've been involved with over the past months. Don't overlook student research; there are a number of periodicals that welcome student research papers. Look at your recent job experiences. Have you found new tools to use or new ways to use them? Have you, or perhaps some of the people you have worked with, developed new procedures, methods, or techniques? Can you suggest a way to do anything cheaper, quicker, or better? Finally, give some thought to your hobbies or avocational interests. Do you know ways to catch more fish or take better pictures or keep from bending your kayak around a rock? Any of these topics might make a good article.

Next, check around a bit to see that no one has beaten you to the punch. For highly technical topics, consult the abstracts and indexes in your field; for topics of a more general nature, consult the *Reader's Guide to Periodical Literature*. Make up a working bibliography and spend an afternoon in the library. Your goal is not to learn more about the topic—you're already an expert. (If you're not, you have no business considering writing an article on the topic.) Your goal is to find out what has already been said so that you can say something new and different. What is more, this background reading will set you up for two crucial later steps.

The next step is to develop a basic plan for what you have to say. Prepare a fairly detailed outline; maybe even write a very rough draft. Consider writing a prospectus or descriptive (actually, *prescriptive*) abstract of your article. Above all, thoroughly analyze the communication context for your article. Specify audience, purpose, scope, and type of information. Give detailed answers to "Who cares?" and "Why?" Then truck all your preliminary material around to at least seven different people. If you can get four of them to encourage you, proceed; if not, pick another topic and start over again. Don't waste any more time.

Identify your first, second, and third choices for journals or magazines you would like your article to appear in. For each, get a copy of the requirements for authors. These may be published in the first or last issue of the year, or you may have to write to the editor or publisher for them. (Or a professor in your major field may have a copy that you can borrow and duplicate.) Also for each of your top three choices, check out or borrow several recent issues and read them from cover to cover. Pay particular attention to the length of articles, the number and type of illustrations used, the tone and style of the writing, and the general outlines or organizational patterns followed. Between the author guidelines or manuscript requirements and your own empirical observations, you should be able to determine what will "fit" in your target periodicals.

Next, write your article. Revise it at least twice on your own. Then ask several co-workers, fellow students, professors, or colleagues to review it informally. If you have chosen your reviewers wisely, this will lead to further revision. Finally, have the manuscript typed according to the format specifications of your first-choice periodical, proofread it scrupulously, and send a copy off to the editor. Be prepared to wait anywhere from several weeks to six months or more for a reply. Also, be prepared for a rejection slip; most periodicals publish only a fraction of the manuscripts they receive. While you're waiting, consider submitting your manuscript to your second and third choices. Some journals and magazines still demand the right of exclusive consideration, but the practice of multiple submission of manuscripts is becoming increasingly more widespread. Unless a publisher specifically forbids multiple submission, feel free.

□ Articles for Scholarly and Professional Journals

If you plan to publish in a scholarly or professional journal, the first thing you should do is get a copy of Robert A. Day's *How to Write and Publish a Scientific Paper* (ISI Press, Philadelphia, 1979). This slim book—hardly an evening's reading—is packed with practical information and presents it in a thoroughly delightful fashion. The next thing you should do is acquire a copy of the appropriate style guide for your discipline or profession. (See Chapter 9 for a list of the common style guides available.) Study both sources diligently; your time will be well spent.

Organization

The most common organizational structure for scientific and technical articles reporting original research is:

1. Title and Author(s)
2. Abstract
3. Introduction
4. Materials and Methods
5. Results
6. Discussion
7. References

Increasingly common, however, is the practice of presenting conclusions immediately after the introduction and relegating methodology and results to a position at the end. Follow the organizational requirements of your target journal. If you have some latitude, consider your readers' needs and desires and consider communication effectiveness. Few, if any, readers really want to follow a chronological narrative of how you came to do what you did. Most will want to know (1) the problem you were addressing, (2) the conclusions you reached, and (3) the implications of your findings. Then if they're still interested, they might look for a presentation and interpretation of the specific results you obtained. Finally, those readers interested in evaluating your research or replicating it will want to see how your work interfaces with previously published work and will want to study your methodology in some detail.

Choose your title and write your abstract with considerable care. More people by far will read your title and abstract than will ever read your article. Moreover, how you word your title and abstract will bear directly on how many people *do* read your article. Avoid empty phrases like "a study of" or "an experimental approach to." Identify the key indexing words for your study and use them in your title. Make your abstract more informative than descriptive (see Chapter 13) but keep it brief. Many journals specify a 200- or 250-word maximum for abstracts.

In your introduction, tell your readers what you have done and why. Let them know the specific problem you have studied, identify your research hypotheses, explain your research strategy or design, and discuss the theoretical aspects of your work. Discuss earlier published works that are directly pertinent to your study, but stop short of an exhaustive literature review; instead refer your readers to general surveys or reviews. It is always good form to acknowledge your intellectual debt to others, but don't go overboard. Finally, if your article is lengthy—say twenty or more manuscript pages—give your readers a brief overview of the main body.

Your materials and methods section should be complete but brief. It is complete if your readers can see how you arrived at your results and can replicate your experiments if they so desire. It is brief if it does that and no more. Include (as appropriate) subsections on materials, equipment and apparatus, experimental and control groups, sampling techniques and statistical methods, and research procedure. If you used standard materials, apparatus, and procedures, just name them and let it go at that. If your methodology involved anything nonstandard—and particularly if it is controversial—describe it in more detail. Just don't get carried away with model numbers of stop watches or manufacturers of Erlenmeyer flasks. Above all, do not include definitions, derivations, or explanations from standard textbooks; if you can find it, so can your readers.

Under "Results," summarize your major measurements and calculated values and present them graphically as tables, charts, etc. (see Chapter 8). Follow the conventions of your discipline with respect to precision, accuracy, and statistical significance, and assume that your readers are familiar with the conventions. Make sure your text and your tables and figures complement one another. Interpret your data at least to the extent of pointing out salient values and significant differences. Give your readers a hint as to what they should look for in your tables and graphs or what they should see in your photos or drawings. Stop short of drawing inferences, qualifying your results, or speculating on implications; save that for the next section.

In your discussion section, you may wax a bit eloquent, evaluating your methods and results in the light of previous research, conflicting theories, your own research hypotheses, and the consequences and implications of what you have found. Start with a clear statement of your major conclusions, however; you're not writing a mystery novel—or even an essay. You may note shortcomings in your study and make suggestions for improvement, but don't dwell on the negative aspects. (Others will do that for you.) Similarly, note inconclusive or negative results, but don't attempt to explain them away. Some speculation can be in order, but always identify it as such. ("It is assumed that . . ." really means "I think . . ."!) Include comments on practical applications of your research, but again be brief. If you have trouble winding up your discussion, close with a simple summary paragraph: problem, conclusions, and implications.

Most technical articles for scholarly and professional journals close with a bibliography of "references cited." This means that you are obliged to list all but only the references that you actually cite in the body of your article. Pay scrupulous attention to bibliographic format requirements for your target journal. Double-check all titles, the spelling of authors' names, and page numbers, volume numbers, dates, etc., all the way back to your original sources. (See Chapter 9.) If you submit your manuscript to several different journals, you may have to reformat your references section (and perhaps the bibliographic citations themselves in the main body of your article). Remember, though, that it's quicker to do it right than it is to do it over. And besides, good form may help you get your article accepted.

Style

Certainly, a technical article for a scholarly or professional journal qualifies as a writing situation that demands excellence. It will have a large audience, it will have an important audience, and your professional reputation and that of your employer are on the line. And there is the further consideration that your article will be around for a long time—probably longer than you will. So everything in Part Four, "Revising for Excellence" is applicable here—in spades and doubled. Pay particular attention to Chapter 21, "Style at the Word and Phrase Level."

Historically, articles for scholarly and professional journals have tended to be formal in style, reserved in tone, objective in point of view, and unbiased in presentation—as indeed they should be. During the first half of the twentieth century, though, we began to see what has been regarded by many as too much of a good thing. Formal style became elaborate and pompous; reservation gave way to stuffiness; precision in word choice led to showing off with arcane and obscure jargon; and authors in a quest for objectivity lost themselves behind clouds of third-person pronouns, passive clauses, and indirect constructions. Since the mid-1950s the opinion of authorities—the authors of textbooks and style guides, the editors of journals, and professional technical writers—has swung in the other direction toward short, simple, and concrete words; straightforward clauses and sentences; and in general an active, personal, direct, and dynamic writing style. Unfortunately, there has been a great deal of inertia. As the writing authorities moved in new directions, writers in some disciplines—notably medicine, engineering, and the physical sciences—continued to move in the old direction. So what we find in the scholarly and professional journals at present is a wide range of styles, with some moving in one direction, some moving in another, and some just marking time.

With the exception of a few journals in limited fields, virtually anything within reason is acceptable, so far as style goes. No editor I know will either accept or reject a manuscript solely on the basis of writing style. So write in a style that you're comfortable with, revise for orderliness, completeness, and brevity (Chapter 5), and give some thought to your paragraphs and sections (Chapter 19), your clauses and sentences (Chapter 20), and your words and phrases (Chapter 21). Above all, be deliberate in your organization and development, your syntax, and your diction. Don't write just from habit.

□ Articles for Trade, Commercial, and Popular Magazines

Generally speaking, technical articles for trade journals and commercial and popular magazines are less rigid in format and freer in style than those for scholarly and professional journals. Also, as a rule, they are more profusely and more professionally illustrated (Figure 17–4). In fact, articles for commercial and popular magazines (the kind you get money for!) may well be accepted as much for the quality of your photographs and drawings as for the quality of your writing. More than anything,

Figure 17–4 Profusely Illustrated Trade Journals and Commercial Magazines

though, your article must be interesting. You must have something interesting to say, and you must present it in an interesting fashion.

Organization

It is impossible to suggest a general outline for trade, commercial, and popular articles. Some observations are in order, however. Many such articles intersperse personal narrative with description, explanation, and instruction as a way of building and maintaining interest. Articles on hunting, fishing, sailing, skiing, scuba diving, drag racing, and mountain climbing—to name just a few—use this technique extensively. At the level of the trade journal, articles on road construction, fabrication, manufacturing, architectural design, and the like often use this technique. The narrative may be either first person singular, referring to the author, or first person plural, referring to the author's project group, firm, or organization.

Another organizational framework that is used extensively in trade journals and commercial magazines—particularly in business fields and the social sciences—is the case study approach. With this technique, the author intersperses actual or slightly fictionalized descriptions of field situations with explanation and commentary on them. The case studies lend an air of reality and immediacy to what otherwise might become some rather dull theory.

The beginnings of trade and popular articles warrant special consideration. Since you are dealing with a much larger and much less specific group of potential readers, you have to take specific steps to arouse interest and lead readers into your article. Such beginnings, in fact, are called "leads" or "grabbers." One technique if you are using narrative in your article is to begin *in medias res* (literally, into the middle of things) at a high point in the narrative and then drop back to either description and explanation or the chronological beginning of your narrative. For an article that doesn't involve narrative, you might consider beginning with a statement of the significance of what you report or a dramatic item of statistical data. A rhetorical question can be useful as well, especially if it involves saving time or money. Actually, most of the techniques you learned for beginning essays are applicable here.

Style

As you move from the scholarly article to the popular article your writing style should become less formal, more conversational, and more personal. You can accomplish this by manipulating some relatively minor variables. Control your technical vocabulary by substituting more general (albeit less precise) synonyms for most of your technical terms. Control your general vocabulary by using more and more words from the familiar list (see Chapter 21). As you move toward the popular end of the spectrum, you can lend an informal flavor to your writing by introducing an occasional colloquial expression or folksy metaphor or "pop" phrase. Use these like seasoning, though. A pinch here and a sprinkle there is fine, but don't dump in the whole box.

You can make your style more conversational and personal by introducing a few contractions (don't, can't, you'll, haven't), employing more first and second person pronouns, including a number of direct quotations in newspaper interview style, and offering here and there some personal observations about what you think, feel, and believe. You can also increase informality (or decrease formality, as it were) by introducing colorful and dynamic examples of some of the points you wish to make.

It is not necessary to use a consistent style throughout your article; in fact, it may be wise not to. You might open with a conversational—even chatty—lead and then drop back to a still personal but somewhat more reserved style for the remainder of your introduction. Then when you get to the meat of your article, you can retain the style appropriate for the professional journal, winding up, finally, with a more conversational and informal summary. You can do the same thing with paragraphs and sections, too. A crisp, direct—even witty—sentence at the beginning or end can add flavor to an otherwise bland passage. Give it a try.

☐ Specimen Excerpts from Articles

INDEXATION: WILL TAXPAYERS GET ITS BENEFITS?

Considering bracket creep's impact, Congress may have second thoughts, the authors say.

by C. W. Bastable and Stephen L. Fogg

The Economic Recovery Tax Act of 1981 provides for implementation, in 1985, of a system of indexation of individual tax brackets. Indexation keys taxation to purchasing power and represents a major change in tax policy designed to eliminate perceived inequities in the present system that stem from inflation-induced bracket creep. It provides, in effect, through governmental action, a mechanism to help taxpayers stem erosion of the purchasing power of their take-home pay.

However, we are concerned that indexation may not become a reality. Statement of Tax Policy no. 6, *Indexation of the Tax Laws for Inflation,* provides an idea of why the federal government may have second thoughts about implementing indexation when the statement observes, "Indexation . . . stops the increase in taxes that is caused solely by inflation, so that overt action by Congress is required to increase taxes."[1] In the ample time that Congress has available to reflect on its current support of indexation, how palatable will it find the need to take the "overt" action from which it heretofore has been shielded by bracket creep? Will a rising deficit cause defections by indexation supporters? The accounting profession must be on guard against specious arguments to abort indexation, for it may be the only hope the beleaguered taxpayer has.

Statement of Tax Policy no. 6 also notes that "the income earner who receives cost-of-living increases is subjected to . . . bracket creep; he pays higher marginal tax rates and a higher average rate. Therefore, a taxpayer's purchasing power drops even if his income rises as fast as the cost of living."[2] This major impact of bracket creep on CPAs' clients—and on all wage earners who are currently under pressure from the ravages of inflation—makes the subject of bracket creep and indexation enormously significant to the accounting profession. Only if accountants better understand the exact impact of bracket creep and indexation on income can they try to foresee whether governmental policy will be equitable and effective. Such knowledge can help accountants to pass meaningful advice along to lawmakers. As professionals, CPAs' responsibilities to help government in such important decisions is paramount.

To enhance the profession's comprehension of the significance of a change to indexation of individual tax brackets, we examined and measured various ramifications of the bracket-creep phenomenon in relationship to the salaries (or wages) of individuals. We also determined the savings that would be attributable to indexation had it been scheduled for implementation in 1981 instead of 1985. The results afford some startling insights con-

C.W. BASTABLE, CPA, Ph.D., is professor of business at the Graduate School of Business, Columbia University, New York. A member of the American Institute of CPAs, the American Accounting Association and the Financial Executives Institute, Dr. Bastable is a former book editor for the *Journal of Accountancy*. **STEPHEN L. FOGG**, CPA, Ph.D, is assistant professor of accounting at the School of Business Administration of Temple University, Philadelphia, Pennsylvania. Dr. Fogg is a member of the AICPA, the AAA, the National Association of Accountants and the Pennsylvania Institute of CPAs.

[1] Statement of Tax Policy no. 6, *Indexation of the Tax Laws for Inflation* (New York: AICPA, 1980), p. 11.
[2] Ibid., p. 6.

Excerpt from a Trade Journal Article. Two pages from "Indexation: Will Taxpayers Get Its Benefits?" by C. W. Bastable and Stephen L. Fogg. *Journal of Accountancy*, January 1982.

cerning bracket creep, which, unfortunately, will persist for at least four more tax years. Furthermore, the insights make us even more skeptical about whether indexation will be implemented, given the enormous loss of revenue if bracket creep is eliminated.

Projections of Salaries, Federal Income Taxes and Aftertax Incomes

To obtain a set of figures for measurement purposes, we made projections for a 1-year interval and a 10-year interval for salaries, federal income taxes and aftertax incomes when salaries were assumed to keep pace with inflation. The projections appear in table 1, page 54.

So that the projections will not be misunderstood and misused, the assumptions employed in making them should be mentioned. We assumed that salaries in general will rise at a rate that matches the inflation rate; of course, they may rise at lower or higher rates. Furthermore, the salaries associated with some of the segments that constitute the total population of salaried personnel may change at different rates than those for other segments. Nonetheless, we believe that the most objective bracket-creep measurements stem from the assumption that salaries keep pace with inflation. To assume otherwise would introduce an additional variable that would cloud measurements.

To enable readers to exercise their own judgments concerning the future course of inflation, we employed 8 percent, 10 percent and 12 percent rates of inflation in making the projections. Given the 12 percent-plus rates that prevailed during much of 1980, the rates are realistic and they encompass commonly mentioned double-digit inflation. To reflect annual compounding for the 10-year interval in part B of table 1, we used respective multipliers of 2.1589250, 2.5937425 and 3.1058482; they are simply 1.08, 1.10 and 1.12, each raised to the tenth power. For example, a $12,000 salary at a compounded annual rate of 8 percent becomes $25,907 ($12,000 × 2.1589250) in 10 years.

The federal income tax figures are for a married taxpayer with four exemptions who files a joint return and does not itemize deductions. We used the 1980 tax table B and schedule Y for all computations. Where appropriate, the 50 percent maximum tax on personal service income was employed; we did not attempt income averaging in those limited instances when it might be applicable.

We recognize that low-income taxpayers ordinarily do not itemize deductions, while many high-income taxpayers do. In not making any allowance for itemized deductions, the federal income tax computations for high-income taxpayers may be overstated. However, we have no universal profiles of high-income taxpayers that permit factoring itemized deductions into the tax computations. Notably, state income taxes qualify as deductions, but there are substantial jurisdictional diversities in state taxation. Rather than increase the subjectivity that would stem from imputing arbitrarily derived deductions, we have accepted the risk of upward bias in the tax computations because of employing the zero-bracket amount (formerly the standard deduction) throughout. We doubt that the bias will have a material effect on any generalizations that stem from the data.

High-income taxpayers often have other income in addition to their salaries, but salaries alone are at issue here. Consequently, we made no attempt to introduce other income. However, each salary in table 1 is, of course, adjusted gross income for federal income tax purposes. For the reader who is so inclined, there is no reason why each salary cannot be

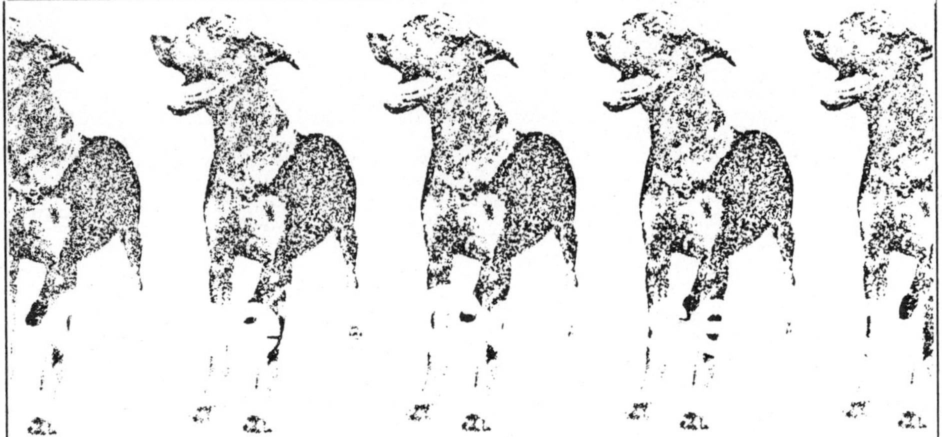

Through genetic counseling, which is an instructional and communicative process, veterinary practitioners can provide a service in the prevention of inherited disease and in improving the outlook of breeders.

Thomas M. Fisher, MS
1150 N. Park Terrace Drive
Horseheads, NY 14845

As a practitioner, you may be faced by a panic-struck breeder who has just discovered a serious genetic defect in the 3rd generation of his champion sire. He has come to you for advice about the dog's breeding future. Among the canine recessive traits you may encounter in this regard are: Collie-eye syndrome, generalized progressive retinal atrophy (PRA or GPRA), most non-PRA-linked cataracts, central retinal atrophy (CPRA) which seems to be recessive in America (contrary to Labrador Retriever studies in England[1]), familial renal failure, and cryptorchidism (in which inbreeding and backcrossing led to severely epidemic proportions in Toy Poodles).

Your role in genetic counseling (GC) is to prevent disease by increasing the dog public's knowledge in the field. To this end you should attempt to improve breeder mentality, which tends to avoid frank discussions of problems.[2] There is a wall of noncommunication in which breeders fear that openness will lead to being traumatized by fellow breeders engaged in a so-called witch hunt. GC is an instructional and communicative process involving both information and support directed toward alleviation of animal suffering. Both the breeder and his or her veterinarian must make 2 important distinctions: (1) genetic responsibility, or how the genetic defect can affect the individual animal, and (2) euthenic responsibility, or how passing on of such defective genes can affect the future of the breed or that variety of the breed.

As applied to breeders, GC involves several important factors that must be considered by both the breeder and you as the consulting veterinarian:

- the need for a precise diagnosis
- the seriousness of the defect
- the availability of therapy and its usual degree of success
- the psychologic nature of the breeder regarding ability to cope with recurrence of the defect and negative feelings other breeders may express
- personal opinions concerning genetic and euthenic responsibilities
- reasonable and general agreement concerning the mode of inheritance of the defect
- calculation of the genetic risk.

The farm animal veterinarian may also be involved with GC: the classic forms of hemophilia (see section on sex-linkage) were discovered in the horse. You should be especially concerned with the principles of the test-cross and their application since the owner of a prize stud may want to prove him a pure dominant for some genetic condition such as disease resistance, milk production or some other desired quality.

The degree of handicap to the animal is impor-

Excerpt from a Special Interest Magazine Article. Two pages from "Genetic Counseling" by Thomas Fisher. *Modern Veterinary Practices*, January 1982.

GENETIC

tant to the success of partial or complete alleviation of the defect. Until environmental management technics are available, GC remains the best method of reducing risk even if the reliability of the mode of inheritance is not absolute. An attempt should be made to alter breeder mentality so that a person is not traumatized by his peers for having produced a defective animal. It is a much greater disgrace to either perpetuate a defect and hope that one's line will not become suspect or to not communicate openly so others may learn from one's experience. Only when that mentality has become popular in animal breeding will there be an honest chance to overcome a host of genetic problems.

GC in Simple Recessive Inheritance

Most of the cases with which you will be working will be in the recessive inheritance mode. There are highly sophisticated formulae for GC, but the simpler and more easily understood the suggestions, the greater will be the willingness of breeders to cooperate.

To the breeder faced with a recessive defect in his animal's pedigree, the most immediate problem is that of genetic risk. Regardless of the risk factor involved, if a planned breeding involves a highly serious defect, you should recommend against it.

FIG 1. "SUM OF DIGITS" METHOD (SEE TABLE A FOR INTERPRETATION).

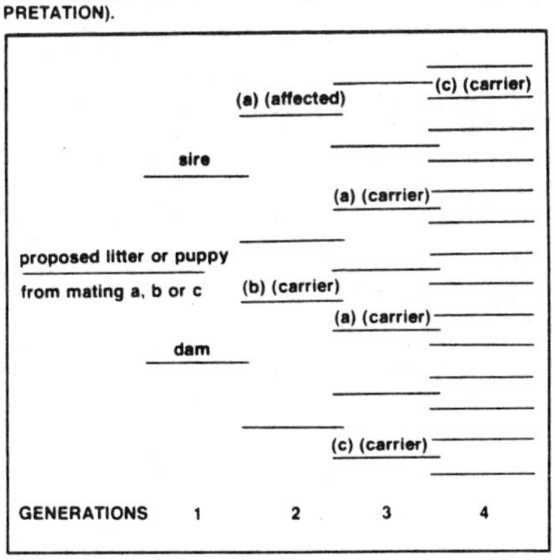

TABLE A. PROBABILITY FOR AN AFFECTED (nn) PUPPY BASED ON SUM OF DIGITS

sum of digits	probability for an affected (nn) puppy		
	AA	AC	CC
2	100%	50%	25%
3	50%	25%	12.5%
4	25%	12.5%	6.25%
5	12.5%	6.25%	3.13%
6	6.25%	3.13%	1.56%
7	3.13%	1.56%	0.78%
8	1.56%	0.78%	0.39%

Example of table interpretation based on Fig 1: for the breeding indicated by (a), there is an affected animal in position 2 and a carrier in position 3 (2 + 3 = 5); since one is affected and the other a carrier, use the AC column to find a probability of 6.25%. For the breeding designated as (b), there are carriers in positions 2 and 3; therefore, from column CC one obtains a probability of 3.13%. Calculate the probability for breeding (c); did you get 0.78%?

Most serious breeders will appreciate such a recommendation.

The crux of the problem with recessive GC is the case in which the defect is not highly serious or life-threatening, and the breeder wants to know the probabilities for that breeding because of other favorable qualities of temperament or conformation. An example is geriatric PRA in Poodles, which is a genocopy (same disease mediated by different genes) of traditional PRA and is manifested at an age when most dogs are nearly blind from various causes. This is one of the more difficult GC situations. For a recessively transmitted defect of a non-life-threatening nature, it is my opinion that, if the breeding is good on all bases save for the defect, the breeding should be permitted if one major condition is met: the risk of producing an affected puppy must be lower than the incidence of that defect in the breed population.

Regardless of the failure to communicate, serious and responsible breeders do know the locations of affected and carrier ancestors of their animals, and most national breed clubs know the incidence of diseases in their breed. The breeder must tell you the location (generation) of the nearest carrier or affected animals on both sides of the pedigree of the proposed breeding pair. Construct a pedigree chart (Fig 1) and use Table A for interpretation: AA indicates that the nearest located ancestors were both affected; AC that one was affected and the other a

Getting Framed: An Inside Job

BY STEVE NETHERBY

It had been a long two weeks chained to the desk. My body and brain were both in need of new scenery. Jackie had the kids downstate to visit their godparents for a few days. I called Mel, my friend the airline pilot, but he was spending the weekend in Kansas City, Louisville, and Philadelphia.

It was trout season. Trout season! I loaded a pack and threw it into the red Jeep and headed for the southern Sierras. The lake I had in mind was in a national forest wilderness area, just above timberline, up with the high peaks. There was a trail to the lake, but I cut cross-country.

It was the perfect office antidote. The air sparkled like diamond dust. The granite crags stood out like shark's teeth. There were steep pulls, brush to bust, knee-high snow in the shaded spots, and, near the lake, boulder slides to climb, then a cliffside to scramble down. I spent the weekend without seeing another person—catching brook trout on nymphs and looking at mountains.

The backpack that helped to give me this mountain freedom was a member of a new generation of packs uniquely suited to rough-country travel. They're called *internal-frame* packs, and right now they're where the excitement is in backpacking equipment.

There was a time, just past, when the trademark of a backpacker was his external-frame pack. The only time you saw those streamlined packs with frames hidden inside, they were overflowing with climbing ropes and crampons and slung on the backs of muscular, intense types who talked about jams, rurps, chocks, snargs, pitons, and other exotic dishes. You knew climbers *had* to use packs like those for some mysterious reason, but you also knew that if you were going to spend more than a couple of days prospecting for mountain trout, or carry more than 35 pounds, you were supposed to do it with a packbag clevis-pinned to a tubular frame.

On my first ski-mountaineering trips—in Colorado's Rockies and Wyoming's Wind Rivers—I used an external-frame pack. It was piled high with winter gear and was quite comfortable when I was *(Please turn the page)*

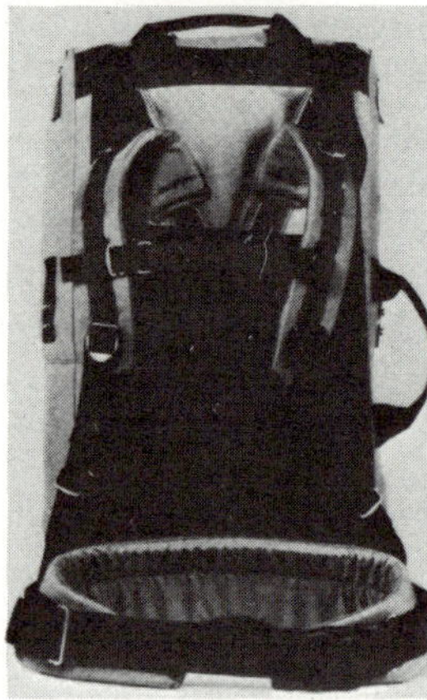

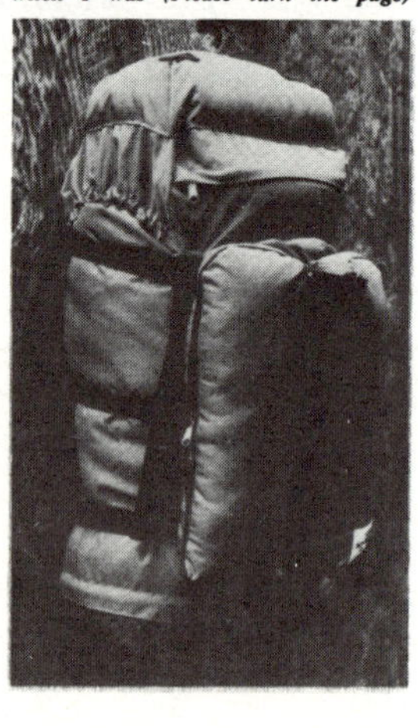

An array of the new internal-frame packs (clockwise from top): Synergy Systems 5.9 FL, North Face Moraine, Grade VI Freighter; Lowe Trek I

26

Excerpt from a Popular Magazine Article. Two pages from "Getting Framed: An Inside Job" by Steve Netherby, *Field & Stream*, February 1982. Reprinted by permission of *Field & Stream*.

moving slowly on the flat or ascending slopes with climbing skins strapped to my skis. But I soon learned that my ski technique had to be precise. To lean too far in a turn, or to slip in any direction, meant a sure fall. The pack's high center of gravity, suspended so far out from my own "CG," would pull me right over. Other external-frame packs threatened similar uncomfortable episodes on snow-free slopes.

CLIMBERS move in a world where a loss of balance can mean disaster. So they bought internal-frame and frameless packs that fit as close as their wool shirt. These packs usually lacked the outside pockets common to outside-framed packbags—to afford the climber more freedom of movement and maneuverability in tight quarters. Cross-country skiers were attracted to similar packs for similar reasons.

But on long approaches to the mountain, climbers often suffered. Their packs really weren't as comfortable on the trail as the backpackers' external-frame models. They couldn't handle the weight as well. Loads heavier than 30 or 40 pounds could turn climbing packs into instruments of torture. And their body-hugging ways made them sweat producers on warm days.

Still, packmakers, many of whom were mountaineers themselves, weighed the advantage of close-fitting packs against the limitations of outside-frame packs—and decided body-huggers held the most promise. They turned their talents to making the body-huggers all-round performers—comfortable on the trail and stable on the mountain. And, as my own recent tests proved, they succeeded brilliantly.

In a rurp-shell, here's the present state of backpack art: Early on, most designers decided that soft, frameless packs—packs that depended on the cut of the fabric for support—could carry a maximum of about 30 or 40 pounds comfortably. So they began to concentrate their efforts on the development of packs with inside frames made of flat metal stays that gave shape and support to the load and concentrated the weight at the bearer's hips, where it seems to belong.

Nowadays, internal-frame packs come in two species: parallel-frame and X-frame. Parallel-frame packs generally fit the back more intimately and are lighter and more stable. X-frame models can handle very large loads comfortably and are cooler to wear, but generally move around more on the back and can sway across the shoulders in rough country or when the wearer leans to the side. Both types, now, are generally considered more competent and comfortable than external-frame packs—both on and off trail—except in the face of the biggest, bulkiest loads. (I still prefer an outside-frame for packing out elk quarters, for instance. It can handle the weight, size, and odd shape better and provides solid lash points. However, some of the big top-loading expedition internal frames will also do the job.)

Aside from carrying qualities, internal-frame packs have other talents that endear them to outdoorspersons. Though they have straps and lash points that allow equipment to be carried on the outside, the design of most of these packs encourages storage inside where gear is better protected. You find yourself ruthlessly editing out nonessentials in order to keep everything inside, and this makes life easier on the trail. Their clean exterior makes internal-frame packs ideal luggage for airline, jeep, packhorse, canoe, or bushplane.

Finally, the best of these packs are very versatile. Compression straps that work like cinch straps on a saddle to snug the bag up to the frame allow the biggest packs to carry everything from expedition loads to day-hiking loads efficiently.

TRAIL TESTS

Almost without realizing it, I started switching several years ago, carrying internal-frame packs on most of my trips. I began with a Synergy Works X-Pedition. Last year I used a Grade VI Freighter. Deciding "inside jobs" were here to stay, I recently gathered two more for testing, along with an updated Synergy, selecting manufacturers and models that would spotlight the exciting technology and diversity available in these new packs. When I had the four together I tested them under heavy load, in up-and-down country, on trail and far off. Here's what I found:

Synergy Systems 5.9 FL. An incredibly useful, extremely comfortable backpack—the handsomest pack I've worn. The 5.9 FL is a big Cordura-nylon box that loads from the front through two zippered "barn doors." Upper and lower compartments are separated by a zippered shelf that can be used to organize and distribute weight, or zipped out to accommodate tall gear. The box can be compressed, to hold small or large loads against the back, with three compression straps. Up to six pockets can be added to the box and attached with "Pull the Dot" snaps that won't let go unless you want them to.

The frame is an aircraft-aluminum X that can be bowed to lift the load away from your back with straps attached to a full mesh back panel. Shoulder straps are shaped to fit your shoulders and attached to lifter straps that pull weight off shoulder tops. A quick-release "sternum strap" connects the shoulder straps. The hip belt is conically cut and shaped to keep the weight stable and well distributed and has a quick-release buckle. Suspension is adjustable in a wide size range.

The workmanship is superior. All inside seams are taped for no-fray. The pack has reinforced stitching and the best hardware. It's great bushplane, airline luggage: it can be stripped of pockets and suspension, or the suspension can be tucked behind the mesh back panel. The pocket can be detached for use as a toiletry kit. All contents are easily accessible. There are nylon tote handles on the top and side.

I FIND the X-frame packs the most comfortable on the trail. With the load lifted off the back, cooling air blowing through, there are times I forget my burden. This pack is the easiest of all to walk upright in. And it's a very quiet pack, unlike the early Synergy I tested.

On the minus side: Compression straps are too stingy; they need more surplus for easy use. Shoulder straps slip through slick metal buckles. The pack shares a problem common to X-frames: it sways sideways at the shoulders with radical use, impairing balance. This is an excellent pack for the trail, but marginal in the rough.

Weight, 5.1 pounds (with pockets). Volume, 6,050 cubic inches (with pockets). Price, about $224.

Grade VI Freighter. I'll be surprised if this pack doesn't last longer than I do. It's made better than I am. The bag is sewn from waterproofed 18-ounce ballistic cloth—the stuff bulletproof vests are made of. All material is hot cut to prevent edge fraying. Threads are extra heavy, self-sealing; stitching is reinforced at stress points. All sliders and buckles are rustproof polycarbonate plastic. External zippers are heavy coil types with locking double sliders. Yoke suspension is sophisticated in design but simple in construction.

Freighter is a top-loading pack, tapered at the top, with a huge bottom compartment accessible through a zipper that bisects the bag. A zippered shelf divides top from bottom or zips out of the way of tall gear. There's a big, full-opening front pocket and an exceptionally roomy top pocket that can float up or down with the size of the load and can be removed for use as a daypack or fanny pack. A waterproof Cordura extension tunnel at the top opening can accommodate big overloads. Side compression straps cinch the pack down for small cargo. The quick-release hip belt has stabilizer straps at the sides to steady the load.

The aircraft-aluminum X frame is separated from the back by an internal panel of firm foam, which also serves to keep gear from rubbing you wrong. On hot days the foam panel helps minimize points of contact with the back to keep you cool.

The teardrop shape of Freighter reduces the X-frame sway, but the pack still shifts more than the parallel frames I tested. Yet its low-CG configuration and streamlined, body-hugging shape make it a good performer in all but the most radical off-trail conditions. On trails, like the Synergy, it becomes as comfortable as an old shirt. While you're walking, the full-access top pocket is easy to use.

(Continued on page 76)

Writing Exercises

1. Consult several scholarly journals in your field. Compare the overall organizational structures of the articles in them. Do you find recurring patterns? Variations? Summarize your findings in a report to the class.
2. Do the same for several trade or commercial magazines in your field.
3. Do the same for several popular magazines in a field you are interested in (photography, hunting or fishing, fashion, personal computing, automobiles, skiing, backpacking).
4. Prepare a list of topics that you might write technical articles on. Make sure the topics are those you have some expertise on and those that a fair number of readers would be interested in.

Writing about Writing

1. Should conclusions precede or follow discussions of methodology and results? Defend your position in some detail and offer specific examples.
2. Should the technical article in a scholarly journal serve as a model for other types of technical writing? Defend your position in some detail and offer specific examples.
3. Write an essay on the relationship of objectivity in research and impersonality in the reporting of research. Aim the essay at faculty and students in your major field or discipline.
4. Develop a questionnaire that could be sent to editors of scholarly, trade, or commercial journals and magazines to determine their attitudes toward questions of writing style appropriate for their journals.

Writing Assignments

1. Write up the results of some research project you have been involved with as if for publication in a specific scholarly journal. Identify the journal and include with your paper a copy of the journal's style guide or requirements for authors.
2. Write an article for a trade journal in your field. Include illustrations as appropriate. Again, identify the journal and include a copy of the journal's style guide or requirements for authors.
3. Write an article on some appropriate topic for a commercial magazine. Check the topic and magazine with your instructor, though, before you get too far into the writing. (If your article is accepted and you get paid for it, you get an automatic A in the class! At least you do in my class.)

□□□□ FOUR □□□□

Revising for Excellence

Many students, it seems, approach a technical writing course wanting to learn the secret of turning out Pulitzer prize-winning prose on the first draft. There is no secret. There is no shortcut to excellence. The only way to achieve excellence in written communication is to start with an adequate paper and keep revising it until it meets the standards of excellence. The aim of Part Four is to help you in the successive stages of revision. Now, not every writing situation you encounter will demand excellence; in fact not even most of them will. But some will. Hence, it behooves you to know what is involved in generating excellent written communication of technical information and to be able to demonstrate that you are capable of doing so—just in case you are ever asked to produce excellent writing on the job.

Chapter 18, "Grammar for Engineers (and Others Who Don't Like Grammar)," offers some definitions and explains some concepts that you will need in order to digest the remaining four chapters of Part Four. Chapter 19, "Style at the Paragraph and Section Level," introduces the theme of style as choice and suggests the range of options that you as writer should be selecting from to achieve excellence at these higher levels of language structure. Chapter 20, "Style at the Clause and Sentence Level," and Chapter 21, "Style at the Word and Phrase Level," discuss further aspects of writing style at lower levels of language structure and present additional stylistic choices that careful writers select among in their pursuit of excellence in technical communication. Each of these three chapters closes with a set of suggested revision techniques that together constitute a procedure you can follow in revising for excellence. Finally, Chapter 22, "A Method for Analyzing Technical Writing Style," suggests an analytic technique you can use to assess the quality of a piece of technical writing—either someone else's or your own.

It is difficult to say how one should proceed through these five chapters. If your background in grammar is strong, you can probably skip Chapter 18. If not, then not. Many students find it useful to study the style-analysis procedure in Chapter 22 before looking in detail at the style chapters—Chapter 19, Chapter 20, and Chapter 21. Others find it useful to skim all five chapters and then refer to specific information when they are in the actual process of revising for excellence. Regardless of how you handle Part Four, though, don't forget to go back and review the three chapters of Part One—Preliminaries. Look especially at Chapter 2, "Adequacy and Excellence." And while you're at it, review Chapter 5, "Report-Writing Procedures," as well.

18

Grammar for Engineers (and Others Who Don't Like Grammar)

It is quite possible to write adequately without knowing a single thing about grammar (or at least without knowing a single thing that is generally taught under the heading of "grammar" in the public schools). Furthermore, knowing a whole lot about grammar is not likely to make you an excellent writer. In fact, one survey after another has shown that formal study of grammar in school has no correlation whatsoever with ability to write effectively. So why bother?

Simple. You can write at a level of adequacy without worrying too much about the structures and processes of language and without knowing anything about the names grammarians give to these structures and processes. But if your goal is excellence in writing, you have got to be concerned with writing style; and no discussion of style can proceed beyond simple emotive responses ("That is neat; I don't like that!") without getting involved with the elements of language and the patterns in which these elements appear. And to be able to communicate effectively about these elements and patterns—these structures and processes—we have got to have a command of certain basic terminology. In short, we've got to know what we're talking about.

That, then, is the goal of this chapter: to provide you with the basic terminology of grammar (and the related concepts) so that you'll know what you're talking about. (Or, more important, so that you'll know what *I'm* talking about in the several chapters on style.) What is here is the bare minimum you have to know in order to read about and discuss writing style with a modicum of understanding. If you want more than this minimum, you'll have to go elsewhere. Some of you may already possess the necessary minimum knowledge of grammatical terms and concepts. If so, proceed directly to the chapters on style. If you find yourself understanding and agreeing with most of the definitions, go forward; if not, go back. Whatever you do, though, keep in mind that this chapter is here to provide you with the necessary grammatical background so that the chapters on style do not have to be cluttered up with interruptive discussions of terminology and concepts.

□ Structure in Language

Language exhibits a hierarchical structure, much like that of a military group or a corporate organization (see Figure 18–1). We can identify various levels of command in the military, or levels of authority in a corporation. Here are some widely recognized levels of structure in language:

- □ Sound level
- □ Word level
- □ Phrase level
- □ Clause level
- □ Sentence level
- □ Paragraph level
- □ Section level
- □ Discourse level

At each level in the language hierarchy (except the lowest), we have units that are made up of one or more units from the next lower level. At the word level, for instance, we have units called *words,* which are made up of one or more sounds (or letters representing sounds). At the phrase level, we have *phrases,* which are made up of one or more words. And so on. A *paragraph* is a unit of language structure that consists of one or more *sentences;* a sentence is a unit consisting of one or more *clauses;* a clause is a unit consisting of one or more *phrases.*

Most of the units at any given level of structure consist of more than one unit from the next lower level, but we do on occasion find such things as one-sentence paragraphs, one-clause sentences, one-word phrases, and so on. These one-unit units (as we shall see) can be a source of confusion for people studying language structure.

□ Formal and Functional Definitions

As we look at varieties of units at different levels of language structure (types of words, types of sentences, and so on), we will want to consider

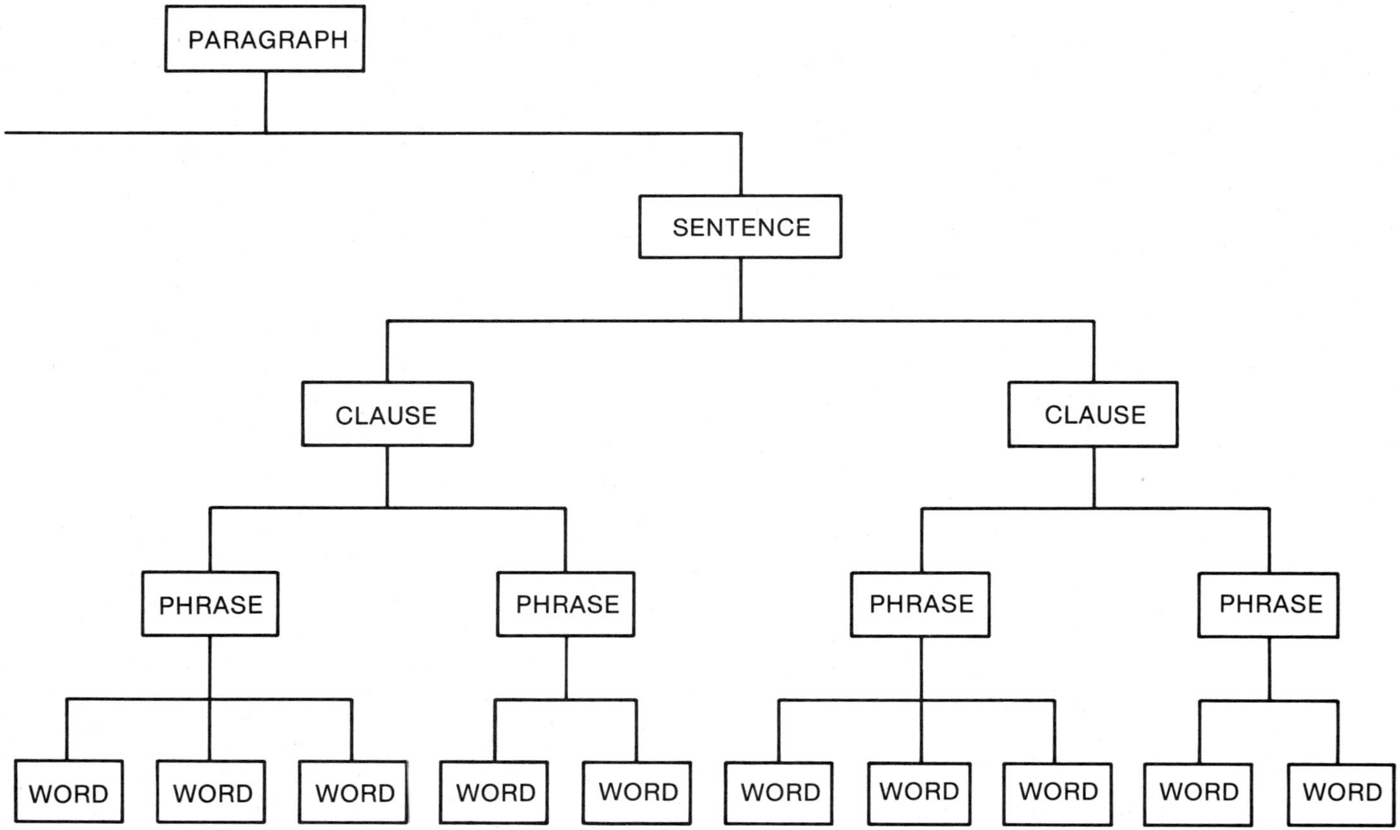

Figure 18–1 Hierarchical Structure in Language

definitions of the various units so that we can attempt to keep them straight and so that we can identify particular types of units when we encounter them. Now, we can define a unit of language (or anything else, for that matter) either in terms of *what it looks like* or in terms of *what it does*. Let's call the definition that focuses on the form, or shape, of a unit a *formal definition;* and let's call the definition that focuses on the function, or purpose, of a unit a *functional definition*. We need both. At times we want to know what a unit looks like so we can distinguish it from other units, and at times we want to know how a unit behaves so that we can more clearly see its role in communicating information.

A source of confusion in many traditional discussions of grammar has been that the definitions offered have constituted a mixed bag of formal and functional definitions. Sometimes we've gotten one and sometimes the other. The following sections will consistently offer both formal and functional definitions. Where appropriate, I will mention both semantic function and syntactic function. Many of the definitions will be familiar to you. Some might be different from what you're used to. As you read the following sections, though, keep in mind that the definitions are those that will prove most useful in our study of style.

□ Words

We'll start with words as the smallest units of language germane to the study of effectiveness in written communication. Sound patterns are interesting, and they are important for the study of other forms of writing, but you really don't have to know anything about them to discuss style in the sense that we are using the term here. Now, everybody knows what a word is, and all of you can recognize words when you see them, so I won't attempt a definition—that is, we will accept *word* as an undefined primitive term. The important categories of words are:

- □ Nouns
- □ Verbs
- □ Modifiers
- □ Function words

In order for you to analyze style—either of your own writing or someone else's—it is crucial that you be able to identify all of the nouns, verbs, modifiers, and function words in a passage of writing. Let's start with nouns.

Nouns

A noun is a word that forms its plural with *-s* or *-es* and functions semantically as the name of a person, place, thing, or concept. (There are exceptions, of course, but let's not get bogged down.) Syntactically, a noun functions as the head of a noun phrase (see "Phrases," below). All of the nouns in the following passage are italicized:

> Fourscore and seven *years* ago our *fathers* brought forth on this *continent* a new *nation*, conceived in *liberty*, and dedicated to the *proposition* that all *men* are created equal. Now we are engaged in a great civil *war*, testing whether that *nation*, or any *nation* so conceived and so dedicated, can long endure. We are met on a great *battlefield* of that *war*. We have come to dedicate a *portion* of that *field*, as a final resting *place* for those who here gave their *lives* that that *nation* might live.

From the standpoint of style analysis, an important subcategory of nouns is those derived from verbs and adjectives—let's call these "derived nouns." Some examples:

Source Verb or Adjective	*Derived Noun*
tempt *(v)*	temptation
explain *(v)*	explanation
communicate *(v)*	communication
involve *(v)*	involvement
place *(v)*	placement
happy *(adj)*	happiness
full *(adj)*	fullness

Another categorical distinction worth knowing is that between "count nouns" and "mass nouns." Count nouns refer to things that can be broken up into discrete, countable units: dogs, chairs, books, bicycles, ideas. Mass nouns refer to things that comprise a mass and cannot be isolated into discrete units: sand, sugar, alcohol, electricity, grammar, bait.

Verbs

A verb is a word that is inflected (that is, has marked distinctions) for tense, number, and person and functions semantically to express an action or state of being. Syntactically, a verb functions as head of a verb phrase (see "Phrases," below). The following passage has all its verbs italicized:

> Fourscore and seven years ago our fathers *brought forth* on this continent a new nation, *conceived* in liberty, and *dedicated* to the proposition that all men *are created* equal. Now we *are engaged* in a great civil war, *testing* whether that nation, or any nation so *conceived* and so *dedicated*, *can* long *endure*. We *are met* on a great battlefield of that war. We *have come to dedicate* a portion of that field, as a final *resting* place for those who here *gave* their lives that that nation *might live*.

Typically, verbs form their past tense with *-ed*, their progressive form with *-ing*, and their third person singular with *-s*, although there are so many exceptions that these forms are only a rough guide at best for identifying the verbs in a passage. Here are some subcategories of verbs that you should know about:

- **Auxiliary Verbs** These are the helping verbs: *be (is, was, am, are, been, being), have (has, had, having), will, shall, should, could, might, can, ought,* and *dare*.
- **Main Verbs** All others.
- **Transitive Verbs** These are verbs that take nouns or noun phrases as complements (see "Clauses," below).
- **Linking Verbs** These are verbs that take modifiers as complements: *be, become, seem, appear, feel, taste, smell, look,* and *sound*.
- **Intransitive Verbs** These are verbs that don't take any complement at all.
- **Verbals** These are words that have the *form* of verbs but often *function* as nouns or modifiers. *Participles, gerunds,* and *infinitives* are all varieties of verbals. Verbals are also called *nonfinite verbs* (for reasons we won't go into here) and serve as the predicates of an extremely important group of clauses called nonfinite clauses (see "Clauses," below).
- **"Weak" Verbs** These are verbs that don't have much semantic or lexical content (that is to say, they don't carry much meaning). Some of the more common weak verbs are *be, do, make, use, give, become,* and *have*.

Modifiers

Modifiers function semantically to limit, change, augment, restrict, or intensify other words. *Adjectives* modify nouns (or pronouns), while *adverbs* modify verbs or other modifiers. Many adjectives form their comparative with *-er* and their superlative with *-est,* and many adverbs end in *-ly;* but as with verbs there are so many exceptions that these forms are only a rough guide for identification. Here's the Gettysburg Address once again, with all the modifiers italicized:

> *Fourscore* and *seven* years ago *our* fathers brought forth on *this* continent *a new* nation, conceived in liberty, and dedicated to *the* proposition that *all* men are created *equal. Now* we are engaged in *a great civil* war, testing whether *that* nation, or *any* nation so conceived and so dedicated, can *long* endure. We are met on *a great* battlefield of *that* war. We have come to dedicate *a* portion of *that* field as *a final resting* place for those who *here* gave *their* lives that *that* nation might live.

(Notice that the word *resting* in the last sentence of the passage was counted both as a verb and as a modifier. In this sentence, the word *resting* is a participle, with the form of a verb and the function of an adjective; hence the dual classification.) Here are some important subcategories of modifiers:

- **Adjectives** As I noted above, adjectives modify nouns (or pronouns).

- □ **Adverbs** Adverbs either modify verbs or modify other modifiers.
- □ **Definite Articles** *the.*
- □ **Indefinite Articles** *a* or *an.*
- □ **Demonstrative Pronouns** *this, that, these, those.*

Function Words

Function words are (typically) short words from closed sets that have relatively little or severely limited semantic content. Syntactically, they function to show relationships between the major lexical (or meaningful) words in a sentence—the nouns, verbs, and modifiers. Among the major categories of functions are *pronouns, conjunctions, interjections* (or exclamations), and *prepositions.* Many grammarians include articles and demonstratives in the list of function words as well, giving them a dual listing like verbals. I will follow this practice, as in the following passage:

> Fourscore *and* seven years *ago our* fathers brought forth *on this* continent *a* new nation, conceived *in* liberty, *and* dedicated *to the* proposition *that* all men are created equal. Now *we* are engaged *in* a great civil war, testing whether *that* nation, *or* any nation *so* conceived and *so* dedicated, can long endure. *We* are met *on a* great battlefield *of that* war. *We* have come to dedicate *a* portion *of that* field, *as a* final resting place *for those who* here gave *their* lives *that that* nation might live.

I won't take the time here to define and give examples of the various types of function words. They're familiar to most of you, anyway, and they don't cause that much of a problem so far as the study of style goes.

□ Phrases

A phrase is a unit of language structure that consists of one or more words and functions syntactically as a major component of a clause (see "Clauses," below). The three categories of phrase that we will be concerned with here are *noun phrases, verb phrases,* and *prepositional phrases.*

Noun Phrases

A noun phrase consists of a head noun or pronoun (called the "nominal") and any associated modifiers or modifying elements. Syntactically, a noun phrase functions either as the subject of a clause or the complement of a clause; semantically, a noun phrase serves to completely identify a person, place, thing, or concept. In noun phrases, some modifiers precede the nominal and some follow it. (Keep in mind, though, that many noun phrases do not have modifiers at all—they are one-word phrases that consist of just a nominal.) Here are some examples:

Prenominal Modifiers	*Nominal*	*Postnominal Modifiers*
	John	
	Sally	
	we	
	man	
big	John	
the	man	in the corner
the little old	man	who lives upstairs
all three of the red brick	houses	that were built before the war
the United Nations General	Assembly	
the plain white aluminum beer	can	second from the left
	all	of us who can do it
	two	of the kittens

Verb Phrases

Verb phrases consist of main verbs together with any associated auxiliary verbs and verb modifiers. The verb phrase functions syntactically as the predicate of a clause and functions semantically to completely express an action or state of being. Here are some representative verb phrases:

go
have gone
sang quietly
might have crept softly
probably will be traveling
resting
to drink beer
are singing loudly
functions
is

Prepositional Phrases

A prepositional phrase consists of a preposition followed by a noun phrase. Prepositions are function words that typically show temporal relationships (before, during, after, since) or spatial relationships (on, under, over, beside, in front of, underneath), and prepositional phrases have pretty much the same semantic function. Structurally, they function (with a couple of minor exceptions) either as postnominal modifiers in noun phrases or as clause adjuncts (see "Clauses," below).

□ Clauses

The clause is the unit of language we use to make a statement or express a thought or idea. It is the syntactic vehicle for a semantic proposition, or predication. The clause has five structural or functional slots:

Connector Subject Predicate Complement Adjunct

Of the five, subject, predicate, and complement are central; these are conceptually required in all clauses. The connector and adjunct slots are peripheral in that they are optional clause elements; not all clauses have connectors or adjuncts.

The subject and complement slots are filled by noun phrases, the predicate slot by verb phrases, the adjunct slot by adverbials of time, place, and/or manner, and the connector slot by conjunctions or sentence adverbials. Some adverbials—either in the adjunct slot or the connector slot—are single words, while others are phrases, typically prepositional phrases. Since the discussion here concerns clauses in isolation and the connector slot comes into play only when we have clauses in multiclause sentences or in connected discourse, I will not show examples of fillers for the connector slot in the following examples. Instead I will show examples of connectors later in the chapter in discussing clauses in complex contexts.

Examples:

Subject	*Predicate*	*Complement*	*Adjunct*
NP	VP	NP	Adv (or PrepP)
John	spilled	the beans	to the police
The three boys	enjoyed	the play	immensely
Judy	hit	the ball	over the fence

As you may have guessed by now, the fillers of the adjunct slot are optional. Each of the following is a well-formed clause: John spilled the beans; The three boys enjoyed the play; Judy hit the ball.

Moreover, clauses with certain kinds of verbs in them (intransitive verbs, if you're curious) do not have complements. For instance:

Subject	*Predicate*	*Complement*	*Adjunct*
NP	VP	Ø	Adv (or PrepP)
John	talked		all night long
The three boys	played		vigorously
Judy	jumped		up and down

Clauses with linking verbs in the predicate verb phrase take modifiers as complements:

Subject	*Predicate*	*Complement*	*Adjunct*
NP	VP	Adj	Adv
Grass	is	green	
The teacher	seems	pleasant	today
The sun	looks	red	just at sunset
Satin	feels	smooth	to the touch

We can say, then, that a clause is a group of two or more syntactically related phrases filling functional slots as follows:

an obligatory subject slot filled by a noun phrase
an obligatory predicate slot filled by a verb phrase

an optional complement slot filled by a noun phrase or a modifier
an optional adjunct slot filled by an adverbial

For purposes of studying writing style, we want to recognize four subcategories of clause:

- □ Independent clauses
- □ Dependent clauses
- □ Nonfinite clauses
- □ Elliptical clauses

Independent clauses are simply clauses that could stand alone as one-clause sentences. Dependent clauses are all others. Dependent clauses, of course, appear only in multiclause sentences and are either subordinate clauses or embedded clauses, depending on the pattern of clause combination in the sentence (see "Sentences," below). Nonfinite and elliptical clauses need a bit more explanation.

Nonfinite Clauses

Finite verb phrases contain main verbs, which are inflected for tense, person, and number.

Tense He *studies* English. He *studied* English.
Person I *am*. You *are*. She *is*.
Number He *reads* the paper. They *read* the paper.

Nonfinite verb phrases contain verbals, which are not inflected for tense, person, or number.

Tense He likes *to eat*. He used to like *to eat*.
Person I like *to eat*. You like *to eat*. She likes *to eat*.
Number He likes *to eat*. They like *to eat*.

There are three varieties of nonfinite verb phrases:

- □ The infinite form (with or without the infinitive marker "to")

  To drive without seat belts is dangerous.
  Rather than May *drive* I'll have Tom drive.

- □ The -*ing* participle form

  Driving without seat belts is dangerous.

- □ The -*ed* participle form

  The trip *finished*, we unbuckled our seat belts.

So much for nonfinite verb phrases. Now, nonfinite clauses are nothing more than clauses with nonfinite verb phrases as predicates. Two special features about nonfinite clauses: (1) they cannot stand alone; they appear only as embedded clauses or subordinate clauses, and (2) they can appear without overt subjects, and usually do. Examples:

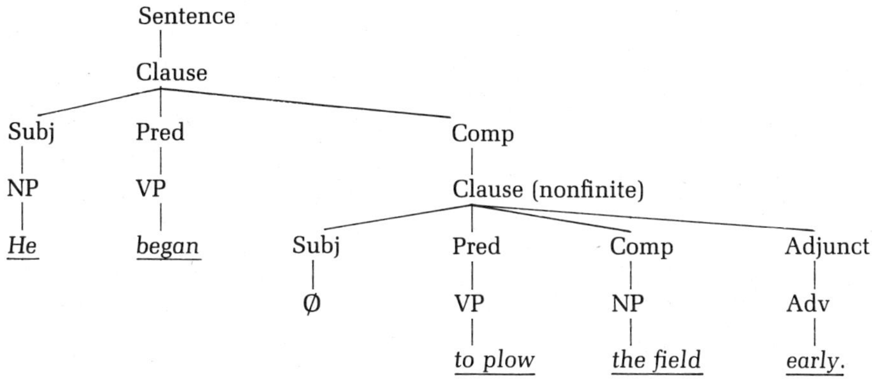

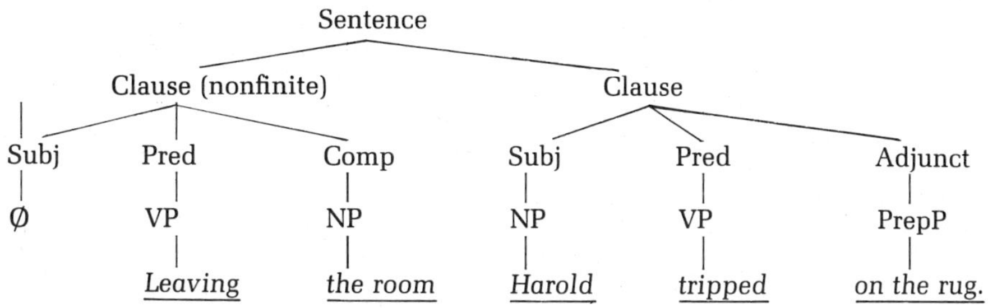

Of course, nonfinite clauses—like all clauses—may themselves have other clauses (either finite or nonfinite) embeded within them or subordinated to them. In fact, it is through the successive embedding and subordinating of nonfinite clauses that mature writers are able to achieve the syntactic compactness that makes for concise writing. And it is this syntactic compactness of the nonfinite clause that makes them so important in consideration of style and effectiveness in technical communication (see Chapter 20, "Style at the Clause and Sentence Level").

Elliptical Clauses

Elliptical clauses are another device mature writers use to gain syntactic compactness in their writing. An elliptical clause is simply a clause that has one or more of its major functional slots missing (or more accurately, deleted from its surface structure). Perhaps the most familiar type of elliptical clause is the imperative clause—the one with the "understood" *you* as the subject. The surface form of such clauses is as follows:

> Close the door.
> Jump in the lake.
> Be quiet!

But, as we have all learned, they have underlying structures like these:

(You) close the door.
(You) jump in the lake.
(You) be quiet!

Another type of elliptical clause can be seen when two (or more) clauses with identical subjects are conjoined and the identical subject of the second (and subsequent) clause(s) is left out of (or deleted from) the surface structure. Here are some examples:

John ran up the stairs and opened the door.
Peter showered, shaved, and changed clothes.
The missile altered its course and evaded the interceptor.

Conceptually, these sentences have underlying structures like the following:

John ran up the stairs, and (John) opened the door.
Peter showered, (Peter) shaved, and (Peter) changed clothes.
The missile altered its course, and (the missile) evaded the interceptor.

Traditional grammar would call these single clauses (or simple sentences) with compound predicates, but for reasons I point out below, it's better (at least in my opinion) to regard these as multiple clauses with elliptical elements.

A third variety of elliptical clause arises when two or more clauses with identical predicates (and perhaps identical complements and adjuncts) are conjoined and the identical elements in the first clause(s) are deleted from the surface structure. Consider the following:

Mildred and Harry walked down town.
John and June are in the living room.
Tom, Dick, Harry, and Jack went to the dance.

We could regard these as having underlying structures like the following:

Mildred (walked down town), and Harry walked down town.
John (is in the living room), and June is in the living room.
Tom (went to the dance), Dick (went to the dance), Harry (went to the dance), and Jack went to the dance.

A sentence like the last one, then, we would say consists of four clauses—three of them elliptical. Now, in traditional grammar we would call these single clauses (or simple sentences, as it were) with compound subjects. From one standpoint, I suppose, it doesn't make much difference what you call constructions like this as long as you know what you're doing. From another standpoint, however, it makes sense to call a construction like:

Mildred and Harry walked down town.

two clauses instead of one because it certainly seems to be making two predications. Besides, if you're counting clauses (as I suggest you do in Chapter 22, "A Method for Analyzing Technical Writing Style"), you'll come up with more clauses if you count elliptical clauses as well as full surface structure clauses. Most importantly, though, if you get in the habit of using elliptical clauses (together with nonfinite clauses) in your writing, you'll improve your writing style and become more effective in your communication of technical information. And that's what this textbook is all about.

A word of caution. Some sentences that look like instances of elliptical clauses with deleted predicates and complements are really instances of conjoined noun phrases. Consider this sentence:

> John and Mary ate all the peanuts.

This can't have an underlying structure of:

> John (ate all the peanuts), and Mary ate all the peanuts.

because there aren't that many peanuts. If John ate them all, there wouldn't be any left for Mary! Watch out for sentences that can't logically be instances of elliptical clauses. These you must regard as representing conjoined subject noun phrases.

A final variety of elliptical clause that some writers employ is the sort that I used to begin the preceding paragraph. Sentences like these, with subject missing, predicate missing, or both missing, are often called "sentence fragements." I would prefer to reserve the term "sentence fragment" to describe those instances where elements are missing accidentally or unintentionally and to refer to sentences like "A word of caution" as another type of elliptical clause. The distinction, I guess, is neither formal nor functional, but more a matter of the writer's intention.

□ Sentences

The sentence is the unit of language that consists of one or more clauses. The one-clause sentence is called a simple sentence; all others are called complex sentences or multiclause sentences. There are two basic ways to combine clauses into a multiclause sentence:

- □ Conjoining
- □ Embedding

Conjoining is a matter of hooking clauses together, like boxcars in a train. Embedding, on the other hand, is a matter of plugging one clause inside another—sort of like loading one of the boxcars. Some examples:

Conjoining:

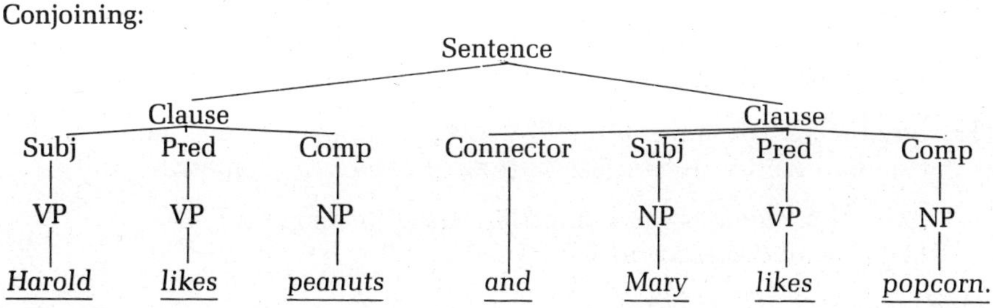

Embedding:

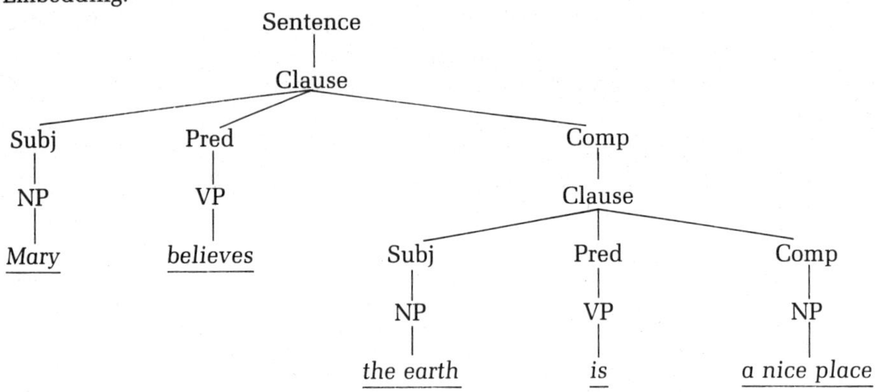

or

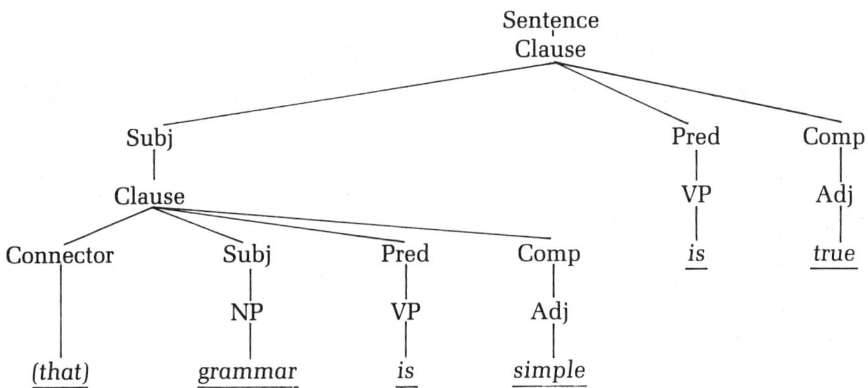

Two major types of conjoining exist: coordinative and subordinative. They are distinguished both by the form of the conjunctive signal used and by semantic emphasis or weight given each of the clauses.

Coordinate Conjunction:

- □ Signals are ", *and*" ", *but*" ", *or*" or ";"
- □ Clauses have equal emphasis or weight.

Subordinate conjunction:

- □ Signals are words like *as, although, since, because* . . .
- □ One clause has less weight or emphasis than the other.

The signal of coordinate conjunction always appears in the connector slot of the second clause, regardless of their order in the sentence:

> Harold ate all the peanuts, and Mary ate all the popcorn.
> Mary ate all the popcorn, and Harold ate all the peanuts.

With subordinate clauses, however, the signal of subordinate conjunction always appears at the beginning of the subordinate clause, in the connector slot of that clause. The subordinate clause typically follows the main clause, but the subordinate clause can be moved up ahead of the main clause. (In such cases, it is conventional to use a comma at the end of the subordinate clause to signal the end of it and the beginning of the main clause.) Examples:

> Jane smiled because she was happy.
> Because she was happy, Jane smiled.

Similarly, two major types of embedding exist: clause-level embedding and deep embedding. With clause-level embedding we have one clause embedded as a major constituent of another clause. That is to say, we have clause as subject, clause as complement, or clause as adjunct. With deep embedding, we have a clause embedded not as a constituent but as a constituent of a constituent. That is, we have a clause embedded all the way down to the phrase level (or even the word level). One example of deep embedding is the relative clause, which functions as a modifier in a noun phrase. For example:

> The little old ladies *who live upstairs* . . .
> The brick house *which is in the middle of the block* . . .
> The car *that gets there first* . . .

When we embed at the word level, we usually use hyphens to signal what we're doing:

> I don't like your *I-don't-give-a-damn* attitude!

Also of some interest is the sentence that consists of a single clause which in turn consists of a single phrase consisting of a single word:

> Jump!
> Fire!
> Shazbat!

Single-word sentences and sentences embedded at the word level don't show up in technical writing very often, though, so I wouldn't give them a great deal of concern if I were you. The key skills so far as sentence analysis goes are being able to tell how many clauses a sentence has and for multiclause sentences being able to identify the various patterns of conjoining and embedding.

□ Paragraphs and Sections

As units of language structure, paragraphs and sections are a little more difficult to pin down than some of the lower-level units. Their forms are simple enough. A paragraph consists of one or more sentences, and a section consists of one or more paragraphs. We have to add the further stipulation that the sentences of a paragraph and the paragraphs of a section be related to one another in some way. We can't just take a set of random sentences and call it a paragraph. But the nature of the required relatedness is not understood all that well. We need some semantic relatedness. The sentences of a paragraph (or the paragraphs of a section) in some way or another have to be "talking about the same thing." We also need some rhetorical relatedness. The sentences of a paragraph have to further some rhetorical purpose: they have to add detail to a description, offer examples of an observation, provide development for an argument, or do something similar. And finally, we need some syntactic and lexical relatedness. The sentences of a paragraph have to cohere—stick together—through the use of devices such as parallel structure, pronoun references, patterns of sentence focus, repetition of identical or related words. (I have developed these concepts more fully in Chapter 19, "Style at the Paragraph and Section Level.") Where we run into trouble is trying to specify the nature of the required semantic, rhetorical, and grammatical relatedness. Much more study is needed in the area of discourse structure before we will be able to discuss paragraphs and sections with the same precision that we can discuss words, clauses, and sentences.

We *can* make a few observations, though. In general, we can recognize four functional categories of paragraph: introductory paragraphs, main paragraphs, summary paragraphs, and transition paragraphs. Main paragraphs tend to be longer and more tightly structured than the other three, with a clearly identifiable topic sentence, often a sentence or two restricting or defining the theme stated in the topic sentence, several sentences developing the theme, and often a summary, or wrap-up, sentence. Introductory paragraphs tend to be short—often only one or two sentences—and for the most part list the items that will be discussed in the paragraph or paragraphs that follow. Similarly, summary paragraphs are short, offering a restatement or recapitulation of what has been said in the paragraph or paragraphs that precede. Transition paragraphs serve as bridges between successive main paragraphs or between successive sections. In various ways, they pick up the theme or train of thought from one paragraph (or section) and redirect it, thereby reorienting the readers so that they may proceed with what follows. Each of the four categories of paragraphs—introductory, main, summary, and transition—has its function, with main paragraphs advancing the discourse and the other three serving as route guides to keep the readers on the track.

Since introductory, summary, and transition paragraphs are short, they are often combined with (or punctuated as part of) main paragraphs. Consider the preceding paragraph. The first sentence is really a one-sentence transition paragraph, picking up on the theme of "lack of knowledge about discourse structure" and redirecting it with the antithetical "though." The second sentence is really a one-sentence introductory paragraph, listing the four functional categories of paragraph. And the last sentence in the preceding paragraph is really a one-sentence summary paragraph that recapitulates the information in the main paragraph. In a real sense, then, what I wrote was four *functionally distinct* paragraphs combined in the *form* of a single paragraph. Now, when most people (myself included) just say "paragraph" they are referring to the formal unit—that which appears between two successive paragraph indentions. But it is good to keep the functional distinctions in mind—especially when we're discussing style and paragraphs.

Sections are similar to paragraphs. In fact, in some senses they can be regarded as superparagraphs. We have the same demands of thematic unity for sections that we have for paragraphs, and we have the same four functional categories. Indeed, introductory sections are often headed "Introduction," and summary sections are often headed "Summary." As for main sections—particularly those in reports, proposals, articles, and other documents with section headings and subheads—the demands for thematic, or topical, unity are perhaps stronger for sections than for paragraphs. When you put a heading on a section—say, "Facilities Available"—you are, in effect, saying two things to your readers:

1. *Everything* in the following paragraph(s) has to do with the topic "Facilities Available." I haven't snuck in any information on any other topic.
2. *All* that I have to say about the topic "Facilities Available" is right here. I haven't scattered information about facilities anywhere else (unless I refer you to it specifically).

In summary, then, we can make the following observations about well-formed paragraphs and sections:

- □ They serve introductory, summary, transition, and main information-transfer functions.
- □ They have clearly identified themes, or topics.
- □ They develop their stated themes with some degree of detail and completeness.
- □ They exhibit thematic unity: they present *all* the information on a given topic and *only* information on that topic.

□ Syntax and Semantics

Language has both a syntactic structure and a semantic structure. For the most part in the preceding discussions of language units, I have been emphasizing the syntactic structure of language (although I have made occasional reference to semantic function as well as syntactic function). Whenever we speak of words, clauses, and sentences or noun phrases and verb phrases or subjects, predicates, and complements, we are speaking of syntactic units and structures. Semantics, on the other hand, has to do with concepts like meaning, content, actions, agents, and statements, or predications. Put simply, a syntactic *word* identifies a semantic "thing" or "idea"; a syntactic *clause* expresses a semantic "statement" or "proposition"; and a syntactic *paragraph* expresses or manifests a semantic "theme" or "topic." The word is not the thing; the clause is not the statement; the paragraph is not the theme. Rather, syntactic words, clauses, and paragraphs bear or carry or express semantic ideas, statements, and themes.

We think semantics, but we write syntax. As we formulate our ideas, marshal our thoughts, organize our data, and plan our approach, we are engaged in semantic activities. But we can't transfer ideas, thoughts, and data directly from mind to mind; if we want to communicate our ideas, thoughts, and data to another human being, we must do so through the medium of language. And that's what writing is all about—giving syntactic form to semantic content, putting our ideas in words, as it were. And the closer we can come to a one-to-one mapping from semantic content to syntactic form, the better our writing is. The more effective our communication is.

Fortunately, we start off with a pretty close mapping between semantics and syntax. Consider the basic syntactic structure of the clause:

Connector Subject Predicate Complement Adjunct

This correlates directly with the structure of a semantic proposition:

Relationship Agent Action Recipient Setting

where the agent is the perpetrator or doer of a deed, the action is the deed itself, the recipient is the receiver of the action, and the setting is the temporal and spatial conditions under which the action takes place. Consider the one-clause sentence:

Harry mowed the grass this morning.

We can label the syntactic and semantic structural components as follows:

Syntactic	Subject	Predicate	Complement	Adjunct
	Harry	mowed	the grass	this morning
Semantic	Agent	Action	Recipient	Setting

If we move away from this one-to-one mapping, we can generate a sentence like:

The grass mowed Harry this morning.

Now this is a syntactically correct sentence, but it doesn't make sense semantically. Or at least the syntactic sentence doesn't map with the semantic proposition we started off with.

The same thing happens at the word level. Suppose we have in mind the semantic idea that John "received and acknowledged receipt of" a package. If we write:

John accepted a package.

everything is fine, but if we write:

John excepted a package.

we're in trouble, because "except" doesn't mean "to receive and acknowledge receipt of"—it means something else. Or to put it another way, the syntactic word *except* does not map one-to-one with the semantic idea we had in mind.

The key to successful, effective communication, then, is to start with sound semantic structures—clear thinking—and then select those syntactic elements—words, sentences, and paragraphs—that allow you to come as close as you can to a one-to-one mapping between your semantics and your syntax.

□ Voice

Of the several grammatical categories of person, number, gender, tense, mood, and voice, the category of voice is the one that causes the most people the most trouble. To be thorough in your stylistic analysis and effective in your revisions for excellence, you must be able to identify instances of active and passive voice when you see them and you must be able to convert at will from active voice to passive and—more important—from passive voice to active. Let's start with identification.

First of all, active voice has nothing at all to do with action verbs. And passive voice has nothing at all to do with past tense. Voice is a characteristic of clauses—not verbs—and it has to do with the mapping of the semantic structure of the clause onto the syntactic structure of the clause. Active voice is the normal, unmarked, nonspecialized situation where the semantic *agent* corresponds (or is mapped onto) the syntactic *subject* of the clause, and the semantic *recipient* corresponds to the syntactic *complement* of the clause:

Syntactic	*Subject*	*Predicate*	*Complement*
	A horse	kicked	Marvin
Semantic	*Agent*	*Action*	*Recipient*

The passive voice is a specialized, restricted structure where the semantic *agent* is mapped onto the syntactic *complement*, and the semantic *recipient* of the action of the clause is mapped onto the syntactic *subject*:

Syntactic	*Subject*	*Predicate*	*Complement*
	Marvin	was kicked	by a horse
Semantic	*Recipient*	*Action*	*Agent*

This is the key identificational characteristic of the passive voice: the syntactic subject of the clause is not the agent, or doer, of the action; rather, the syntactic subject is the recipient of the action. You can use a couple of other identificational characteristics, too, to spot clauses in the passive voice (see Figure 18–2, page 396). The agent will be marked with

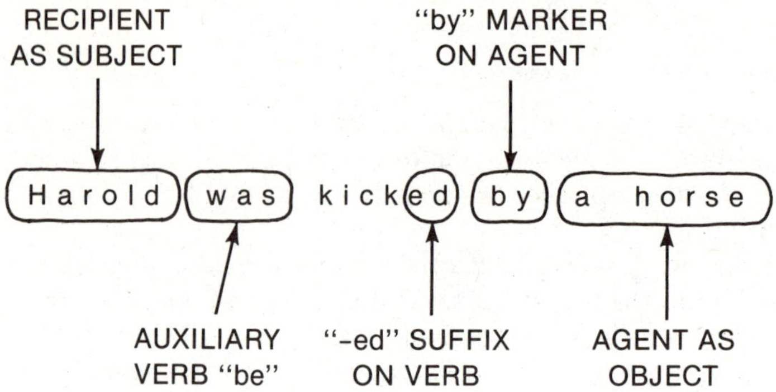

Figure 18–2 Characteristics of the Passive Voice

the word *by,* and the verb in the predicate will be preceded by a form of the auxiliary verb *be (is, was, were, will be)* and will have (in most cases) an *-ed* or *-en* suffix on it. Here are some clauses in the active voice, together with their passive voice counterparts:

Active John spilled the beans.
Passive The beans were spilled by John.

Active A dog is biting Harold.
Passive Harold is being bitten by a dog.

Active The principal investigator will conduct the experiments.
Passive The experiments will be conducted by the principal investigator.

Active The satellite relays the signal back to earth.
Passive The signal is relayed back to earth by the satellite.

Active Mary brought the cookies and hot chocolate.
Passive The cookies and hot chocolate were brought by Mary.

The passive voice is used in situations where, for reasons of clause emphasis or paragraph cohesion, the semantic agent wants to be in a position of focus at the beginning of a clause. Careful writers do not use the passive voice in any other situations.

There is a further identificational problem with the passive voice. Sometimes, the semantic agent is deleted from the surface structure of the clause, yielding what we call in the trade "The Passive Irresponsible":

The beans were spilled.
The window got broken.
The Teton Dam was designed poorly.
The data were misinterpreted.
Marvin was kicked.

Of course, in such clauses you can't spot the *by* marker on the agent—there's no agent—but you still can see that the syntactic subject of the clause is the recipient of the action rather than the agent of the action.

Converting from active to passive voice is simple (see Figure 18–3):

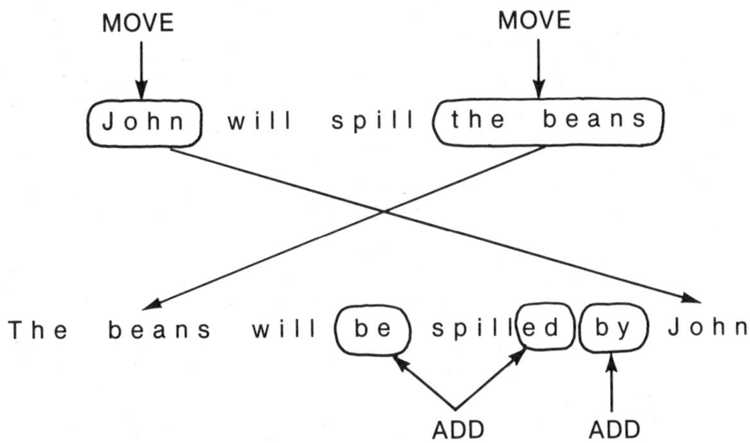

Figure 18–3 Converting from Active Voice to Passive Voice

1. Move the agent to the complement slot.
2. Move the recipient to the subject slot.
3. Add a *by* marker to the agent.
4. Add the appropriate markers to the verb in the predicate slot.

If the clause has an adjunct, most often the passive voice version will read more smoothly if you switch the positions of the complement and adjunct so that the agent-as-complement comes at the end of the sentence:

Active John spilled the beans on the floor.
Passive The beans were spilled by John on the floor.
Switched The beans were spilled on the floor by John.

Active A horse kicked Marvin in the face.
Passive Marvin was kicked by a horse in the face.
Switched Marvin was kicked in the face by a horse.

Active The satellite relays the signal back to earth.
Passive The signal is relayed by the satellite back to earth.
Switched The signal is relayed back to earth by the satellite.

Converting from passive voice back to active is somewhat more involved (see Figure 18–4, page 398)—especially from the passive irresponsible:

1. If the agent is missing, supply one.
2. Move the agent forward to the subject slot.
3. Move the recipient back to the complement slot.
4. Remove the *by* marker and the passive-voice markers from the verb.

For more (a good deal more, actually) on why you should convert passive voice back to active voice—or why you should avoid the passive voice in the first place—see Chapter 20, "Style at the Clause and Sentence Level."

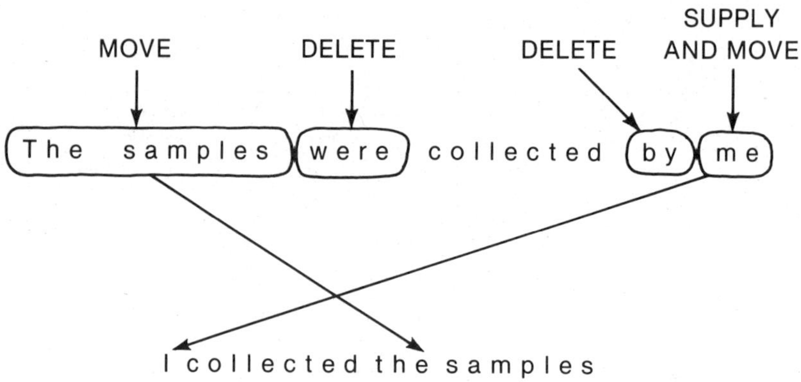

Figure 18–4 Converting from Passive Voice to Active Voice

Writing Exercises

1. Select a 250-word sample of technical writing from your field or discipline. Underline all the nouns in the sample once and all of the verbs twice.
2. Select a 250-word sample of technical writing from your field or discipline. Underline all the modifiers once and all of the function words twice.
3. Select a 250-word sample of technical writing from your field or discipline. Underline all of the noun phrases once and all of the verb phrases twice.
4. Select a 250-word sample of technical writing from your field or discipline. Underline all of the prepositional phrases. What percentage of the total number of words in the sample are contained in prepositional phrases?
5. Select a 250-word sample of technical writing from your field or discipline. Identify all the clauses. Classify each clause as: (a) independent, (b) dependent, (c) nonfinite, or (d) elliptical.
6. Find or construct five examples each of: (a) clause embedded as subject, (b) clause embedded as complement, (c) clause embedded as adjunct, and (d) clause embedded at phrase or word level.
7. Select a sample of technical writing from your field or discipline that contains at least 20 clauses. Prepare a table that shows the complete subject, predicate, complement, and adjunct, as appropriate, for each of the first 20 clauses.

Writing about Writing

1. Write an essay on the topic "The relationship of studying grammar to being able to write well." Aim the essay either at your classmates or at faculty and students in your major field.

2. The concepts reflected by the words *grammar, usage,* and *mechanics* are related but yet distinct. Write an essay exploring the similarities and differences among the three concepts and their relative importance to technical writing.

19

Style at the Paragraph and Section Level

Paragraphs and sections are, of course, the largest units of language structure that make up the text of a report, proposal, manual, or other piece of technical discourse. In syntactic form, a section is that portion of a text that appears between two successive headings or subheads, while a paragraph is that portion of a text that appears or *can appear* between two successive paragraph indentions (see "Punctuating Paragraphs," below). A paragraph consists of one or more sentences, and a section consists of one or more paragraphs. Paragraphs and sections serve four syntactic functions. They can:

- Introduce a theme or topic
- Provide for the main development of the theme or topic
- Summarize or recapitulate what has been said previously
- Serve as bridges between successive main paragraphs or sections

We call these, respectively, introductory, main, summary, and transition paragraphs or sections. (See "Paragraphs and Sections" in Chapter 18 for additional discussion and examples.)

Semantically, paragraphs—especially main paragraphs—serve to develop and extend themes and topics and generally carry the discourse toward the achievement of its rhetorical aim or purpose. In a well-formed discourse, each paragraph furthers the rhetorical cause, as it were, and carries the readers of the discourse firmly and significantly toward the conclusion. Excellent writing consists, first of all, of effective paragraphs (and sections)—paragraphs that are developed with some degree of detail and completeness, that exhibit thematic unity, and that are structurally coherent. Let's look in detail at each of these characteristics of excellence.

□ Paragraph and Section Development

A key characteristic of excellent technical communication is longish (hence well-developed) paragraphs. A first step, then, in revising for effective style is to scan your copy for one-sentence or two-sentence paragraphs, or even one- or two-sentence sections. Such short units generally indicate a lack of development. Introductory, summary, or transition paragraphs may (and often do) consist of only a sentence or two, but your main paragraphs and all one-paragraph sections should exhibit a fair degree of development: five or more sentences; a hundred or more words. You need an awfully good reason for less. Now, much of the time paragraph development just isn't a problem: we have plenty to say and we say it, and well-developed paragraphs are the natural result. But if in revising a report you find yourself with too many short, underdeveloped paragraphs, here are some things you might consider:

- □ If you have several short paragraphs in a row, often you can just combine them into a single paragraph. The demands of unity (see below) will probably require a new topic sentence to tie the separate elements together, and you may need as well a summary sentence at the end.
- □ Try developing a short expository or argumentative paragraph by adding examples or illustrations of the point you're trying to make. Get in the habit of using phrases like "for instance" and "for example."
- □ You can develop a thin descriptive or narrative paragraph by adding specific details at a lower level of abstraction. For instance, if you have described someone's home as a typical mid-sixties suburban ranch house, you might go on to say that it has an attached two-car garage, two and a half baths, and a sliding glass door to a patio off the dining room.
- □ Consider restating your topic sentence for clarity or for emphasis. The restatement can come either before or after illustrations or details. Remember that redundancy is *unnecessary* repetition; if repeating or restating an idea contributes to paragraph development by making ideas clearer or giving emphasis to them, the repetition is not redundant.
- □ Don't be afraid to attach one-sentence introductory, summary, or transitional paragraphs to a preceding or following main paragraph—especially if the main paragraph in question is rather short anyway.
- □ Finally, you can use a combination of these techniques to provide as full a development as you want: combine short para-

graphs, add examples and illustrations, add detail at a lower level of abstraction, restate for clarity or emphasis, and attach short functional paragraphs to adjacent main paragraphs.

With paragraph development as with any other aspect of writing style, too much of a good thing can be bad. Don't overlook the value of a short paragraph for emphasis. Keep in mind that a little variety is in order, too. If you are *averaging* 100 words or so with your paragraphs, you'll want some paragraphs of 40 or 50 words (or fewer) and some of 150 words (or more). Don't just mechanically develop a string of five-sentence, hundred-word paragraphs all of a similar syntactic structure. Develop where development is necessary, but don't get carried away.

□ Paragraph and Section Unity

Achieving and maintaining thematic unity with paragraphs and sections is simply a matter of talking about one thing at a time—sticking to a single topic or theme within a given paragraph or section. With most kinds of technical writing, thematic unity with sections is not much of a problem—especially if you're using headings and subheads. Maintaining unity is just a matter of making sure that each sentence in each paragraph relates to the topic expressed by the heading or subhead. With paragraphs, your topic sentence serves as a sort of "heading." Make sure that each of your paragraphs (your main paragraphs, that is) *has* a topic sentence, and then scan each of the sentences in the paragraph to see that it relates to the idea expressed in your topic sentence. Admittedly, the term *relates to* is somewhat subjective and requires a certain amount of individual judgment, but it is an area of subjectivity that most writers are able to operate within once they begin to inspect their sentences for relatedness to headings and subheads and to topic sentences.

There are some problems, however. One problem is deciding whether to include tangentially related material in a given section or paragraph or to cross-reference the reader to another section. Suppose, for instance, you are explaining how to estimate the volume of lumber in a tree. You state that one of the first steps is to measure DBH (diameter, breast height). You could go on to define DBH, explain why the diameter is not measured at ground level, and add a few cautions about knots or burls and about measuring the DBH of trees growing on side hills. In fact, you could work up a pretty good little paragraph on DBH. Or you could just insert a single sentence: *See "Measuring DBH" on page 6.* Similarly, suppose you are writing a functional description of an automobile carburetor. You mention the vacuum line that goes to the vacuum advance mechanism on the distributor. You could mention that the vacuum advance mechanism advances the distributor timing on acceleration when there is a strong vacuum in the intake manifold. Or you could refer your readers to your section on "Ignition Timing" and "Distributor Operation."

If you can explain the tangentially related material in a sentence or two, I suggest putting it in. A minor digression like this is not a serious violation of thematic unity. This is especially true if your readers would have to refer to two or more other sections to get the complete picture. On the other hand, if the digression amounts to several sentences or a whole paragraph, you're probably better off including the information in a section of its own. Be sure to cross-reference, though. Leave a trail of bread crumbs so that your readers can find their way through your document. Also, keep in mind that while an occasionally minor digression is acceptable, a whole series of minor digressions can destroy your thematic unity and make your writing very difficult to follow.

A second and related problem is the habit that many writers have of writing a statement that is fraught with implications and raises more questions than it answers. This phenomenon is common when you have simplified a rather complex topic for a lay audience and then start to worry about some other technical expert who might read what you have written. So you include a sort of "cover-your-tail" statement. You might say that water freezes at 32° F *or sometimes a little lower.* Or you might say that by convention electric current flows from positive to negative *except that the electrons really flow from negative to positive.* Or perhaps you have explained how to use a pH meter and conclude by saying "*Of course, there are other ways of measuring the acidity of a solution.*" Now, each of these statements is true and in a sense relevant to the discussion, but they tend to leave your readers hanging. You suggest there's more to the story than what you've told them, but then you fail to tell them the rest of the story. Under what conditions does water freeze at lower temperatures? Are these important? What is the relationship between current flow and electron flow? Does it make a difference? What are the other ways of measuring acidity? Should I know what they are and how to perform them?

The solution is simple. Avoid all such statements that lead nowhere. Either develop the idea fully and answer all the implied questions or eliminate the qualifying statement altogether and don't worry about being accused of oversimplifying or being inaccurate. The essence of clear explanation is avoiding unnecessary complications, so write at the level of your intended audience and resist the temptation to try to look good for the experts.

A third problem with thematic unity is drifting off the track. Suppose you're explaining how to sail a boat. You discuss coming about to a new compass heading. Then you explain that the new compass heading is not really the boat's course but the direction of a tack. Then you explain the relationship between magnetic bearings and true bearings and how to set the magnetic declination on a compass. Pretty soon you've written two and a half pages on celestial navigation, while your poor readers are wondering how to tie off the sheets and set the tiller. You can draw a direct connection between each sentence and the preceding one, but somehow you've gone way wide of what you wanted to say.

The key is "what you wanted to say." Here's where a tight outline and a clear purpose, or communicative goal, is important. If you keep your goal clearly before you (just as in sailing), each tack or change in direction should take you closer to your goal. If you begin to fall off your mark or fail to make way, it's time to come about.

□ Paragraph and Section Coherence

A well-written paragraph sticks together. It is coherent. Furthermore, it is smooth—it flows without lumps or bumps. In some respects, a well-formed paragraph is like good bread dough: it hangs together and doesn't crumble when you pick it up, it is smooth and even-textured, and (this is unity) it doesn't have any pieces of funny stuff in it. To push the analogy (perhaps further than it ought to be pushed), a good paragraph is combined from quality ingredients; it is mixed and kneaded with care, and it yields a product that is a delight to consume. Good dough makes good bread.

Coherence is a semantic concept, like the concept of *relatedness* discussed in the previous section. (See also "Syntax and Semantics" in Chapter 18). In well-formed discourse, it is the semantic propositions and themes that must cohere. This semantic coherence is manifested by or reflected through syntactic *cohesion*. If the clauses and sentences of a paragraph are cohesive, the ideas and thoughts carried by these structural units will be coherent. If not, then not. (This distinction between *coherence* and *cohesion* can be confusing at first, but as with *form* versus *function* and *syntax* versus *semantics*, it is important—even crucial—to maintain the distinction. A great deal of confusion has arisen by confounding the two concepts and failing to keep them separate and distinct. Coherence has to do with the semantic function of a discourse, while cohesion is a matter of syntactic structure.) Semantic coherence in paragraphs and sections is manifested chiefly through three cohesive factors:

- □ The frequent signaling of semantic relationships—particularly with sentence adverbials
- □ The strategic repetition of thematic tags, the key words that carry and advance the theme of a discourse
- □ The careful distribution of known and new information

Signaling Semantic Relationships

For a paragraph or section to be well-structured, its statements and propositions must stand in some meaningful relationships to one another and exhibit some degree of orderliness and completeness. Writing that fails to achieve this is less than adequate. But a common failing of mediocre writing is that these semantic relationships are left implicit rather than being signaled overtly. It is up to the reader to *figure out* what the relationships are. Excellent writing, however, doesn't force the reader to guess at the semantic relationships between and among its statements and propositions. The relationships are signaled—clearly and frequently—relationships like cause-effect, reason-result, temporal pattern, spatial pattern, antithetical relationships, cumulative relationships, and the like.

The signals excellent writers use to indicate semantic relationships are a rather loose set of words and phrases I will call sentence adverbials. Such words and phrases fill the connector slot in the clause or sentence (see "Clauses" in Chapter 18), and they serve to signal relationships between full, independent clauses—often in different sentences. Some examples of sentence adverbials are *therefore, moreover, nevertheless, furthermore, for instance, for example, on the other hand, in spite of, yet, first, second, last, inasmuch as, namely, so, now, actually, confidentially, occasionally,* and *finally*. This list is not exhaustive, by any means; there are hundreds of other sentence adverbials—like *by any means,* for instance. Consider some of the semantic relationships that are signaled by sentence adverbials:

however, nevertheless, though, on the other hand, yet, still, in any case, after all, anyway, besides	ideas are related concessively
furthermore, moreover, in addition, what is more, then, above all	later ideas reinforce earlier ideas
first(ly), second(ly), third(ly), last(ly), finally, next, then, in the first place, to begin with, a, b, c, (1), (2), (3)	ideas are related enumeratively
therefore, consequently, hence, so, thus, as a result	ideas are related causally
likewise, equally, similarly, in a like manner	ideas are related equatively
for example, for instance, namely, that is, that is to say, e.g., i.e.	later ideas expand upon earlier ideas
in summary, in conclusion, to sum up	ideas are summative
obviously, of course, to be sure, surprisingly, understandably, remarkably	these adverbials express the writer's attitude toward what he is saying
frankly, seriously, confidentially, honestly, personally, generally	these adverbials express the style or manner in which the writer speaks

□ Success Strategy for Student Writers □

A key characteristic of excellent technical writing is a fairly high frequency of sentence adverbials. Cultivate the habit of employing sentence adverbials to signal semantic relationships. Try doing this as a step, or pass, in your revision process. Go through your entire initial draft with the sole purpose of finding places where you can incorporate sentence adverbials. Don't worry about getting too many; if you go overboard your instructor will let you know, and you can cut back a bit.

Thematic Tags

The theme of the text in Figure 19–1 (page 407) is "the Sun." We know this not only because *the Sun* is the object of the first sentence in the text but also because the phrase *the Sun* is repeated throughout the text (see Figure 19–2). Subthemes are "energy" (light, heat, radiation) and "the Earth"; you can see how these words and phrases are introduced early in the text and repeated throughout. *The Sun, energy,* and *the Earth* are the thematic tags of the passage, and the frequent repetition of the thematic tags contributes to the coherence of the text. Contrast our "Sun" text to the text in Figure 19–3 (page 409). The text here is incoherent—largely because there is no identifiable theme; there are no repeated thematic tags.

There's more to achieving cohesion through the use of thematic tags than just repeating a given word. You can achieve cohesion in your writing by repeating cognates, or different grammatical forms of your theme word, as in Figure 19–4 (page 410). Also, you can achieve cohesion through the use of pronoun reference, as in Figure 19–5. And finally, you can achieve cohesion through repetition of words that in some sense "have to do with one another"—words that are, as we say, from the same lexical set. Consider the paragraph in Figure 19–6 (page 411). The theme of this paragraph is "atoms." We know this in spite of the fact that *atom(s)* is only mentioned twice, because all of the circled words and phrases have to do with atoms, or are members of the lexical set that includes *atoms*. All of these repetitions of thematic tags contribute to cohesion—repetition of the same word, repetition of cognate words, repetition of reference with pronouns, and repetition of words from a lexical set.

In fact, you should be able to pick up any piece of discourse and map out the chains of lexical and thematic repetition, as in Figure 19–7 (page 412). If the sample you choose is well-structured and coherent, all parts of it will be laced together with these cross-sentence ties. Any gaps in thematic linkage indicate discontinuities in the train of thought of a paragraph.

Known and New Information

Patterns of known and new information also contribute to cohesion in discourse. When an item or idea has been mentioned already in a stretch of discourse, we say that it is *known*, but when an item or idea is mentioned for the first time, we say that it is *new*. Generally, we reserve the front half of a sentence or clause for known information and present new information in the back half. That way, we're always beginning on familiar ground, so to speak, and then stepping out into unknown, or new, territory. By always moving from known information to new information,

Consider the Sun. The most obvious characteristic of that body is the quantity of light and heat it delivers despite the fact that it is 93,000,000 miles from us. It lights and warms the Earth and has done so constantly through all of history.

The energy in the form of light and heat pouring down from the noonday Sun upon a single square centimeter of the Earth's surface in a single minute is 1.97 calories. This quantity, 1.97 cal/cm^2/min, is called the solar constant.

A cross section of the Earth in a plane perpendicular to the radiation reaching it from the Sun is about 1,280,000,000,000,000,000 or 1.28 x 10^{18} square centimeters in area. Therefore, the total radiation striking the Earth each minute is about 2,510,000,000,000,000,000 or 2.51 x 10^{18} calories.

Even this by no means expresses all the radiation of the Sun. The Sun radiates energy in all directions and only very little of it strikes the Earth.

Imagine a huge, hollow sphere with the Sun enclosed at its center, with every part of the sphere 93,000,000 miles from the Sun. The Sun would light and heat every part of that sphere just as it does the Earth, and the area of this huge sphere would be over two billion times the cross-sectional area of the Earth. That means that the Sun radiates more than two billion times as much energy as the Earth manages to intercept.

Here, then, is the crucial question: Where is all that energy coming from? If the law of conservation of energy applies to the Sun as well as to the Earth, then the incredibly vast supply of energy being poured into space by the Sun cannot be created out of nothing. Energy can only be changed from one form to another, and therefore the Sun's radiation must be at the expense of another form of energy. But what other form?

Figure 19–1 Sample Text (from Isaac Asimov, *The Nutrino* [Garden City, NY: Doubleday, 1966], pp. 37–38)

Consider the Sun. The most obvious characteristic of that body is the quantity of light and heat it delivers despite the fact that it is 93,000,000 miles from us. It lights and warms the Earth and has done so constantly through all of history.

The energy in the form of light and heat pouring down from the noonday Sun upon a single square centimeter of the Earth's surface in a single minute is 1.97 calories. This quantity, 1.97 cal/cm^2/min, is called the solar constant.

A cross section of the Earth in a plane perpendicular to the radiation reaching it from the Sun is about 1,280,000,000,000,000,000 or 1.28 x 10^{18} square centimeters in area. Therefore, the total radiation striking the Earth each minute is about 2,510,000,000,000,000,000 or 2.51 x 10^{18} calories.

Even this by no means expresses all the radiation of the Sun. The Sun radiates energy in all directions and only very little of it strikes the Earth.

Imagine a huge, hollow sphere with the Sun enclosed at its center, with every part of the sphere 93,000,000 miles from the Sun. The Sun would light and heat every part of that sphere just as it does the Earth, and the area of this huge sphere would be over two billion times the cross-sectional area of the Earth. That means that the Sun radiates more than two billion times as much energy as the Earth manages to intercept.

Here, then, is the crucial question: Where is all that energy coming from? If the law of conservation of energy applies to the Sun as well as to the Earth, then the incredibly vast supply of energy being poured into space by the Sun cannot be created out of nothing. Energy can only be changed from one form to another, and therefore the Sun's radiation must be at the expense of another form of energy. But what other form?

Figure 19–2 Sample Text with Thematic Tags Noted

It pervades the galleries of Renaissance art, floods the corridors of Greek and Roman sculpture and settles with a reverent calm before the mysteries of ancient Egyptian art. Unfortunately, arguments for trade protection ignore fundamental political and economic forces which severely limit their validity. In the 1930s (and earlier), those who troubled to examine the technical facts know that rockets could lift people to the moon. An eight-million-dollar renovation, building and expansion program extending over the last four years has resulted in the completion of a new educational and orientation center.

In the next decades, commemorative oratory welled up again on each July 4, and still more on Washington's birthday, February 22, which was a national holiday of almost equal importance. These innovative systems from Bell break with the past by bringing the latest in advanced communications to a new world of critical applications--from reviewing the diets and health records of hospital patients to monitoring room status in hotels.

Furthermore, the coati is not the sort of animal that inspires official research. Twinings, teamen to the royal family now as earlier, remains devoted to trading quality teas, though its fragrant Earl Greys or Lapsang Souchongs are more likely to be sold in tea bags to supermarket patrons than dispensed in heavy wooden chests to English country houses.

Figure 19–3 Sample of Incoherent Text

the careful writer keeps his readers on track. Now, in normal clause order the subject slot is at the front and the complement slot is at the back, so a pattern of known-item-as-subject and new-item-as-complement is common. In fact, by introducing a new item in the complement slot of one sentence and then repeating it in the subject slot of the following sentence, you can achieve further continuity and coherence in your writing. If these two sentences happen to span a paragraph or section break, as in Figure 19–8 (page 412), this kind of lexical cohesion can serve to knit your paragraphs together and make for a smooth transition from one to the other.

CHAPTER 10 --DOCUMENTATION

As a student writer you documented or "footnoted" your term papers largely to avoid allegations of plagiarism and to provide your instructor with a handy list of your sources so that he could check to see if you were thinking for yourself or just stringing together other people's ideas. There was mention of giving credit where it was due and perhaps some mention as well of providing your readers with a list of sources in case they wanted to pursue your subject matter in greater detail. All in all, though, documenting your papers was a frustrating and time-consuming chore that seemed to have little relevance to your actual writing.

As a writer on the job, however, you will find all of this turned around. First of all, you won't do all that much documentation. Most of the things you write will not require it. When you do document, you will do so primarily to help your readers find additional information on the subject you are writing about. Questions of proper credit and formal recognition will be present, but they will take a secondary role. You include references to other sources because you are convinced that it is in your readers' best interests to consult these other sources and you want to make it as easy as possible for your readers to find these sources. In short, you document as an act of communication.

Figure 19–4 Different Grammatical Forms of the Theme Word as Thematic Tags

Consider the Sun. The most obvious characteristic of that body is the quantity of light and heat it delivers despite the fact that it is 93,000,000 miles from us. It lights and warms the Earth and had done so constantly through all of history.

Figure 19–5 Pronoun Reference to the Theme Word as Thematic Tags

One might try, out of a spirit of neatness, to divide the atom cleanly between the two major physical sciences, awarding the electrons to the chemist and the nucleus to the physicist.

To attempt such a cleancut division would, however, be a violation of the spirit of science, which is all-one-piece. The structure of the nucleus, however remote that might seem from the world of ordinary chemical reactions, must nevertheless be of keen interest to the chemist, if only because of its effect upon that fundamental chemical datum, the atomic weight.

By the time the nineteenth century had come to its end, the matter of the atomic weight seemed settled. Each element had a characteristic atomic weight, chemists believed, and the only future in that respect was the greater precision with which the fourth and fifth decimal places might be determined.

Then came the discovery of isotopes and all that had seemed certain about atomic weights immediately went into discard. The notion, dating from Dalton, that all atoms of a single element possessed identical mass and that the atomic weight expressed this mass was seen to be false. Instead, most elements were made up of two or more varieties of atoms differing in mass. The atomic weight was merely the weighted average of the masses of these isotopes.

If the term "atomic weight" is reserved for this weighted average of the isotope mass as found in their natural distribution within an element, then one ought not to speak of the atomic weight of an individual isotope (as I have been doing so far in this book). It is better to use a different phrase and speak of the relative mass of an individual isotope as its mass number.

Figure 19–6 Words from a Lexical Set as Thematic Tags (from Isaac Asimov, *Understanding Physics: The Electron, Proton, and Neutron* [New York: Walker, 1966], pp. 145–146)

□ Information Density

Suppose we could identify a semantic entity called "Minimal Unit of Information." (We can't, but suppose we could.) This unit of information might correspond to a single fact or a single idea. By counting up these units in, say, a paragraph, we could determine the quantity of information in that paragraph. We could say that paragraph has so many units of information in it, and that paragraph B contains more, or fewer, units of information. This minimal unit of information would give us, then, a measure of the amount of information contained in a stretch of discourse.

By counting both the number of words and the number of informa-

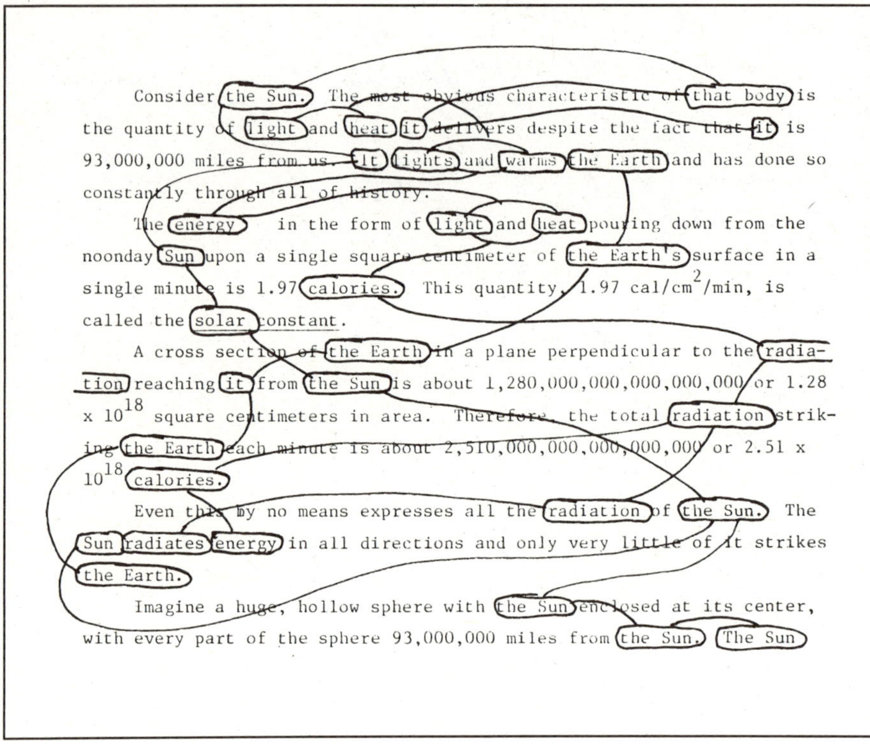

Consider the Sun. The most obvious characteristic of that body is the quantity of light and heat it delivers despite the fact that it is 93,000,000 miles from us. It lights and warms the Earth and has done so constantly through all of history.

The energy in the form of light and heat pouring down from the noonday Sun upon a single square centimeter of the Earth's surface in a single minute is 1.97 calories. This quantity, 1.97 cal/cm^2/min, is called the solar constant.

A cross section of the Earth in a plane perpendicular to the radiation reaching it from the Sun is about 1,280,000,000,000,000,000 or 1.28 x 10^{18} square centimeters in area. Therefore, the total radiation striking the Earth each minute is about 2,510,000,000,000,000,000 or 2.51 x 10^{18} calories.

Even this by no means expresses all the radiation of the Sun. The Sun radiates energy in all directions and only very little of it strikes the Earth.

Imagine a huge, hollow sphere with the Sun enclosed at its center, with every part of the sphere 93,000,000 miles from the Sun. The Sun

Figure 19–7 Complete Mapping of Lexical Chains and Thematic Repetition

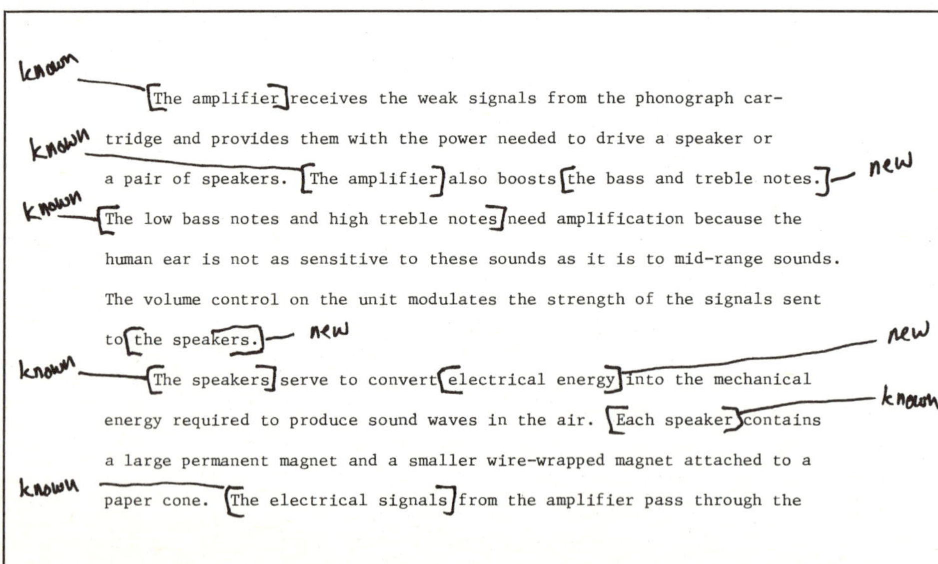

The amplifier receives the weak signals from the phonograph cartridge and provides them with the power needed to drive a speaker or a pair of speakers. The amplifier also boosts the bass and treble notes. The low bass notes and high treble notes need amplification because the human ear is not as sensitive to these sounds as it is to mid-range sounds. The volume control on the unit modulates the strength of the signals sent to the speakers.

The speakers serve to convert electrical energy into the mechanical energy required to produce sound waves in the air. Each speaker contains a large permanent magnet and a smaller wire-wrapped magnet attached to a paper cone. The electrical signals from the amplifier pass through the

Figure 19–8 Patterns of Known and New Information in a Text

tion units in a given paragraph and dividing the former by the latter, we could calculate the number of words per information unit for the paragraph:

$$(1)\ \text{Words per Information Unit} = \frac{\text{Number of Words}}{\text{Number of Information Units}}$$

By taking the reciprocal of words per information unit, we can calculate a value that we could call "Information Density":

$$(2)\ \text{Information Density} = \frac{\text{Number of Information Units}}{\text{Number of Words}}$$

This calculated value would give us a measure of the relative density of information in a given text. We could see which writers are packing lots of information into what they write, and which writers are using lots of words to say very little.

Now, we can't really measure information density, but I am sure that if we could we would find many suspicions confirmed. We would find that our calculus text and that philosophical treatise by Wittgenstein had very high information densities indeed; we would find that newspapers and magazine articles had somewhat lower information densities; and we would find that political speeches had information densities approaching zero—all words and no information.

Here's another interesting observation we could make. If we measured the information density in different parts of some well-written technical reports, proposals, or articles, we would find different information densities in these parts. We would find that careful writers are controlling their information densities, allowing the pace of information transfer to increase in introductory, summary, and other general sections and slowing the pace a bit (by decreasing the information density) in those areas where the concepts being presented are new and somewhat difficult to grasp. A careful writer controls the pace of his presentation by manipulating his information density: he moves right along with introductory and background material and keeps a fairly high pace (high information density) when he's dealing with easy ideas and concepts. But when he gets into the difficult stuff, he slows down and lets his readers catch their breath, so to speak. We have all had the experience of having to read a passage two or three times in order to understand it. The problem more often than not is that the passage has too high an information density—the facts and ideas are coming at us too fast—for us to grasp them on a single reading.

So how can you control the information density in the reports that you write? The key to lowering the information density in a given paragraph or section is providing more development. Once you've stated an idea, restate it. Offer an example. If you have one example, offer another. Summarize your ideas periodically, and restate your conclusions. The place to do this, of course, is in the main body of your report—particularly in those areas where the concepts and ideas you are discussing are difficult, important, or both. Don't be afraid to slow the pace a bit when the going gets tough; your writing will be more readable and easier to understand if you do. On the other hand, if you find yourself going on and on with a topic, you may wish to increase the information density and make your writing more compact. You can do this by eliminating restatements and redundancies, deleting examples and summaries, and boiling some of the extra words out of what you have written. To increase your information density, edit for brevity and conciseness.

The key word with information density is *control*. Be conscious and aware of the ratio between the number of words you are using and the amount of information you are presenting. Manipulate this ratio by lowering information density in the difficult portions of your report and raising it in the introductory, summary, and general portions. Lower your information density by developing your ideas more fully, and raise your information density by editing for brevity and conciseness. Above all, control your writing.

□ Punctuating Paragraphs

The word *paragraph* is ambiguous. It can refer to the functional unit (introductory, main, transitional, or summary) or it can refer to the formal unit that appears between two successive paragraph indentions. Most of the time, this ambiguity doesn't cause any difficulty because more often than not the functional unit and the formal unit are identical. But the two meanings of the word can get us into trouble when we talk about paragraph structure (see "Paragraphs and Sections" in Chapter 18) and when we talk about punctuating paragraphs. For this discussion, then, let us distinguish between functional paragraphs and formal paragraphs.

Any stretch of discourse that spans several functional paragraphs can be broken up into a number of formal paragraphs by inserting paragraph indentions (or by skipping a line or by using some other device to indicate a paragraph break). We can't just put paragraph indentions in any old place, however. We can put them in only at the junctures between functional paragraphs. But we don't have to put in a paragraph indention at *every* juncture; we have a choice as to whether or not we want to exercise the option of starting a new formal paragraph every time we start a new functional paragraph. We can, for instance, choose to attach a short introductory, summary, or transition paragraph to the main paragraph preceding it or following it. Similarly, we have some choice with successive main paragraphs. Suppose, for instance, that you were presenting an argument that consisted of an introduction, three main points, and a conclusion. You could, if you wished, punctuate these five functional paragraphs as a single formal paragraph. Or you could punctuate them as five separate formal paragraphs. Or you could combine the three main points into a single unit and have three formal paragraphs—introduction, body, and conclusion. You could even combine your introduction with the first main point or combine your conclusion with your last main point (although to do both would be odd). What you can't do is make a formal paragraph out of your first point and half the second or put in a paragraph indention halfway through the development of your third main point.

What factors govern the choice of paragraph punctuation? Length, of course, is one. If you have a five- or six-sentence introductory unit, you would probably let it stand alone as a separate formal paragraph. If your introductory unit is only a single sentence (as with the present paragraph), you may well punctuate it as part of the following formal paragraph. Similarly, if your three main points are rather short you may combine them into a single formal paragraph; on the other hand if by combining them you would run on for a page or a page and a half, you're better off keeping them separate.

The pattern of paragraphing in the nearby portions of your text may influence your choice as well. If you have just completed a series of rather short paragraphs, you may be inclined to combine two or more functional paragraphs into a single longish formal paragraph. Conversely, if you have a series of long paragraphs, you may choose to punctuate a short summary or transitional paragraph as a formal unit by itself—just to break up your pattern a little bit and introduce some variety.

Finally, you should consider the physical form of your page—especially as it will appear in final reproduction form. The key factors to consider are type size and column width. For most of your main paragraphs the width and length (in inches) want to be approximately equal, with neither dimension more than about twice the other. A paragraph that is attractive to the eye is about square—or perhaps a rectangle that is a little shorter than it is wide. If you have a wide column—say, a typewritten page—you will need 125 to 150 words to get a square paragraph. With a handwritten draft, 50 or 60 words may give you a square paragraph—especially if your handwriting is rather large. With a narrow column, as in a newspaper or a journal or magazine with two-column pages, 30 or 35 words may give you a square paragraph. This is one of the reasons that journalists write one- and two-sentence paragraphs.

□ Revising at the Paragraph and Section Level

If you are following the program I suggested under "Achieving Adequacy and Achieving Excellence" in Chapter 2, then revising at the paragraph and section level will be the first major step in revising for excellence. It will not, however, be your first revision. Before you reach this point in your revisions, you have checked for the utility, accuracy, completeness, orderliness, and correctness that make for adequate writing, and you have revised for orderliness, completeness, and brevity, as outlined in Chapter 5. This has given you a more-than-adequate—in fact an above-average—document. Now you are ready to begin to make it an excellent document.

- □ Perhaps the most important revision is to eliminate the underdevelopment that is characteristic of mediocre writing. Scan your draft for short paragraphs. Drastically reduce the number of short paragraphs by combining short paragraphs; adding examples, illustrations, and details; and restating for clarity or emphasis.
- □ Check the thematic unity of each of your main paragraphs and sections. Make sure you haven't digressed too much into unrelated or only tangentially related material. Eliminate qualifying statements that raise more questions than they answer. Make sure your paragraphs don't drift off course.
- □ See to it that you have sufficient cohesive devices in your writing so that your paragraphs and sections are coherent. Pay particular attention to signaling your semantic relationships with sentence adverbials. Don't ignore thematic tags and known and new information, however.
- □ Control your information density so that you have a fairly rapid pace with familiar information and a much slower pace with new and difficult information.
- □ Finally, adjust your paragraph punctuation to your type size and column width so that your page is pleasing to the eye as well as communicatively functional.

Writing Exercises

1. Find a sample of technical writing from your field or discipline that consists of a series of short paragraphs (under 50 words in length). Combine several of these into longer paragraphs. Attach a copy of the original.
2. Find a sample of technical writing from your field or discipline that consists of one or more long paragraphs (250 words or more). Break the long paragraph(s) up into several shorter paragraphs. Attach a copy of the original.
3. Revise a 250-word sample of technical writing (either yours or someone else's) by adding sentence adverbials or other signals of cohesion to *every* sentence. Go overboard—this is just an exercise. Attach a copy of the original.
4. Find a sample of technical writing from your field or discipline that exhibits a high information density. Revise it to lower the information density significantly. Attach a Xerox copy of the original.

5. Select a 250-word sample of technical writing (either yours or someone else's). Map out the patterns of cohesion in the sample in the manner illustrated in Figure 19–7. Comment briefly on your findings.

Writing about Writing

1. Write a library research paper on the topic "Coherence and cohesion in English." Consult some linguistics textbooks and textbooks in technical writing, business writing, and freshman composition. Consult some primary sources, too, if you can justify the time. Aim your research paper at your classmates.
2. Write a library research paper on the topic "Known and new information." Consult the same kinds of sources as suggested in the previous assignment. Again, aim your research paper at your classmates.
3. Write an essay on the topic "Paragraph development" and/or "Paragraph unity" as applicable to technical writing. Aim the essay either at your classmates or at faculty and students in your major field.

20

Style at the Clause and Sentence Level

Clauses are the fundamental building blocks of discourse. We use words and phrases to name and identify, but it is clauses we use to make statements, show relationships, and express ideas. Statements, relationships, and ideas, of course, are semantic concepts or semantic units. They are the stuff of which thoughts are made; they constitute the substance of communication—the meanings that are being communicated. Clauses are the syntactic units that express or manifest semantic propositions. Sentences are carriers of clauses; they are the syntactic units that consist of one or more clauses and provide a matrix for clauses to exist in several structural relationships to one another.

As writer, you have a considerably wider range of choice with clauses and sentences than you do with paragraphs and sections. You can control the lengths of your sentences and the variety of their beginnings. You can control the voice of your clauses. You can control the complexity of your sentences—both in terms of number of clauses and in terms of types of clauses used. Finally, you can give greater or lesser emphasis to certain of your semantic propositions by choosing to express them as independent, conjoined, embedded, or reduced clauses. The choices you exercise at the clause and sentence level affect the smoothness and clarity of your writing more than those at any other level.

□ Controlling Sentence Length

Mature expository prose averages about 19 words per sentence, regardless of whether it is excellent or barely adequate. There is no significant difference in average sentence length between adequate and excellent writing. There is, however, a significant difference in the *range* of sentence length, or in the variation from the average. Mediocre writing tends to show patterns of sentence length like those in Figure 20–1, while excellent writing tends more toward sentence-length patterns like those in Figure 20–2. And it stands to reason, doesn't it, that a plodding sameness of sentence length will lead to judgments of dullness, while a varying pattern of short, middle-sized, and long sentences will at least be a little more interesting to read? Moreover, the short sentence is emphatic. You can lend a certain punch to your major ideas by expressing them in short, one-clause sentences—so long as these occasional short sentences appear against a background of midlength and longish sentences. And finally, some ideas and patterns of ideas are just too complex to be expressed clearly in a series of short, one- or two-clause sentences (see "Combining Clauses," below).

Figure 20–1 Patterns of Sentence Length in Mediocre Writing

Figure 20–2 Patterns of Sentence Length in Excellent Writing

Your goal is variety. If your sentences average 19 words in length, you don't want them all between 15 and 25; you want a fair number of 10-to-15-word sentences and a fair number of 25-to-30-word sentences, and you want a few 5-to-7-word sentences and a few 35-to-45-word sentences. In fact, in any 200-word sample of your writing, your longest sentences should have five to eight times as many words in them as your shortest sentences. For most of you, this will mean learning how to develop and sustain the longer sentence. Student writers tend to err in the direction of too many short sentences rather than too many long sentences.

□ Controlling Sentence Beginnings

Another characteristic of mediocre writing is an inordinate number of sentences that begin with the same words. Perhaps the chief violators here

This proposal for analysis of the Bloomington water distribution system can be broken down into three phases. These phases are: Preliminary Investigation, Computer Analysis, and Interpretation.

The first phase, Preliminary Investigation, consists of inspecting all water sources such as pumps, wells, and storage tanks. The next step is to identify all residential, commercial and industrial water users. This information will help determine your maximum water usage including water for fighting fires. The data will be used by our technical staff to develop a system map of water usage.

The next phase is the Computer Analysis phase. This phase involves coding the pipe system and developing the computer data. The University of Illinois' computer system will be used to analyze the data and determine the pressures and headlosses in each pipe.

The Interpretation phase involves drafting pressure contour maps and recommending any changes in the system. You will receive copies of the contour maps, together with our suggested changes. The contour maps will be printed on 8½ by 14 inch paper and bound together with our report. The entire package will be presented at the May city council meeting.

Figure 20–3 Passage with Excessive Th- Word Beginnings

are the "*th-* words": the definite article *the* and the demonstrative pronouns *this, that, these,* and *those.* It is not uncommon in mediocre writing to find 50, 60, 70, even 80 percent of the sentences in a passage beginning with one of these *th-* words. Such sameness of beginning does contribute to dullness. Now this is not to suggest that there is something wrong with *th-* words; quite the contrary, they are among the most frequently used words in the English language. (In fact, *the* is the *most* frequently occurring word in the language.) Furthermore, this is not to suggest that you should never begin a sentence with *the, this, that, these,* or *those.* Just don't go overboard. Excellent writing may have up to 25 percent of its sentences beginning with *th-* words; it's when you get over 40 percent that you begin to get into trouble. Notice the passage in Figure 20–3. Rather dull, don't you think?

Another kind of sameness of beginning that you should be on guard against is starting almost every sentence in a personal narrative with *I.* This shows up quite often in the mediocre job application letter or resume cover letter, as shown in Figure 20–4.

I am writing to inquire about the possibility of employment as an accountant with your company. I will graduate this June with a B.S. in Business Administration and Accounting. I have worked three summers with Chase and Harding, CPAs.

I feel that my educational background and work experience would qualify me for a beginning position with your firm. I am particularly interested in the opportunity for overseas travel, and I know you have several branches in South America. I am willing to travel anywhere, however.

I am available for interview during spring break (March 15 to 20), and I would appreciate an opportunity to tour your office facilities. I can be reached at the above address or by phone at (906) 786-6165.

I appreciate your taking the time to read my letter, and I hope you find my qualifications suitable.

Figure 20–4 Resume Cover Letter with Too Many Sentences Beginning with "I"

Finally, keep an eye out for an inordinate number of sentences beginning with the words *it is* or *there are.* Such sentence beginnings are called "indirect beginnings" and they are weak for two reasons. First, of course, they get a little boring when they show up every third or fourth sentence. Second—and more important—they are weak because of their very indirectness. Consider two sentences:

(1) There were 27 people who contributed to the success of the project.
(2) Twenty-seven people contributed to the success of the project.

Let's look at the structure of sentence (1).

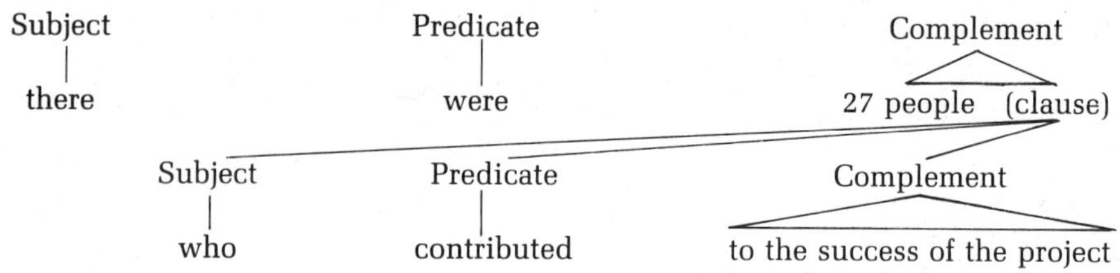

The key predication of the sentence is that people contributed to success. Yet we have to dig all the way down to the predicate and complement of an embedded clause to find the core of the key predication. In contrast, let's look at the structure of sentence (2):

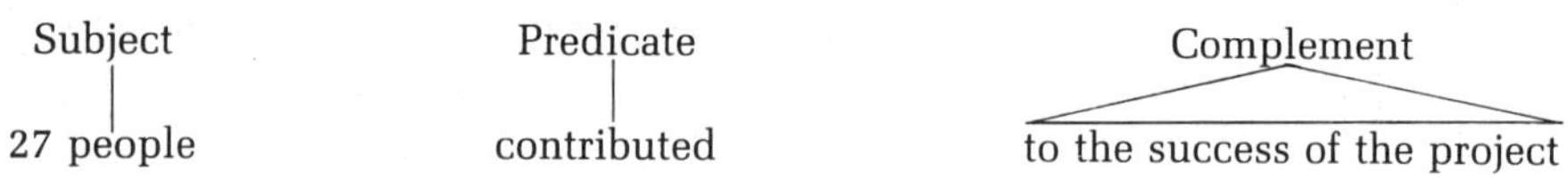

Here, the key predication is readily available in the topmost clause of the sentence. Directness in syntax contributed to clarity in communicating.

□ Controlling the Passive Voice

Excessive and unwarranted use of passive clauses is at once the most common and the most serious stylistic weakness in the writing of technical and scientific people. If you write like the great majority of science and technology students, there is nothing you can do that will make a more significant and noticeable improvement in your writing style than learning how to control the passive voice. There was a time, 50 or 75 years ago, when stylists encouraged scientists to write in the passive voice because some felt that it promoted objectivity in reporting. For at least the past three decades, however, the overwhelming opinion of editors, stylists, and teachers of technical writing has favored the direct communication of the active voice. Let's sample what some of the major scientific and technical style guides have to say on the matter:

> Authors should not always use verbs in the third person, passive voice. When a verb concerns the interaction of inanimate objects, the active voice is usually preferable because it is more direct and concise. When a verb concerns an author's belief or conjecture, use of the impersonal passive is highly inappropriate.
>
> *The American National Standard for the Preparation of Scientific Papers* (1972)

> The Society encourages authors to write in the active voice. . . .
>
> *Authors' Guide to the Publications of ASCE* (American Society of Civil Engineers, 1979)

Use the active voice except where you have a good reason to use the passive. The active voice is the natural voice, the one in which people usually speak and write, and its use is less likely to lead to wordiness and ambiguity. Avoid the "passive of modesty," a device of writers who shun the first-person singular.

Council of Biology Editors Style Manual (1978)

When directness is desired, use the active voice. Avoid passive constructions.

Handbook for Authors (American Chemical Society, 1978)

The passive voice is generally less emphatic than the active and should be avoided where emphasis is desired. . . . In many sentences the passive form is wordy and weak, and it may mean that the author is trying to convince himself.

USGS Suggestions to Authors (United States Geological Survey, 1958)

Absolute insistence on the third person and the passive voice has been a strong tradition in scientific writing. Authorities on style and readability have clearly shown that this practice results in the deadliness and pomposity they call "scientificese." Some scientists maintain that this style preserves objectivity, but the validity of this assertion is suspect. Now, reputable journals are breaking the tradition with notable success, and writing manuals are recommending a more personal style. . . . An experienced writer can use the first person and the active voice without dominating the communication and without sacrificing the objectivity of the research. If any discipline should appreciate the value of personal communication, it should be psychology.

Publication Manual of the American Psychological Association (1974)

In any type of writing, the active voice is usually more precise and less wordy than the passive voice. Why, then, do scientists insist on using the passive voice? Perhaps this bad habit is the result of the erroneous idea that it is somehow impolite to use first-person pronouns. . . . I herewith ask all young scientists to renounce the false modesty of previous generations of scientists. Do not be afraid to name the agent of the action in a sentence, even when it is "I" or "we."

How to Write and Publish a Scientific Paper, Robert A. Day, 1979. (Robert A. Day is managing editor, Publications Board of the American Society for Microbiology.)

Scientific writing has suffered much from editors who mistakenly demand impersonality in the name of objectivity, but many journals now encourage a more direct and personal style.

The Scientist as Editor, Maeve O'Connor, 1979. (Maeve O'Connor is Senior Editor, Ciba Foundation, London.)

No undesirable feature of style is more common in the writing of profes-

> sional men than the use of the passive voice where other speakers and writers would use the active voice. . . . To write entirely in the passive voice would seem not just unwise but impossible; yet some misled scientists and engineers attempt it.
>
> *Effective Writing for Engineers, Managers, Scientists*, H. J. Tichy, 1966. (H. J. Tichy is a professor of English, communications consultant, and Fellow of the Society for Technical Communication.)

The first step in controlling the passive voice is being able to recognize it when you see it. Next, you must be able to convert from passive to active. See "Voice" in Chapter 18 for some help in this department.

The active voice is the natural, normal, unmarked, standard form for a clause. It is the one you should start with; that is to say, conceptually, all of your clauses start off as active. Some you change to passive *if and only if* you have a good reason to do so. All others remain active. So, cultivate the habit of writing your initial drafts in the active voice. If you cannot bring yourself to do that, then as one of your first revisions at the clause and sentence level, identify each instance of passive voice and convert it back to active. Start with a clean slate, as it were.

Then, look for situations where the passive voice might be appropriate. One such situation might be when the demands of paragraph unity (see "Paragraph and Section Unity" in Chapter 19) require that the agent of a clause be put in the position of known information at the front. Consider the following:

> Ladies and gentlemen, we have a problem. This problem was called to my attention by John Hubert in purchasing. John was told of it by one of his field representatives.

Here, the passive voice of the second sentence allows "the problem" to be in the known position at the front of the sentence, where—among other things—it serves as a lexical link with "a problem" in the first sentence. Similarly, the passive voice of the third sentence allows "John" to link with "John Hubert" in the second sentence.

Another situation where the passive voice can be appropriate is when no one knows and no one *cares* who the agent is. Consider these:

> Over 7000 imported cars were purchased last month.
> If you break the law, you will be arrested.
> The parts were shipped yesterday.

Do not use the passive voice to "give a little variety to your sentence structure." Variety is fine but not at the expense of clarity. Do not use the passive voice to "achieve objectivity" in your writing. Objectivity is more a characteristic of your research than your writing, anyway. Do not use the passive voice to hide the fact that you don't know who the agent of a particular action is. If it's important to your readers, *find out* who did it. Finding out is always preferable to copping out. Above all, don't use the passive voice just out of habit. Be able to give a specific reason for every clause you write in the passive voice. Excessive and unwarranted passive voice contributes to the dullness of the passage in Figure 20–5.

Since this is the first report, my overall status is shown by the work completed in this first period. A chart is shown below, indicating work attendance and participation for the time period. The textbooks were purchased from the bookstore as well as the suggested material on computer text editing. All assignments for the period were completed and outside examples of writing were analyzed from several magazines. A topic for a recommendation report was selected, and the background of the topic was totally defined.

The topic for the recommendation report is being researched, and a rough outline must be organized. The report is then to be completed and handed in by the end of January. The material on computer text editing has to be studied, and time must be taken for work on the computer.

No major problems were encountered during this period, and no difficulties are expected during the next reporting period. It is assumed that the recommendation report and an instruction manual will be completed.

Figure 20–5 Passage with Excessive and Unwarranted Passive Voice

□ Combining Clauses into Longer Sentences

When the need arises to generate longer sentences in your writing, you have two choices open to you: (1) you can gain your sentence length by adding modifiers and prepositional phrases to existing short sentences, or (2) you can combine short one-clause sentences into longer multiclause sentences. Of the two choices, the second is far more effective than the first. As you add modifiers and prepositional phrases, all you are doing is increasing the length of your noun phrases (see "Phrases" in Chapter 18)—in effect, giving longer and longer names for things. Such a "naming" or "nominal" style tends to be rather static and dull. In order for your writing to be more dynamic and lively, you want to develop an "acting" or "verbal" style. You achieve this by using more verbs and verb phrases as opposed to longer noun phrases; and since verbs and verb phrases lie at the heart of clauses, what you are doing in fact is packing more clauses into your sentences. This you do by combining clauses into longer sentences—either by conjoining or by embedding.

Conjoining Clauses

We conjoin clauses by using a set of function words called *conjunctions*. Conjunctions come in two varieties: coordinating and subordinating. As you develop sentence sophistication and employ more and more patterns of coordination and subordination, you will perforce use more conjunctions. I urge you to use them wisely. The coordinating conjunctions are *and, but,* and *or*. The presence of any of these between two clauses signals, first of all, a semantic relationship of equivalence, or equality of emphasis. That's what *coordination* means. In addition, though, each of these coordinating conjunctions signals a further semantic relationship as follows:

and	ideas are cumulative or additive
but	ideas are antithetical
or	ideas are disjunctive and mutually exclusive

The subordinating conjunctions are a little more involved. There's quite a long list of subordinating conjunctions; the list includes such words as *as, after, although, because, if, once, since, until, when, while,* and *yet*. The presence of a subordinating conjunction at the beginning of a clause signals, as you might suspect, a relationship of subordination of that clause and the superordinate (or main) clause with which it is associated. Moreover, each subordinate conjunction signals a further semantic relationship, as suggested by the following:

after, before, since, until, when	ideas are related temporally
where, wherever	ideas are related spatially
if, although, whereas	ideas are related conditionally or concessively
because, since, as	ideas are related causally

(There are, of course, a good many more subordinating conjunctions and a good many more semantic relationships, as well. For a thorough treatment of this topic, see any good grammar of English.)

We can see, then, that conjunctions signal two things about the clauses they conjoin: (1) an emphasis relationship of either coordination or subordination, and (2) a semantic relationship of some kind or other. Keep in mind that this semantic relationship is not *created* by the conjunction; it exists beforehand as a relationship between the ideas or semantic propositions you're trying to communicate.

Be wary of too much coordination and not enough subordination in your writing. The easiest way to conjoin clauses is with *and*, but *and* doesn't say much about the relationship between the clauses you are conjoining. Moreover, if you use too much coordination, your style begins to remind people of the run-on sentences characteristic of the second- and third-grader, as illustrated in Figure 20–6. Be sensitive to the logical relationships between and among your clauses. Be sensitive as well to the

```
     I got up in the morning and I washed my face and I combed my
hair and I ate breakfast and Daddy got the car out and we drove to
the fair grounds and we parked the car and we got out and we went
to the gate and we got some tickets and we went on the Ferris wheel
and the Ferris wheel went up in the air and the chair rocked when
it stopped at the top and we could see the whole fair grounds and
I was so scared and . . .
```

Figure 20–6 Example of Immature Run-on Sentences

relative importance of your clauses (see "Relative Emphasis of Clause Patterns," below). If these relationships call for subordinate conjunctions, don't cop out with an all-purpose *and*. Get in the habit of using subordinate conjunctions.

Embedding Clauses

The second method of combining clauses is embedding one clause within another. As you recall from Chapter 18, you can embed one clause as a major component of another clause. Here are some examples:

Clause as Subject:	*That aspirin is an effective pain reliever* has long been known.
Clause as Complement:	Harold thought *that Mildred would go to the dance.*
Clause as Adjunct:	Sam was a failure *as far as grammar was concerned.*

You can embed clauses at the phrase level. The most common instance of embedding at the phrase level is the relative clause:

We analyzed several samples of lunar rock, *which we borrowed from NASA.*

The research assistants, *who are all graduate students in wildlife biology,* will install the radio tracking collars.

We will use only those statistical measures *which are applicable to nominal or ordinal data.*

As you begin to combine clauses into longer sentences, look for chances to embed one clause as the subject, complement, or adjunct of another clause or as a relative clause within another clause.

□ Using Nonfinite and Elliptical Clauses

As writers gain sophistication with their sentence structures, they tend to use more and more reduced clauses—either nonfinite clauses or elliptical clauses (see "Clauses" in Chapter 18). Using reduced clauses has several stylistic advantages: it lets you express ideas with fewer words, since reduced clauses have certain elements "understood"; it makes for a more verbal writing style; and it allows you to use a greater range of clause emphasis, since the reduced clause is less emphatic than the full clause in a comparable position (see "Relative Emphasis of Clause Patterns," below). The passages in Figure 20–7 employ a good number of nonfinite clauses and several elliptical clauses.

□ Relative Emphasis of Clause Patterns

Careful writers exercise choice in the arrangement of their clauses not only to provide cohesion but also to give relative emphasis to different

1. It is difficult to overestimate the force with which this pessimistic prophecy grips the popular mind, especially among young people. Hammered into their heads by a stream of movies, plays and books, fed by a prestigious line of authors from Kafka and Orwell to Whyte, Marcuse and Ellul, the fear of bureaucracy permeates their thought. In the United States everyone "knows" that it is just such faceless bureaucrats who invent all-digit telephone numbers, who send out cards marked "do not fold, spindle, or mutilate," who ruthlessly dehumanize students, and whom you cannot fight at City Hall. The fear of being swallowed up by this mechanized beast drives executives

2. Brainstorming sessions go something like this: Some professors are dawdling over coffee in the faculty dining room. They are complaining about the quality of the students these days, the difficulty of finding parking spaces immediately adjacent to the buildings in which they teach occasionally, the stupidity of the dean, and the difficulty of promotion. One of them adds that he can't stand to teach in room 322 because the blackboard is at the back of the room behind the students, and when he wants to draw some cabalistic symbols, he has to thread his way through the desks. In another age, this wouldn't have been all that bad, but nowadays the girls are wearing denim work shirts

Figure 20–7 Passages with a Good Number of Nonfinite and Elliptical Clauses (*top,* from Alvin Toffler, *Future Shock* [New York: Random House, 1970], p. 112; *bottom,* from Richard Mitchell, *Less Than Words Can Say* [Boston: Little, Brown, 1979], p. 117)

clauses according to the importance of the ideas the clauses express. Some clause patterns carry greater relative emphasis than others. Now, the whole question of comparative stress on ideas in terms of the clause patterns the ideas are expressed in is one that needs more hard research. We just don't have enough quantitative data about how writers make such choices and how readers react to the choices that are made. We can start with the suggestion you find in every freshman composition text: "Put main ideas in main clauses and subordinate ideas in subordinate clauses." This is sound advice and good as far as it goes. It doesn't go far, though. Beyond that, we can establish a rank-ordered list of clause patterns in descending order of emphasis value. Let me suggest the following:

First Order Emphasis
- □ One-clause sentence

Second Order Emphasis
- □ Member of a coordinate pair (or series)
- □ Main clause of a main-subordinate pair

Third Order Emphasis
- □ Subordinate member of a main-subordinate pair
- □ Embedded relative clause

Fourth Order Emphasis
- □ Clause embedded as subject, complement, or adjunct
- □ Nonfinite clause

Relative emphasis is also affected by linear position in a series of clauses within a single sentence. In general, the first clause in a sentence bears the greatest emphasis, the last clause the next greatest emphasis, and the medial clauses the least emphasis. Both of these factors are at work, then, in showing the comparative stress you as writer place on the ideas expressed in a pair of clauses: the "distance" between them in emphasis value, and their positional relationship in the clause string. You can emphasize certain of your ideas by: (1) couching them in first- or second-order clause patterns, (2) placing them first in the sentence, and (3) de-emphasizing (or subordinating) your remaining ideas.

□ Variety in Sentence Patterns

As we have seen in the preceding sections, varying your sentence patterns makes your written communication effective by allowing you to signal semantic relationships, achieve cohesion and continuity, and emphasize important ideas. But beyond this, variety in sentence patterns has value in and of itself in that it keeps your writing from becoming dull and boring. A couple of decades ago, highway engineers discovered that interstate expressways are safer and more efficient conduits for cars and trucks if they have a few hills and curves than if they are straight and flat. The New

York State Thruway is a good example of such a highway. Its gentle curves and varying profile tend to keep drivers alert and prevent the "highway hypnosis" that is common on the arrow-straight highways of some Western states. The same is true with writing. An endless succession of independent-clause sentences all with a sameness of length and a sameness of syntactic structure make it easy to fall asleep reading. A little variety is valuable just for its own sake. If you find your writing following a dull pattern, try couching some of your ideas as introductory subordinate clauses. Make greater use of nonfinite and elliptical clauses, reserving your independent clauses for main ideas. And finally, insert a judicious sentence adverbial now and then to achieve variety in your sentence beginnings. Don't be afraid of a little variety just for the sake of variety; I've never heard anyone complain that a technical report was too interesting to read.

□ Revising at the Clause and Sentence Level

The ability to write mature, sophisticated sentences is a skill that anyone can acquire, but like any skill it can be acquired only through developing good habits and practicing them. A crucial step in acquiring good writing habits is critical reading. Whenever you encounter something that is, in your judgment, well-written, take a few extra minutes to discover what stylistic choices the writer has made in combining his clauses into sentences. First, identify each of the clauses. Second, note the patterns of coordination, subordination, and embedding—and note particularly the use of nonfinite and elliptical clauses. Finally, note how the writer uses semantic signals and cohesive devices. Next, try applying some of these techniques in your own writing. Practice combining your ideas into different clause patterns. Write out various alternatives. Don't assume that the first way you happen to write something is the only way it can be written. Or the best way. Try different patterns and then select the ones that seem to work the best.

- □ Control your sentence length. Compute the average sentence length in your initial draft. If your sentences are too short, combine these short sentences into longer ones. Get your average sentence length up to around 18 to 20 words.
- □ Scan your draft and circle all the sentences that begin with *th-* words. While you're at it, look for other repetitious beginnings as well. If you have problem with sameness of sentence beginnings, introduce some variety.
- □ Don't overlook the value of an occasional one-clause sentence. Use short, one-clause sentences to emphasize the main ideas of paragraphs or the key concepts of sections.
- □ Strive for variety in the length of your sentences. Balance the oc-

casional short sentence with an occasional long sentence—five or six clauses and 35 to 45 words.

- □ Scan your draft and circle all instances of passive voice. Make sure you can recognize the passive voice when you see it. Convert all back to active voice and then employ the passive *only* where you can justify it. Controlling the passive voice is a key step in revision.
- □ Scan your initial draft for sentences that begin with *It is* or *There are*. In most cases, you can improve your style by eliminating these indirect beginnings and freeing up the embedded clause.
- □ Keep pushing yourself in the direction of more sentence sophistication. If your writing runs long to coordination, try subordinating some clauses. When you get comfortable with subordination, try embedding.
- □ As you embed clauses—particularly as subjects and complements—convert a fair number of these to nonfinite clauses. The ability to use nonfinite clauses comfortably is one of the hallmarks of a mature writer.
- □ Get in the habit of expressing your ideas with the most useful syntactic tool. Don't worry about going too far. If you get carried away with a particular technique, your instructor will let you know, and you can back off a little bit.
- □ Use your writing class assignments as opportunities for practicing new skills and testing new techniques. Doing so is a sound strategy and one that will pay off in improved writing style. And that's what you want out of a writing class, isn't it?
- □ Pay attention to the techniques used by writers whose style you admire. You'll probably find that they use one-clause sentences sparingly (but emphatically), they use nonfinite clauses lavishly, and they use quite a few subordinating conjunctions and sentence adverbials.
- □ Develop the ability to write multiclause sentences. Get your average number of clauses per sentence up to three or four. Remember that as you raise your average number of clauses per sentence, your sentences don't necessarily get longer and longer. There are two reasons for this. First, as you employ more and more nonfinite and elliptical clauses, you take advantage of the syntactic compactness that is inherent in these types of clauses. Second, as you use subordination and embedding more and more, you move away from long noun phrases with loads of prenominal modifiers and strings of prepositional phrases and toward short noun phrases that combine with direct verbs to lend conciseness to your writing.
- □ Keep in mind that sophisiticated sentence structure is not a virtue in itself, nor do long, involved, complex sentences invari-

ably make for clear writing and effective communication. It's just that most student writers need to move in the direction of more syntactic sophistication if they want their writing to be judged to be excellent.

Writing Exercises

1. Select a 250-word sample of technical writing (yours or someone else's). For each sentence in the sample, identify the constituent clauses and classify them according to the list on the first page of this chapter.
2. Read an article in *Scientific American, Nature, Omni, Science,* or another journal that has an editorial policy of excellence in writing. After reading the article, identify and list what you feel to be: (a) the major ideas, (b) the secondary ideas, and (c) the minor ideas. After listing these, return to the article and locate the clauses in which the ideas are first mentioned. Classify each such clause according to the list on the first page of this chapter. Then compare your findings to the discussion in the section on "Relative Emphasis of Clause Patterns," above. Report your findings to the class.
3. Select three 250-word samples each from your corpus of excellent writing and your corpus of minimally adequate writing. Analyze each for average sentence length, sentence length ratio, and average number of clauses per sentence. Do you find any significant differences? Report your findings to the class, together with a brief discussion of your findings.
4. Select three 250-word samples each from your corpus of excellent writing and your corpus of minimally adequate writing. For each sample, calculate the frequency of occurrence of clauses in the passive voice and clauses with indirect beginnings. Do you find any significant differences? Report your findings to the class, together with a brief discussion.

Writing about Writing

1. Write a research paper on the topic "Passive voice." Consult various style guides, technical writing textbooks, business writing textbooks, and freshman composition textbooks. Include a discussion of your findings as they apply to technical writing. Aim your research paper either at your classmates or at faculty and students in your major field.
2. Proceed as in the previous assignment on the topic "Personalness versus impersonalness in technical writing."
3. Write an essay on some aspect of "Style at the clause and sentence level" that you are particularly interested in. Aim the essay either at your classmates or at faculty and students in your major field.

21

Style at the Word and Phrase Level

Words and phrases are the lowest-level units of language that we are concerned about when considering technical writing style. A *word* consists of one or more sounds (or letters representing sounds), while a *phrase* consists of one or more words. Words and phrases function semantically to name and identify things, ideas, actions, and concepts. They function syntactically to fill the major functional slots of clauses (see Chapter 18). The way you choose your words and phrases will have more effect than any other aspect of your writing on your tone and your register, or level of formality. The English language is incredibly rich and varied at the level of words and phrases, with a wide range of synonyms and near-synonyms and expressions that are roughly interchangeable. Mastery of effective style at the word and phrase level is largely a matter of becoming aware of the choices you have and the ways in which your choices affect such aspects of style as personalness, level of technicality, pomposity, precision and vagueness, accuracy, clarity, and general readability.

General and Technical Vocabularies

As you write about any scientific or technical topic, you will use both general words and phrases—the kind you learned in high school—and technical words and phrases—the kind that are peculiar to your profession or discipline or perhaps even the firm or organization that you work for. As you consider your choice of words and phrases, it is wise to keep your general vocabulary and your technical vocabulary separate and distinct. By far the majority of your words and phrases—even in a highly technical piece of writing—are from your general vocabulary. It is as if your technical expressions are projected against a background of your general vocabulary. Most stylistic problems at the word and phrase level are problems with the *general* vocabulary rather than the technical. If you keep a close control of your general vocabulary, your technical words and phrases will pretty much take care of themselves.

Within your technical vocabulary, we can identify two levels—which for convenience we will call simply level-one and level-two. Level-one technical words and phrases are those that are peculiar to your particular profession or discipline, while level-two words and phrases are those that are common to all fields of science and technology. Level-two technical terms are those that you can expect to be familiar to a reasonably bright college graduate who has majored in some field of science or technology. The level-one technical terms are those that are likely to be familiar only to those who are up-to-date experts in your field. In a recommendation report on minicomputers, for instance, words like *program, storage, memory,* and *hardware* would be level-two technical terms; you could expect experts in all fields of science and technology to be familiar with them. On the other hand, expressions like *CRT, peripherals, shared-logic, floppy disks,* and *byte* would be level-one technical terms. *Software,* also, would be a level-one technical term even though it is fairly common, because many people outside the computing field think the term refers to the paper associated with computers—the program writeups, the use manuals, and even punch cards and printer paper. They don't realize that by "software" a computer expert means only the various programs that are available for the particular computer system in question. Or in an article on volcanoes, expressions like *lava, fault, plate, vent, crater,* and *cinder cone* would be level-two terms, while expressions like *scoria, pyroplast, magma, anticline, laterals,* and *amorphous silica* would be level-one terms.

□ Avoiding Pomposity

Of the seven deadly sins of technical writing style, the first five are pomposity. If you can avoid the pretentious, inflated, show-off kind of writing that characterizes much of what is written in scientific and technical circles, you will be well on the way to becoming an excellent writer. Pomposity is contagious. It begins to infect students; it is endemic with scientists and other technical experts; and it reaches epidemic proportions with administrators and bureaucrats. Pomposity is technical writing's worst social disease. Fortunately, however, it is curable.

Using Familiar Words

A giant step toward avoiding pomposity is to use words that will be familiar to the majority of your readers. This is particularly true of your general vocabulary. Figure 21–1 shows the first 100 entries from a list of familiar words. If you select your general or background vocabulary from such a list, you will find it virtually impossible to be pompous. Notice that many of the familiar words are also short. Most readability formulas take note of this fact and urge you to use short words. This is not bad advice as far as it goes, but familiarity rather than shortness is what you want. Multisyllable words like *butterfly, everyone, however,* and *neighborhood* are on the familiar list, while short words like *adz, bleb, emir,* and *hadj* are not. You will also see from such lists that *use* is familiar, *utilize* is not; *most* is familiar, *majority* is not; *size* is familiar, *magnitude* is not; *make* is familiar, *fabricate* is not; *after* is familiar, *subsequent to* is not; *end* is familiar, *terminate* is not; and so forth.

a
able
aboard
about
above
absent
accept
accident
account
ache(ing)
acorn
acre
across
act(s)
add
address
admire
adventure
afar
afraid
after
afternoon
afterward(s)
again
against
age
aged
ago
agree
ah
ahead
aid
aim
air
airfield
airplane
airport
airship
airy
alarm
alike
alive
all
alley
alligator
allow
almost
alone
along
aloud
already
also
always
arm
America
American
among
amount
an
and
angel
anger
angry
animal
another
answer
ant
any
anybody
anyhow
anyone
anything
anyway
anywhere
apart
apartment
ape
apiece
appear
apple
April
apron
are
aren't
arise
arithmetic
arm
armful
army
arose
around
arrange
arrive(d)
arrow
art
artist
as
ash(es)
aside
ask

Figure 21–1 First 100 Entries from a List of Familiar Words

Break the habit of using a four-bit word when a two-bit word will do. Use technical terms as you need them, but make sure your general vocabulary is also familiar. Or at least make sure that familiar words are your first choice and that you can justify every use of a nonfamiliar word in your general vocabulary.

Controlling Jargon

Jargon is the specialized technical terminology that is characteristic of a particular trade, profession, or discipline. When medical doctors speak of *myocardial infarction* and *spontaneous pneumothorax,* or lawyers speak of *habeas corpus* and *laches and estoppel,* or waste treatment biologists speak of *floculation* and *biotic mass,* they are using jargon. Jargon isn't limited to strange and wonderful words, though. The special use of a common word also constitutes jargon. For instance, around the race track a *horse* is a mature virile male animal, as opposed to a mare, or a gelding, or a colt. To call a mare or a gelding a horse is to confess your ignorance of equestrian jargon. Similarly, around the kennels, *bitch* is a polite term. And sailors get grumpy when you call their *ship* a boat.

There is nothing wrong with jargon in communication—even written communication—between members of the same profession or discipline. In fact, it is beneficial; it lends a certain precision and economy to the communication. Problems can arise, however, when jargon is used in communication between members of different disciplines. And since most correspondence and reports and *all* management decision-making documents cross disciplinary boundaries, the excessive and unwarranted use of jargon becomes a barrier to communication in most of the categories of technical communication we are considering. Careful writers control their jargon in all communications that go outside their immediate circle of peers. They refer to "the first hill on the blacktop road" rather than "the first crest vertical curve on the bituminous asphaltic roadway." They refer to "the forehead" rather than "the medial anterior cranium." They refer to "tone of voice" rather than "intonational supersegmental phenomena."

A particularly noxious form of jargon is the myriad abbreviations and acronyms that seem to grow like fungus in technical fields. A sanitary engineer can talk all day about SVI, MLSS, MGD, and BOD, and most people won't have the foggiest idea what he's talking about. Foresters fill their reports with DBHs, MBFs, and CFIs; programmers fill their reports with BYTEs, CPUs, and CRTs; police officers fill their reports with DUIs, MTVs, and SOs; and when the USC&GS talks to the BLM about the conflict between the NPS and the USFS, all is lost. It's worse than listening to your CB radio. Don't use abbreviations and acronyms unless you absolutely have to.

A final observation about jargon. As Richard Mitchell points out, whenever jargon leaks from one discipline to another, it always flows downhill. Less prestigious disciplines borrow jargon from the more prestigious disciplines; never the other way around. Lawyers don't borrow terms from bookkeepers; psychiatrists don't borrow jargon from social workers; licensed amateur radio operators don't use the jargon of truck drivers; and doctors don't borrow terms from ambulance attendants. So when the school superintendent talks about physcial plant, productivity, and efficiency, he is saying something about the relationship between education and business. Or when the health and welfare official talks about holding patterns and final approaches, he is acknowledging the superiority of air traffic controllers. Keep this in mind whenever you are tempted to borrow jargon from other fields.

Control your use of jargon in your writing whenever your audience might include people from outside your discipline (and that means 90 percent of the time). If you need the precision, use the technical term—but define it. If you don't need the precision, use a less specialized term. Avoid abbreviations, initials, and acronyms; the space they save isn't worth the confusion they may introduce. Above all, avoid borrowing jargon from other disciplines. Don't write to impress your readers; write to inform them. Keep in mind the classic story about the plumber who wrote to the Bureau of Standards to ask if he should use hydrochloric acid to clean clogged drains. One of the bureau's specialists replied "The efficacy of hydrochloric acid is indisputable, but the corrosive residue is not compatible with metalic permanence." The plumber wrote again, saying he was happy the bureau agreed with him. Another technical expert answered as follows: "We cannot assume responsibility for the production of toxic and noxious residues with hydrochloric acid and advise that you implement an alternative procedure." The plumber again thanked the bureau for supporting his idea. In desperation, the bureaucrats turned to the building janitor, who wrote "Don't use hydrochloric acid. It eats the hell out of the pipes."

Controlling Abstract Nouns

Those abstract nouns derived from verbs and adjectives—the ones ending in *-tion* or *-ment*—can go a long way toward constipating your prose and making it stuffy and pretentious. The examination and evaluation of communication by those who have education in computation and measuration give rise to the speculation that considerable cogitation and cerebration are required for identification of designation and denotation. Moreover, the commencement of infringement of the environment has led to the abatement by the establishment of advancement in containment achievement measurement assessment. Get the picture? These *-tion* and *-ment* nouns make for pretty pompous prose—especially when you have two or three (not to mention 10 or 12) in a row. What is more, nearly every one of these derived abstract nouns covers up a perfectly good verb: *examine, evaluate, measure, identify, commence, establish, achieve,* and *assess.*

Learning to control abstract nouns ranks right with controlling the passive voice as an aspect of technical writing style that science editors and authors of style guides stress. Here are some of the things they have to say:

> The frequent use of nouns formed from verbs and ending in "-tion" produces unnecessarily long sentences and dull prose. . . . The excessive length results in part from the length of the "-tion" nouns and in part from the need to use unnecessary verbs with them. The dullness results from the abstractness of these nouns and the usually passive, weak verbs that accompany them. In reviewing your manuscript, look for the "-tion" nouns that can be replaced in rewritten sentences by their more direct and forceful verb equivalents.
>
> *Council of Biology Editors Style Manual* (1978)

> Another frequent problem in scientific writing is the verbosity that results from use of abstract nouns. This malady is corrected by turning the nouns into verbs.
>
> *How to Write and Publish a Scientific Paper*, Robert A. Day, 1979. (Robert A. Day is managing editor, Publications Board of the American Society for Microbiology.)

> What is it that makes most scientific writing so preeminently dull? I believe it is the failure to use expressive verbs. . . . Yet scientists . . . show an inexplicable urge to use gerunds, abstract nouns derived from verbs, or noun phrases—anything *but* a verb, in fact—to do the work in a sentence. . . . Your first task is, then, to . . . recognize and then *release the hidden verb*, that sleeping beauty so often locked up in a bland ivory tower of an abstract noun.
>
> *Scientific Writing for Graduate Students*, F. Peter Woodford, ed., 1968. (F. Peter Woodford is chairman of the CBE Committee on Graduate Training in Scientific Writing.)

Abstract nouns have a habit of teaming up with weak verbs, like *be*, *do*, *make*, and *use*. The pompous writer doesn't *examine* things, he *makes an examination of* them; he doesn't *compute* answers, he *does computations of* them; he doesn't *evaluate* alternatives, he uses *evaluation instruments for their assessment*. Watch out for this combination of anemic verb and abstract noun. Almost without exception, your writing will be less pompous—and more direct and concise as well—if you avoid the abstract noun and use the underlying verb instead. Now, as the last two sentences illustrate, an occasional "combination" or "exception" doesn't clutter up your writing all that much. And then we have those nouns such as *environment* or *nation* or *portion* for which the language has lost the underlying verbs. Finally, there are times when you prefer the abstract noun and weak verb for reasons of rhythm or emphasis or variety. The key word is *control*. Limit your use of abstract nouns to those cases where they enhance your communication. (Or enhance the way you communicate.) If you have no strong feelings one way or the other, choose the underlying verb. Above all, avoid strings of *-tion* and *-ment* nouns. Strive for a verbal rather than nominal writing style.

□ Controlling Your Level of Formality

As you recall from Chapter 3, register in language—like fashion in men's wear—ranges in almost infinite gradation from extremely formal to extremely casual. For convenience of discussion, we can block the continuum into three more or less discrete categories—formal, standard, and casual—with subdivision as follows:

- Formal
  - Ceremonial
  - Contractual
- Standard
  - Cultured
  - Consultative
  - Conversational
- Casual
  - Colloquial
  - Crude

The extremes of register—ceremonial and crude—are rarely found in technical writing; hence controlling your level of formality is a matter of letting your register move up and down from conversational to cultured, with occasional excursions into contractual and colloquial language. Level of formality is, or course, related to pomposity (see above) and tone (see below). It is hard to sustain a contractual or cultured register if your writing is familiar and free from jargon. It is hard to sustain a conversational or colloquial register if your writing is impersonal, detached, exclusive, or highly technical. If fact, so far as words and phrases go, register is established largely through contractions, plain and fancy words, and colloquial or colorful expressions.

Controlling Contractions

Interestingly enough, the simple matter of contractions is a significant index of register. Consider the passage in Figure 21–2 (page 240). Rather formal, isn't it? Suppose we do nothing to this passage except plug in contractions wherever we can, as in Figure 21–3. This change alone drops the register somewhere down around the conversational or colloquial levels, doesn't it? Here are some guidelines for contractions as they relate to register:

- □ *Ceremonial and Contractual Registers*
  No contractions whatever.
- □ *Cultured Register*
  Occasional contractions with *not:* can't, don't, won't, isn't.

The API Engine Service Classification System, as it is designated, characterizes five levels of gasoline engine oils (SA-SE) and four levels of diesel engine oils (CA-CD). The S denotes spark ignition, while the C denotes compression ignition. Let us look briefly at the S series to determine what those letters on a can of oil mean.

SA is the lowest grade; it is additive free, with the possible exception of a pour-point depressant and a foam inhibitor. It has no performance requirements whatsoever. SB is typical of oils available since the Thirties. It has some antioxidant and antiwear capabilities. SC characterizes oil designed to meet 1964-1967 engine warranties. It is beginning to show control of high- and low-temperature deposits, rust and corrosion. SD, 1968-1971 warranty oil, exceeds SC in all categories. Finally, SE meets engine warranties of cars built since 1972. Its base stock blending and complete additive package give high-temperature oxidation protection, low-temperature anti-sludge and all the other benefits we have come to expect from a can of oil.

Let us suppose that two oils, one refined and the other synthetic, are to be compounded for equivalent performance. (As we will see shortly, economic considerations imply that this is rarely the case.) Also, we will not assume precisely the same materials as additive components: What is best for one base may not be optimal for another. Instead we will discuss the lubricants in terms of functional categories.

Figure 21–2 Passage in Formal Register

□ *Consultative Register*
More frequent contractions with *not,* including: didn't, wouldn't, shouldn't.
Occasional contractions with *will:* I'll, he'll, you'll.
Occasional contractions with *us*: let's, or *is:* it's, there's.

□ *Conversational Register*
Frequent contractions with *not,* including: oughtn't, mightn't.
Frequent contractions with *will.*
Frequent contractions with *us* and *is.*

> The API Engine Service Classification System, as it's designated, characterizes five levels of gasoline engine oils (SA-SE) and four levels of diesel engine oils (CA-CS). The S denotes spark ignition, while the C denotes compression ignition. Let's look briefly at the S series to determine what those letters on a can of oil mean.
>
> SA is the lowest grade; it's additive-free, with the possible exception of a pour-point depressant and a foam inhibitor. It's got no performance requirements whatsoever. SB is typical of oils available since the Thirties. It's got some antioxidant and antiwear capabilities. SC characterizes oil designed to meet 1964-1967 engine warranties. It's beginning to show control of high- and low-temperature deposits, rust and corrosion. SD, 1968-1971 warranty oil, exceeds SC in all categories. Finally, SE meets engine warranties of cars built since 1972. Its base stock blending and complete additive package give high-temperature oxidation protection, low-temperature anti-sludge and all the other benefits we've come to expect from a can of oil.
>
> Let's suppose that two oils, one refined and the other synthetic, are to be compounded for equivalent performance. (As we'll see shortly, economic considerations imply that this is rarely the case.) Also, we'll not assume precisely the same materials as additive components: What's best for one base may not be optimal for another. Instead, we'll discuss the lubricants in terms of functional categories.

Figure 21–3 Formal Passage with Contractions Introduced

Occasional contractions with *have*: I've, you've.
Occasional contractions with *are*: you're, they're, we're.

- *Colloquial Register*
  Frequent contractions with *not* and *will*.
  Frequent contractions with *us* and *is*.
  Frequent contractions with *have* and *are*, including: should've, might've.
  Frequent contractions with *had*: we'd, they'd, you'd.

- *Crude Register*
  No limit on contractions.

Plain and Fancy Words

Another important index of register is the use of eloquent diction, or fancy words. Writers operating in the ceremonial, contractual, and cultured registers tend to show off a little bit with their vocabularies. They tend to say things like "eloquent diction" rather than "fancy words." But when communication situations call for language in the consultative or conversational registers, careful writers tend to favor plain, simple, unadorned words. The distinction between plain and fancy words is closely related to the distinction between familiar and unfamiliar words that we considered under "Avoiding Pomposity," above. Even within the ranks of familiar words, however, some are just a bit more elegant than others. And some, too, are a little on the crude side.

As you revise, you can control your register by selecting words and phrases that play up and down the levels of formality. Consider the simple matter of going to the bathroom. To say that "Johnny voided urine in his trousers" is contractual, or perhaps even ceremonial in the medical circles. To say that he "urinated in his pants" is cultured, while "Johnny wet his pants" is consultative. I'm sure you can think of examples yourself at the colloquial and crude levels of register. Or consider the devices in Figure 21–4. We can refer to these as "spectacles" (ceremonial), "corrective lenses" (contractual), "eyeglasses" (cultured), "glasses" (con-

Figure 21–4 What You Call These Depends on Your Register

The API Engine Service Classification System, as it's called, characterizes five levels of gasoline engine oils (SA-SE) and four for diesels (CA-CD). The S refers to spark ignition; the C to compression ignition. Let's take a brief look at the S series, just to see what those letters on a can of oil mean.

SA is the lowest grade; it's additive-free, with the possible exception of a pour-point depressant and a foam inhibitor. It's got no performance requirements at all. SB is typical of the oils available since the Thirties. It's got some antioxidant and antiwear capabilities. SC characterizes oil designed to meet 1964-1967 engine warranties. It's starting to show control of high-and low-temperature deposits, rust and corrosion. SD, 1968-1971 warranty oil, does everything SC does, only better. Finally, SE meets engine warranties of cars built since 1972. Its base stock blending and complete additive package give high-temperature oxidation protection, low-temperature anti-sludge and all the other benefits we've come to expect from a can of oil.

Let's suppose that two oils, one refined and the other synthetic, are to be compounded for equivalent performance. (As we'll see shortly, economic considerations imply that this is rarely the case.) Also, we'll not assume precisely the same materials as additive components: What's best for one base may not be best for another. Instead, we'll talk about the oils in terms of functional categories.

Figure 21–5 Formal Passage with Plain Words Introduced

sultative), "specs" (conversational), or "cheaters" (colloquial). Select the register that is appropriate for your situation.

We can return to our synthetic lubrication example (see Figure 21–2). By substituting a few plain words for some of the fancy words, as in Figure 21–5, we further decrease the formality of the register.

Colorful Expressions

One final aspect of register deserves mention. Virtually all fields of endeavor develop over the years a collection of vivid, colorful expressions that practitioners of the field use to describe objects and actions connected with the field. Often, these expressions involve imagery and metaphor that can contribute to one's understanding of what is being discussed. For the most part, though, such expressions are labeled "jargon," "colloquial," or "slang." In the construction business, for instance, a motor grader becomes a "blade," a scraper becomes a "can," wet concrete becomes "mud," a sledge hammer becomes a "beater," and a cutting torch becomes a "gas axe." Or around the automotive repair shop, engines become "mills," carburetors become "jugs," manual transmissions become "stick shifts," and an exhaust system becomes a "set of pipes."

The use or nonuse of such expressions is not a question of right or wrong; it is a matter of register. The judicious inclusion of a few such expressions can give an otherwise consultative passage a distinctly conversational or colloquial flavor. Likewise, the strict avoidance of colorful expressions makes for a rather more contractual or cultured register. Let's look once again at the lubrication example and see the impact of adding a few colorful expressions from the field of motor racing (Figure 21–6).

□ Controlling Tone

The tone of your writing, like tone of voice, is not so much a matter of what you say but how you say it. The adjectives we use to describe the tone of a written passage are much the same as the ones we use to describe people, or personalities: lively or dull, friendly or hostile, personal or impersonal, positive or negative, aggressive or meek, open or reserved, and so on. Actually, tone varies continuously along each of these parameters (and others as well); the labels we choose to describe tone tend to be the endpoints of the scales and show the extremes. This is not to suggest, however, that pairs of labels like "personal" and "impersonal" set up distinct, mutually exclusive categories of tone. Tone, like register, varies up and down the scale.

For most types of technical writing we want a neutral or slightly positive tone. Effective technical communication is not coldly impersonal, but it is not chummy, either. It is confident but not cocky, lively but not exuberant, authoritative but not dogmatic, courteous but not condescending. Here are some things to keep in mind as you revise to control the tone of your writing.

Avoiding Emotional Words and Phrases

A good many years ago, Bertrand Russell offered several of what he called "emotive conjugations" to illustrate the range of connotations with synonyms and near-synonyms. Here are some examples:

> I am firm; you are obstinate; he is pig-headed.

The API Engine Service Classification System, as it's called, characterizes five levels of gasoline engine oils (SA-SE) and four for diesels (CA-CD). As you might guess, the S refers to spark ignition; the C to compression ignition. Let's take a brief look at the S series just to see what those letters on a can of oil mean to your modified 455.

SA is the lowest grade; it's additive-free, with the possible exception of a pour-point depressant and a foam inhibitor. It's got no performance requirements, and you get the feeling your old lawnmower deserves better. SB is typical of oils available since the Thirties. It's got some antioxidant and antiwear capabilities. SC characterizes oil designed to meet 1964-1967 engine warranties and can handle the high-compression, high-rev engines of that era. SD, 1968-1971 warranty oil, does everything SC does, only better. Finally, SE meets engine warranties of cars built since 1972. Its base stock blending and complete additive package give high-temperature oxidation protection, low-temperature antisludge and all the other goodies we've come to expect from a can of oil.

Let's suppose that two oils, one refined and the other synthetic, are to be compounded for equivalent performance. (As we'll see shortly, economic considerations imply the game is played somewhat differently.) Also, we'll not assume precisely the same materials as additive components: What's best for one base may not be best for another. Instead we'll talk about the oils in terms of protection for your power mill.

Figure 21–6 Formal Passage with Colorful Expressions Introduced

I have reconsidered; you have changed your mind; he has gone back on his word.

I am stocky; you are overweight; he is fat.

I have missed a payment; you are delinquent; he is a deadbeat.

I am a maintenance engineer; you are a mechanic; he is a grease-monkey.

In each case, all three terms mean more or less the same thing, but the "second-person" and "third-person" terms conjure up negative connotations, or bad vibrations, in the minds of the readers. Such words have a high emotional content. They are similar to the "red flag" words in correspondence (see Chapter 10) and they have much the same effect on readers—they tend to alienate your readers, or turn them off. This you want to avoid, because if you lose your audience—for whatever reason—you can no longer communicate with them. Be direct in your writing (see below) but stop short of being blunt. Exercise a little delicacy.

Using Personal Words

Mediocre technical writing suffers from (among other things) being too impersonal. A good many of the reports, proposals, and articles churned out these days present data and conclusions, describe objects, and portray events as if there were no human involvement whatsoever in scientific and technical pursuits. This phenomenon is mainly a historical one, having its roots in the first third or so of the twentieth century. Eighteenth- and nineteenth-century scientists were quite personal in their writing—read Lavoisier, for instance, or Faraday or Darwin or Mendel or Edison. During the 1910s and 1920s, scientific writing became much more detached and impersonal, largely because of the influence of the empirical method in science and the demands of objectivity. By the 1930s and early 1940s even the textbooks and style guides were calling for impersonal writing and advising writers to avoid first- and second-person pronouns and names of people. The trend shifted in the late 1940s, however, at least so far as the textbooks and style guides were concerned. Authorities on scientific and technical writing style recognized the excesses of previous decades and began to recommend the more natural writing style and tone that had been common for two centuries prior to 1900. In fact, not a single major technical writing textbook or style guide published after 1948 advocates an impersonal tone for scientific and technical documents. On the contrary, the overwhelming majority of them recommend a natural, personal tone.

But there has been a great deal of inertia in this aspect of technical writing style. In the 1940s, World War II and the beginnings of the aerospace programs gave rise to a tenfold increase in technical publications—all written by people schooled in the 1930s. Even into the 1950s and 1960s, publications managers, dissertation directors, and even some journal editors hung tenaciously to the writing styles they had learned and insisted that young writers conform. Style guides, textbooks, and English teachers to the contrary, stark impersonality continued to be the dominant tone for scientific and technical writing. Even today, many student writers believe as a matter of faith that technical writing ought to be impersonal and that somehow it is wrong to say *I* or *me* or *you* in a technical report.

There is nothing wrong with a personal tone. In fact, a little personalness in your writing makes it smoother, more interesting to read, and easier to remember. As we enter the last decades of the twentieth century, I urge you to throw off the fifty-year-old constraints of artificial impersonalness and trust your own instincts—if not to follow the advice of the experts. Be personal and be natural in your writing—especially in your correspondence and in those portions of your report where you are offering interpretations of data and conclusions. Use first- and second-person pronouns; use names of people; and use personal words like *father, daughter, brother, child, family*. Your writing will be more effective if you let your tone hit at least a midpoint on the scale of personalness.

□ Achieving Precision and Directness

Technical writing is not a matter of putting words on paper; it is a matter of transferring technical information from the mind of a writer who has it to the minds of readers who do not. To be effective, the writing should transfer precisely the information that the writer has in mind and furthermore should do so in the most direct manner possible. The way you choose your words and phrases will have a great deal to do with how precise and direct your writing is. You can revise your writing for precision and directness by using concrete and specific nouns and modifiers, by preferring dynamic verb phrases to strings of modifiers and prepositional phrases, and by avoiding redundancy and circumlocution.

Using Concrete and Specific Words

Concreteness and specificity go hand in hand. Words that are concrete and specific stand in opposition to words that are abstract and general, although abstraction and generality are not quite the samc thing. Words range along the scale of concreteness/abstractness according to how far what they refer to is removed from the real world. Concrete words refer to physical, tangible objects, while abstract words refer to intangible ideas and concepts. *Concrete*, the stuff your foundation is made of, is quite concrete; *concreteness*, on the other hand, is an abstract idea. *Concrete* is also specific (although *5/8–minus aggregate Portland cement concrete* would be more specific); *stuff* is quite general. The distinctions tend to blur, however.

It is interesting to see how the words you can choose to refer to a given object can range up and down what S. I. Hayakawa calls the ladder of abstraction (or generality, if you will). Let's take the family car. Here are a set of words and phrases you can use—ranging from concrete and specific to abstract and general:

My 1978 Datsun B210	The blue one, serial number 123456789, with 55,436 miles on it and a dent in the left rear fender.

A 1978 Datsun B210	Any of several tens of thousands manufactured that year.
A Datsun	Any of several hundred thousand manufactured by Datsun.
A compact car	Millions.
A passenger car	Includes Cadillacs and Rolls-Royces.
An automobile	Includes station wagons, trucks, and buses.
A vehicle	Includes trains, boats, and airplanes.
A means of transportation	Includes horses, bicycles, and kayaks.

Or how about tools:

An S-K 9/16″ combination wrench

A 9/16″ combination wrench

A 9/16″ wrench

A wrench

A tool

Something to take this nut off with

Or how about working from abstract and general to concrete and specific with something like an air-conditioned office:

Comfortable

Cool and well-ventilated

70–75 degrees F., 50% relative humidity, and 1750 cfm per person ventilation

The biggest problem that novice writers have in this area is using words and phrases that are much less concrete and specific than the ideas they have in mind. They say "inexpensive" when they have in mind "less than $2.00 a pound"; they say "behind schedule" when they mean "eight working days behind schedule"; they specify a "suitable vehicle" when they want a "3/4–ton four-wheel drive pickup"; or they tell you to use a "container" when they want you to use a "250-ml Erlenmeyer flask." It's pretty hard to communicate any information when you keep your words abstract and general. Consider the fellow who just came back from a vacation on the West Coast. His neighbor asked him where he went surfing. "In the ocean," he replied. And what are they wearing on the California beaches? "The latest fashions." And what do they grow in the Imperial Valley? "Crops." And the factories—what do they make in them? "Manufactured goods." And what do they drive on the freeways? You guessed it: "Cars." Lots of words—no communication.

Make sure your words are as concrete and specific as the ideas you have in your head. Don't draw big circles around fine points. Say precisely what you mean—even if you have to use a few unfamiliar words to do so. As with most aspects of writing style, there are some trade-offs at work here.

Don't go overboard, though. When your words are more precise than your thoughts, you are getting pompous. Besides, communciation may suffer. If what you want is something to catch oil in, you don't worry about whether it's a pan or a pail, plastic or metal, shallow or deep—just so long as it will hold five quarts.

Controlling Modifiers

Another barrier to precision and directness is the stacking of three or more modifiers in front of a noun. With a single modifier, we know that the modifier modifies the noun:

electric typewriter
coffee cup
anaerobic bacteria

With two modifiers in front of a noun, we can usually figure out what's going on:

modern electric typewriter
plastic coffee cup
beneficial anaerobic bacteria

although there can be cases of ambiguity with two modifiers:

plastic tape dispenser
green beer can
hydraulic piston puller

Whenever we find three or more modifiers ahead of a noun, however, we are likely to find confusion as well:

retarded child study team
ABF pump discharge control valve
supersonic wind tunnel air storage tank
engineering college advance placement information
diesel engine truck driver training school instructor

The reason for this is simple: each time we add a modifier to a string, we multiply the possibilities for structural interpretation. For *n* modifiers in front of a noun, there are *n!* structural interpretations:

for 1 modifier, 1 interpretation
for 2 modifiers, 2 interpretations
for 3 modifiers, 6 interpretations
for 4 modifiers, 24 interpretations
for 5 modifiers, 120 interpretations
for 6 modifiers, 720 interpreations
for 7 modifiers, 5040 interpretations

As you can see, the number of possible structural interpretations climbs pretty fast once you get past three. Now, context will eliminate some interpretations and logic some others. But why force your readers to figure out the patterns of modification when it is just as easy to break up strings of modifiers. Just move some of the modifiers around *behind* the noun:

dispenser of plastic tape
air storage tank for the supersonic wind tunnel
information for advance placement in the college of engineering
instructor in the training school for drivers of diesel engine trucks

By doing so, you increase the length of the noun phrase somewhat by adding a preposition or two; but this is a small price to pay for increasing the effectiveness of your communication.

As you revise your initial draft for precision and directness, scan for instances of three or more modifiers ahead of a noun. Be particularly wary of strings of noun modifiers (or nouns functioning as modifiers); these seem to cause the most trouble. If the phrase is a standard expression in your discipline or if there is *no* possibility of misinterpretation, leave it alone. Otherwise, break it up by moving one or more of the modifiers to a position behind the noun:

front engine mounts = mounts for the front engine

hydraulic piston puller = puller of hydraulic pistons

ABF pump discharge control valve = control valve for the ABF pump discharge

retarded child study team = study team for the retarded child

Controlling Prepositional Phrases

Verb-oriented writing is direct and straightforward. Writing that gains its sentence length through long noun phrases, however, tends to be roundabout, static, and dull. A chief source of length in noun phrases is prepositional phrases. Consider the passage in Figure 21–7. As you can see, somewhat over half the words in the passage are in prepositional phrases. Compare this passage to the one in Figure 21–8, where fewer than 20 percent of the words are in prepositional phrases.

Run a spot check on your own writing by selecting some samples and underlining all the prepositional phrases. Then count to see what percentage of your words appear in prepositional phrases. If you are running much more than 40 percent or so, you may want to convert some of the prepositional phrases to clauses—particularly nonfinite clauses.

Avoiding Redundancy and Circumlocution

Redundancy is unnecessary repetition. Circumlocution is good old beating around the bush. There are no firm rules for deciding when repetition is necessary (or useful) or when extra words make your communication more effective (see "Information Density" in Chapter 19). Rather, avoiding redundancy and circumlocution is largely a matter of developing a

This is a report of the status of the radio tracking project during the month of July. In the beginning of the project, we looked into the possibility of using collars from the manufacturers of other equipment in our possession, but later in the project, we went ahead with the use of collars for radio tracking from other manufacturers in different areas of the overall investigation. The status of the project for the month of July is one of being more or less satisfactory, with work on the design of collars and maintenance of the radio receivers progressing with a reasonable degree of success in these areas. The problem of locating herds of elk and other big game in the areas near the foothills of the Sawtooth Mountains in the central portion of the state is in a state of nearing completion.

Figure 21–7 Passage with Too Many Prepositional Phrases

Those who have the habit of correctness and precision can do things by design; those who don't usually have to depend on luck. And when we fly in airplanes or undergo surgery or file our tax forms, we feel better if we can depend on something more than luck. It isn't luck that rings the right phone in Honolulu--it's simply correct dialing. (Getting a dial tone is luck.) We are a notably superstitious people, but we aren't superstitious enough to believe that our keys fit our locks by a marvelous stroke of good fortune. Our world is more and more crowded with things that will work only if lots of people have been correct and precise. That they seem to work less often and

Figure 21–8 Passage with Relatively Few Prepositional Phrases

sensitivity to wordy expressions. Figure 21–9 lists a number of common wordy expressions. Study the list. Then scan your own writing for these or similar expressions. Get in the habit of boiling the fat out of your writing and employing single words or direct phrases in place of redundant or roundabout expressions. If you do so, your writing style will improve.

future plans	plans
general consensus of opinion	consensus
end result	result
at this point in time	now
due to the fact that	because
in the event that	if
have need for	need
at a rapid rate	rapidly
in close proximity to	near
aware of the fact that	know
an overwhelming majority of	most
it is believed that	I think
purple in color	purple

Figure 21–9 Some Wordy Expressions to Avoid

□ Writing to Enhance Memory and Recall

Excellent writing is not only easy to read and interesting to read, it is easy to remember; and there are a number of techniques you can employ to make your writing easier to remember. Some of these techniques operate at higher levels of language structure. Using headings and subheads on your sections and subsections and maintaining thematic unity throughout your sections and subsections go a long way toward making your writing easier to remember (see Chapter 19). Using short, one-clause sentences to emphasize major ideas also enhances memory and recall (see Chapter 20). Similarly, using sentence adverbials to signal logical relationships (*first, second, last; therefore, however, on the other hand*) makes your arguments and line of reasoning easier to follow and hence easier to remember.

At the word and phrase level, one thing you can do to enhance memory and recall is to break out sequences of words and phrases as vertical lists rather than presenting them in paragraph form. Which of the following is easier to remember?

> The alternatives considered for this report were the Black & Decker Model 7190, the Skil Model 2002, the Skil Model 2022, the Wen Model PR53, and the Black & Decker Model 7127. These were evaluated according to cost, speed range, motor amperage, safety features, and availability of parts and service.

(1) The alternatives considered for this report were:

- □ Black & Decker Model 7190
- □ Skil Model 2002
- □ Skil Model 2022
- □ Wen Model PR53
- □ Black & Decker Model 7127

(2) These alternatives were evaluated according to the following criteria:

- □ Cost
- □ Speed range
- □ Motor amperage
- □ Safety features
- □ Availability of parts and service

Another thing that helps memory considerably is referring to alternatives by manufacturer's name rather than by model number. People can remember and keep straight designations like *Vivitar, Sunpak,* and *Minolta* much more easily than designations like *Model 225, Model 120C,* and *Model 128*. If you have several models by the same manufacturer, use both manufacturer's name and model number: *the Vivitar Model 225, the Sunpak Model 120C, the Vivitar Model 283, the Minolta Model 128,* and *the Sunpak Model 121C.*

As you employ successive headings and subheads or develop lists of items, maintain rigorous grammatical parallelism. Similarly, as you mention main points in successive paragraphs, present each main point in the same grammatical form: a noun phrase, a participial clause, or a full sentence.

Figures of speech can make your writing easier to remember. We usually associate figures of speech with creative or expressive writing, but you can use them profitably in communicative writing. Some of the more common figures of speech in technical writing are shape images:

- □ U-bolt
- □ A-frame
- □ cloverleaf (intersection)
- □ square wave

Many writers of scientific and technical information use figurative comparisons as well—analogies, similies, and metaphors. One writer might compare a lunar landing with hitting a moving Ping-Pong ball at 1000 meters with a BB gun. Another writer might suggest that reading environmental impact statements is about as interesting as watching paint dry. Still another may liken writing a computer program to drafting a report. An occasional vivid image or figurative comparison can have the effect not only of perking up your writing and making it more interesting to read but also making certain points easier for your readers to remember. Be wary of worn-out, tired figures of speech, though. If you find yourself writing *tried and true, many and varied, tight as a drum, old as the hills,* and the like, develop some fresh images or eliminate the figurative language altogether. Remember: If you've heard it before, chances are that your readers have heard it before. Don't bore them.

Repeated sound patterns can both help and hinder your writing. Intentional repetition of beginning sounds—alliteration—or repetition of ending sounds—rhyme—can aid your efforts to be remembered. This is a legacy from the days of oral traditions with language. Alliterative, rhythmic, rhyming prose tends to be easy to remember. On the other hand, unintentional alliteration and rhyme can call attention to themselves and distract your readers: *the speed of the spacecraft supplies sufficient scooping; we must stop the drain of brainpower from the West, lest we fall behind in the space race.* Once again, what you must do is develop a sensitivity to sound patterns. Read your writing aloud—or better yet, have someone else read it in your presence. Listen for sound patterns. Retain those that help, and remove those that hinder.

□ Revising at the Word and Phrase Level

Attending to diction, or your choice of words and phrases, is the third wave in your revision process. If you are following the game plan, you have already revised for excellence at the paragraph and section level (Chapter 19) and at the clause and sentence level (Chapter 20). As with the other levels, begin with your executive summary (if you have one) and your introduction. Then work on your summary. Finally, if the communication situation warrants it, revise the diction in your main body and your appendixes. Here is a procedure you can follow to help you eliminate pomposity, control your tone and register, achieve precision and directness, and write to enhance memory and recall.

- □ Review your analysis of the communication situation. Are your ten adjectives for tone still appropriate (see Figure 21–10)? Your level of register? Your level of technicality? Your degree of personalness?
- □ Isolate your technical vocabulary by underlining all technical words and phrases. Do you have too many level-one technical terms? Can you substitute level-two technical terms or general terms? Have you defined all remaining level-one technical terms?
- □ Consider your general vocabulary. Are all your general words and phrases from the familiar list? If not, can you justify the exceptions? In as many instances as you can, substitute familiar words for unfamiliar words; then see if you like your revised version better than your preliminary version.
- □ Circle all instances of jargon—especially jargon borrowed from other fields. Do you have too many circles? Can you substitute shorter general words for longer jargon phrases?

Adjectives for Tone and Approach		
abstract	detached	obscure
ambiguous	devious	orderly
authoritative	dogmatic	personal
balanced	direct	positive
blunt	dull	precise
boring	emotional	pompous
breathless	general	remote
brisk	humorous	rude
bureaucratic	interesting	scholarly
chaotic	"it"-centered	serious
chummy	"I"-centered	simplistic
clear	impressive	slow
cocky	informative	specific
concrete	messy	straightforward
condescending	muddled	stuffy
confident	negative	subjective
confusing	neutral	superficial
convincing	natural	"you"-centered
courteous	objective	wishy-washy
dense	obsequious	

Figure 21–10 Adjectives for Describing Tone and Approach

- □ Scan for weak verbs and derived nouns. Wherever you can, eliminate the weak verb and substitute the muscular verb that underlies the derived noun.
- □ Flag in red all your red-flag words. This takes a little empathy: remember you're not looking for words that offend *you;* you're looking for words that might offend one of your readers. Eliminate all such—unless you are deliberately trying to write something to be belligerent, provocating, or antagonistic.
- □ Underline all your prepositional phrases. Is close to half of your total word count buried in prepositional phrases? If so, convert as many as you can to single-word modifiers or nonfinite clauses (see Chapter 20).
- □ Underline all your auxiliary verbs. How many have you used with contractions? Adjust the ratio, if necessary, to match your desired tone and register level.
- □ Do the same for first- and second-person pronouns and other personal words: count, assess, and adjust as necessary. Then count, assess, and adjust your colloquial expressions. With these two categories you may well want to look for opportunities to insert personal words or colorful expressions if your analysis of the communication context calls for language that is more per-

sonal and conversational than what you have written in your preliminary draft.

- □ Scan for redundancy and circumlocution. Do away with all instances of unnecessary repetition and beating around the bush.
- □ Underline all instances of three or more modifiers stacked ahead of a noun. Break these up by converting one or more of the prenominal modifiers to a prepositional phrase.
- □ Look for other unnecessary or weak modifiers—particularly intensifiers that don't intensify and limiters that don't limit. Pay particular attention to "cover-your-tail" statements. Do you really have to cover your tail?
- □ Run a few of your key nouns up and down the scale of abstraction. Can you profit from nouns that are more concrete or specific? More abstract or general? Do the same for some of your key modifiers.
- □ Scan your copy for series of three or more elements—words or phrases. Can you break these out as vertical lists rather than horizontal strings?
- □ Look for such series as "the three main points," "the major alternatives," "the five criteria," "the key characteristics," or "the important conclusions." Maintain rigorous grammatical parallelism as you list and develop such items.
- □ Scan for images and figures of speech. Can you profit from more shape images? Alliteration? Simile and metaphor? Are there devices that help *you* remember important points? If so, share these with your readers. Are your images predictable? If so, substitute fresh expressions for clichés or just eliminate tired, worn-out expressions altogether.
- □ If you are writing instructions or procedural descriptions check for telegraphic writing style. If you find yourself guilty, insert all necessary function words to achieve normal English.
- □ Run a final check for smoothness. Have someone read your report aloud, and note all places where the reader hesitates, stumbles, or backs up. Scan for annoying or distracting alliteration or rhyme or repetition of identical or similar words. Eliminate these as necessary. Pay attention to the rhythm and meter of your prose and adjust where necessary by expanding, contracting, or altering your sentence structure.

Writing Exercises

1. Select a 250-word sample of technical writing from your field or discipline. Circle all words that you judge to be technical vocabulary rather than general vocabulary. For the remaining general vocabulary words, underline all those that are not on the familiar list. For as many of these as you can, substitute a familiar word or phrase.

2. Select a 250-word sample of technical writing from your field or discipline. Circle all instances of what you judge to be: (a) unwarranted jargon, and (b) pompous words or phrases. For as many of these as you can, substitute clear, straightforward expressions.
3. Select a 250-word sample of technical writing from your field or discipline. Circle all abstract derived nouns. Underline all of these that are teamed up with weak verbs. For as many of these as you can, eliminate the derived noun and weak verb by freeing up the verb that underlies the derived noun.
4. Find a sample of technical writing from your field or discipline that is written in a ceremonial or contractual register. Revise it by changing the register to consultative or conversational. Attach a copy of the original.
5. Find a sample of technical writing from your field or discipline that is written in a conversational or colloquial register. Revise it by changing the register to cultured or contractual. Attach a copy of the original.

Writing about Writing

1. Write a library research paper on the topic "Abstract nouns." Consult various style guides, technical writing textbooks, business writing textbooks, and freshman composition textbooks. Include a discussion of your findings as they apply to technical writing. Aim your research paper either at your classmates or at faculty and students in your major field.
2. Proceed as in the previous assignment on the topic "Jargon in technical writing."
3. Proceed as in the previous assignment on the topic "Pomposity in technical writing."
4. Write an essay on some aspect of "Style at the word and phrase level" that you are particularly interested in. Aim your essay at your classmates or at faculty and students in your major field.
5. Write an essay on the topic "The scholarly journal article as style guide for management decision-making documents."

22

A Method for Analyzing Technical Writing Style

Here is a technique you can use to analyze the style of a report or other piece of technical writing according to the major characteristics of style presented in the preceding three chapters. Try it out on some of the samples of technical writing presented in Part Three. Then try it out on samples from textbooks, reports, and articles you encounter in your major field or in your casual reading. Finally—and most importantly—analyze several samples of your own writing—particularly those writing assignments that your instructor has judged to be adequate but less than excellent. It will give you a good assessment of your strengths and weaknesses and provide specific answers to the question of what you should do to improve your writing style. The analysis technique is described in detail here and summarized later in the chapter.

Procedures for Analyzing Style

Sample Selection

1. Select a 200- to 250-word sample of the writing you wish to analyze. When selecting a sample, keep in mind that writing style often varies in different portions of a report, article, or chapter. You may wish to select a sample from the main body of a work and then choose another from a more general section—the introduction or conclusion, for instance.

2. Number each sentence consecutively for convenience of reference.

Sentence Analysis (see Figure 22–1)

1. Count and record the number of words in each sentence. Count as a word any set of alphanumeric characters bounded by spaces. Use a style analysis worksheet like the one shown in Figure 22–2 to record your counts and calculations.

NO. OF WORDS

32 (1)One might try, out of a spirit of neatness, to divide the atom clean-
ly between the two major physical sciences, awarding the electrons to the
chemist and the nucleus to the physicist.
19 (2)To attempt such a cleancut division would, however, be a violation of
40 the spirit of science, which is all-one-piece. (3)The structure of the nucleus,
however remote that might seem from the world of ordinary chemical reactions,
must nevertheless be of keen interest to the chemist, if only because of its
effect upon that fundamental chemical datum, the atomic weight.
19 (4)By the time the nineteenth century had come to its end, the matter of
31 the atomic weight seemed settled. (5)Each element had a characteristic atomic
weight, chemists believed, and the only future in that respect was the
greater precision with which the fourth and fifth decimal places might be
determined.
19 (6)Then came the discovery of isotopes and all that had seemed certain
28 about atomic weights immediately went into discard. (7)The notion, dating from
Dalton, that all atoms of a single element possessed identical mass and
16 that the atomic weight expressed this mass was seen to be false. (8)Instead,
most elements were made up of two or more varieties of atoms differing in
14 mass. (9)The atomic weight was merely the weighted average of the masses of
these isotopes.
48 (10)If the term "atomic weight" is reserved for this weighted average of
the isotope mass as found in their natural distribution within an element,
then one ought not to speak of the atomic weight of an individual isotope
22 (as I have been doing so far in this book). (11)It is better to use a different
288 phrase and speak of the relative mass of an individual isotope as its <u>mass
number</u>.

Avg. Words per Sentence: 26.2 Sentence Length Ratio: 3.4

Th-Word Percentage: 36.7

Figure 22–1 Sample Passage with Sentence Analysis Completed (from Isaac Asimov, *Understanding Physics: The Electron, Proton, and Neutron* [New York: New American Library, 1969], pp. 145–146.)

Sentence No.	No. of Words	No. of Clauses	Th-Word Beginning
1.			
2.			
3.			
4.			
5.			
6.			
7.			
8.			
9.			
10.			
11.			
12.			

Totals

Plus Items
1st & 2nd Pers. Pronouns:
Names of People:
Sentence Adverbials:
Subord. Conjunctions:
Nonfinite Clauses:

Minus Items
Passive Voice:
Indirect Beginnings:
Stacked Modifiers:
Jargon or Pompous Phrases:
Abstract Nouns:

Calculations
Avg. Words per Sentence:
Sentence Length Ratio
Th-Word Percentage:

Avg. Words per Paragraph:
Avg. Clauses per Sentence:
+ Items per 100 Words:
− Items per 100 Words:

Figure 22–2 Style Analysis Worksheet

2. Calculate and record the average sentence length. Mature prose averages 18 to 20 words per sentence. If your writing averages much less than this (and especially if your average number of clauses per sentence is low), you can improve your writing style by generating a greater number of longish, multiclause sentences. If your average sentence length is 12 or under, you are writing in what is called the "primer style" or the "Dick and Jane book style." This is a serious shortcoming in writing style and calls for practice on your part in combining one-clause sentences into longer ones. See "Sentences" in Chapter 18 and "Combining Clauses into Longer Sentences" in Chapter 20. If your average sentence length is 25 words or more, you should consider breaking up some of your longer sentences.

3. Calculate and record the sentence length ratio. Sentence length ratio is a rough measure of the range in variation of sentence length.

Divide the number of words in the longest sentence in the passage by the number of words in the shortest sentence. Your sentence length ratio should be in the 5.0- to-8.0 range. If it is down around the 2.0-to-3.0 range, you can improve your style by generating more short, one-clause sentences and more 35- to 40-word, multiclause sentences as well.

4. Calculate and record the "*Th-* word" beginning percentage. Count the number of sentences in the passage beginning with *the, this, that, these,* or *those*. Express this count as a percentage of the total. Your "*Th-* word" percentage should be in the 20 percent to 25 percent range. If it is more than this—especially if it is over 50 percent—you can improve your style by varying your sentence beginnings. (While you are looking for "*Th-* word" beginnings, note any other oft-repeated beginnings as well.)

Paragraph Analysis (see Figure 22–3, page 466)

1. Calculate and record the average number of words per paragraph. Count as paragraphs units punctuated as paragraphs. Include vertical lists as parts of paragraphs. Well-developed technical prose averages 85 to 90 words per paragraph (exclusive of short introductory, summary, or transitional paragraphs). If your average is significantly less, you can improve your style by developing your paragraphs more fully (see "Paragraph and Section Development" in Chapter 19).

2. Isolate and identify each functional paragraph in the sample. Disregard paragraph punctuation and concentrate on function. Identify each functional paragraph as (a) introductory, (b) main, (c) summary, or (d) transitional. There are no calculations to be made based on this distribution; the count is for your information only.

Clause Analysis (see Figure 22–4, page 467)

Note: This portion of the style analysis procedure presupposes the ability to recognize clauses of all types—independent, conjoined, embedded, nonfinite, and elliptical. If you're not comfortable with clause analysis yet, you may wish to skip this for now and go on with "Plus-Item" Analysis and "Minus-Item" Analysis.

1. Count and record the number of clauses in each sentence.

A. Underline (lightly in pencil) all of the finite main verbs in each sentence. Count auxiliary verbs (*be, have, can, could, may, should,* etc.) as parts of the main verbs.

B. Underline (lightly in pencil) all participles (*-ing* verbs and *-en* verbs) and all infinitives (*to* verbs). Count each as a clause.

C. Identify all instances of two (or more) conjoined subjects, conjoined predicates, or conjoined adjuncts. Count each as two (or more) clauses.

NO. OF WORDS

32 (1) One might try, out of a spirit of neatness, to divide the atom cleanly between the two major physical sciences, awarding the electrons to the chemist and the nucleus to the physicist. INTRO

59 (2) To attempt such a cleancut division would, however, be a violation of the spirit of science, which is all-one-piece. The structure of the nucleus, however remote that might seem from the world of ordinary chemical reactions, must nevertheless be of keen interest to the chemist, if only because of its effect upon that fundamental chemical datum, the atomic weight. MAIN

50 (3) By the time the nineteenth century had come to its end, the matter of the atomic weight seemed settled. Each element had a characteristic atomic weight, chemists believed, and the only future in that respect was the greater precision with which the fourth and fifth decimal places might be determined. MAIN

77 (4) Then came the discovery of isotopes and all that had seemed certain about atomic weights immediately went into discard. The notion, dating from Dalton, that all atoms of a single element possessed identical mass and that the atomic weight expressed this mass was seen to be false. Instead, most elements were made up of two or more varieties of atoms differing in mass. The atomic weight was merely the weighted average of the masses of these isotopes. MAIN

70 (5) If the term "atomic weight" is reserved for this weighted average of the isotope mass as found in their natural distribution within an element, then one ought not to speak of the atomic weight of an individual isotope (as I have been doing so far in this book). It is better to use a different phrase and speak of the relative mass of an individual isotope as its <u>mass number</u>. MAIN

Avg. Words per Paragraph: 57.6

Paragraph Types:

Introductory: I
Main: IIII
Transitional:
Summary:

Figure 22–3 Sample Passage with Paragraph Analysis Completed

D. Identify all other elliptical clauses—absolute constructions or "sentence fragments." Count each of these as a clause.

Note: A and B together will yield 60 to 70 percent of the clauses in most sample texts. C will yield most of the remaining clauses. In most technical writing, the clauses counted in D will be less than 2 percent of the total.

2. Calculate and record the average number of clauses per sentence. Excellent technical prose averages 3.5 to 4.5 clauses per sentence, when we include nonfinite clauses and elliptical clauses. If your average

NO. OF CLAUSES

4 (1) One might try, out of a spirit of neatness, to divide the atom clean-
ly between the two major physical sciences, awarding the electrons to the conj.
chemist and the nucleus to the physicist.
3 (2) To attempt such a cleancut division would, however, be a violation of
2 the spirit of science, which is all-one-piece. (3) The structure of the nucleus,
however remote that might seem from the world of ordinary chemical reactions,
must nevertheless be of keen interest to the chemist, if only because of its
effect upon that fundamental chemical datum, the atomic weight.
2 (4) By the time the nineteenth century had come to its end, the matter of
5 the atomic weight seemed settled. (5) Each element had a characteristic atomic
weight, chemists believed, and the only future in that respect was the
greater precision with which the fourth and fifth decimal places might be conj.
determined.
3 (6) Then came the discovery of isotopes and all that had seemed certain
5 about atomic weights immediately went into discard. (7) The notion, dating from
Dalton, that all atoms of a single element possessed identical mass and
2 that the atomic weight expressed this mass was seen to be false. (8) Instead,
most elements were made up of two or more varieties of atoms differing in
1 mass. (9) The atomic weight was merely the weighted average of the masses of
these isotopes.
5 (10) If the term "atomic weight" is reserved for this weighted average of
the isotope mass as found in their natural distribution within an element,
then one ought not to speak of the atomic weight of an individual isotope
3 (as I have been doing so far in this book). (11) It is better to use a different
phrase and speak of the relative mass of an individual isotope as its mass
number.

Avg. Clauses per Sentence: 3.2

No. of Indirect Beginnings: 1

Level of Emphasis:

First: |
Second: 卌 卌 ||
Third: 卌
Fourth: 卌 卌 卌 ||

Figure 22–4 Sample Passage with Clause Analysis Completed

number of clauses per sentence is between 1.0 and 3.0 (you must have at least *one* clause per sentence), you can improve your writing style by generating more multiclause sentences—particularly nonfinite and elliptical clauses. If your average is 5.0 or more, you ought to consider simplifying your sentence structure somewhat.

3. Count and record the number of clauses at each level of emphasis. See "Relative Emphasis of Clause Patterns" in Chapter 20. There are no calculations to be made based on this distribution; the count is for your information only.

4. Count and record the number of clauses with indirect beginnings. These are clauses that begin with "it is" or "there are." Record all indirect beginnings under "Minus Item" Analysis, below.

"Plus-Item" Analysis (see Figure 22–5)

1. Underline, count, and record each of the following:
a. Sentence adverbials

One might try, out of a spirit of neatness, to divide the atom cleanly between the two major physical sciences, awarding the electrons to the chemist and the nucleus to the physicist.

To attempt such a cleancut division would, however, be a violation of the spirit of science, which is all-one-piece. The structure of the nucleus, however remote that might seem from the world of ordinary chemical reactions, must nevertheless be of keen interest to the chemist, if only because of its effect upon that fundamental chemical datum, the atomic weight.

By the time the nineteenth century had come to its end, the matter of the atomic weight seemed settled. Each element had a characteristic atomic weight, chemists believed, and the only future in that respect was the greater precision with which the fourth and fifth decimal places might be determined.

Then came the discovery of isotopes and all that had seemed certain about atomic weights immediately went into discard. The notion, dating from Dalton, that all atoms of a single element possessed identical mass and that the atomic weight expressed this mass was seen to be false. Instead, most elements were made up of two or more varieties of atoms differing in mass. The atomic weight was merely the weighted average of the masses of these isotopes.

If the term "atomic weight" is reserved for this weighted average of the isotope mass as found in their natural distribution within an element, then one ought not to speak of the atomic weight of an individual isotope (as I have been doing so far in this book). It is better to use a different phrase and speak of the relative mass of an individual isotope as its mass number.

Plus Items		Minus Items	
1st & 2nd Pers. Pronouns:	I	Passive Voice:	II
Names of People:	I	Indirect Beginnings:	I
Sentence Adverbials:	III	Stacked Modifiers:	—
Subord. Conjunctions:	~~IIII~~ I	Jargon or Pompous Phrases:	—
Nonfinite Clauses:	~~IIII~~ IIII	Abstract Nouns:	~~IIII~~ III
Plus Items per 100 Words:	6.9	Minus Items per 100 Words:	3.8

Figure 22–5 Sample Passage with Plus-and-Minus-Item Analysis Completed

b. Subordinating conjunctions
c. Nonfinite verbs
d. First- and second-person pronouns
e. Names of people.

Sentence adverbials and subordinating conjunctions give you an indication of cohesion present in the sample. Nonfinite verbs (as heads of clauses) indicate a fair level of sentence sophistication. The presence of first- and second-person pronouns and names of people indicate a degree of personalness in tone and approach.

2. Calculate the number of "plus items" per 100 words. The "plus item" percentage is a measure of the extent to which you are "accentuating the positive" with your writing style. A "plus item" count of 7.0 to 10.0 per hundred words is pretty good. If you fall short of this, you can improve your writing style by incorporating more of these "plus items" into your writing.

"Minus-Item" Analysis (see Figure 22–5)

1. Circle, count, and record each of the following:
a. Instances of passive voice
b. Abstract nouns derived from verbs or modifiers
c. Instances of three or more pre-noun modifiers
d. Clauses that begin with indirect constructions
e. Level-one technical terms and instances of jargon
f. Pompous words or phrases

Unwarranted passive voice and abstract nouns are the two biggest causes of dullness and indirectness in the writing of scientific and technical people. Count *all* instances of passive voice and abstract nouns—don't stop to judge whether or not they are warranted. A few instances of either passive voice or abstract nouns won't hurt your style evaluation that much, and if you have a good many of either, chances are most of them are unwarranted. Stacked pre-noun modifiers constipate your writing. Indirect beginnings are, of course, indirect, but they also affect your clause emphasis. Technical terms, jargon, and pompous words and phrases are somewhat judgmental. Moreover, level-one technical terms and some jargon may be warranted in certain communication contexts. Count them all, though, and be critical rather than defensive in your judgments. Your goal is to improve your writing style—not to justify the style you entered the class with.

2. Calculate the number of "minus items" per 100 words. The "minus item" percentage is a measure of the extent to which are you "eliminating the negative" (or accentuating the negative, as it were) with your writing style. A "minus item" count of less than 3.0 per hundred words is acceptable. If your count is much higher than this, you can improve your writing style by learning to control your use of these "minus items" in your writing.

□ Comments on the Style Analysis Procedure

Over the years, a good many formulas and yardsticks have been developed to assess "readability" or "communicative effectiveness" or "style." All of them are useful; none of them is perfect. The procedure I have outlined here is no exception. It fails to deal at all with some very important aspects of excellent writing style—thematic unity, for instance, or precision in word choice, or patterns of known and new information, or general smoothness and grace in writing, or appropriateness for given audiences and purposes. Moreover, it is quite cumbersome to apply—compared to those that deal only with word length and sentence length.

On the other hand, it has the value as a tool to help you improve your writing style. The chief value lies in encouraging you to look critically at aspects of style in technical writing—both your writing and the writing of others. As you become increasingly aware of the elements of style, you will become more sensitive to those characteristics of writing that account for (or at least correlate with) judgments of excellence. And as you look in detail at paragraphs, sentences, and words and focus on the various "plus items" and "minus items" listed here, you will begin to see other elements and patterns of style—good or bad—that you may wish to incorporate in or eliminate from your own writing.

□ Summary of the Style Analysis Procedure

This summary of the procedure covers the same ground as above but without the notes and comments. Also, the order of steps is changed somewhat to make for quicker counting and calculating. Refer to Figure 22–6 for a completed style analysis sheet.

1. Select a 200- to 250-word sample of the writing you wish to analyze.
2. Count and record the number of words in each sentence.
3. Record each sentence that begins with a "*Th*- word."
4. *Underline*, count, and record each of the following "plus items":
   a. Sentence adverbials
   b. Subordinating conjunctions
   c. Nonfinite verbs
   d. First- and second-person pronouns
   e. Names of people
5. *Circle*, count, and record each of the following "minus items":
   a. Instances of passive voice
   b. Abstract nouns derived from verbs or modifiers

Sentence No.	No. of Words	No. of Clauses	Th-Word Beginning
1.	32	4	
2.	19	3	
3.	40	2	X
4.	19	2	
5.	31	5	
6.	19	3	
7.	28	5	X
8.	16	2	
9.	14	1	X
10.	48	5	
11.	22	3	
12.			
Totals	288	35	3

Plus Items		*Minus Items*	
1st & 2nd Pers. Pronouns:	I	Passive Voice:	II
Names of People:	I	Indirect Beginnings:	I
Sentence Adverbials:	III	Stacked Modifiers:	—
Subord. Conjunctions:	~~IIII~~ I	Jargon or Pompous Phrases:	—
Nonfinite Clauses:	~~IIII~~ IIII	Abstract Nouns:	~~IIII~~ III
	20		11

Calculations			
		Avg. Words per Paragraph:	
Avg. Words per Sentence:	26.2	Avg. Clauses per Sentence:	3.2
Sentence Length Ratio	3.4	+ Items per 100 Words:	6.9
Th-Word Percentage:	36.7	– Items per 100 Words:	3.8

Figure 22–6 Style Analysis Worksheet Completed for Sample Passage

c. Instances of three or more pre-noun modifiers
d. Clauses that begin with indirect constructions
e. Level-one technical terms and instances of jargon
f. Pompous words or phrases

6. Make a quick check: do you have more circles than underlines?
7. Calculate and record your average sentence length.
8. Calculate and record your sentence length ratio.
9. Calculate and record your "Th- word" beginning percentage.
10. Calculate and record your average number of words per paragraph.
11. (Optional) Isolate and identify each functional paragraph in the sample.
12. Calculate and record the number of "plus items" per 100 words.
13. Calculate and record the number of "minus items" per 100 words.
14. Count and record the number of clauses in each sentence.

15. Calculate and record your average number of clauses per sentence.
16. (Optional) Count and record the number of clauses at each level of emphasis.
17. Compare your results with these benchmarks for excellence:

Average sentence length	18 to 20 words
Sentence length ratio	5.0 to 8.0
"*Th*- word" beginning ratio	20% to 25%
Average number of words per paragraph	85 to 95 words
Number of "plus items" per 100 words	7.0 to 10.0
Number of "minus items" per 100 words	Less than 3.0
Average number of clauses per sentence	3.5 to 4.5

Writing Exercises

1. Team up with four or five classmates and attempt to establish a corpus of excellent writing and a corpus of (minimally) adequate writing. Exchange the examples of excellent and adequate writing that each of you has been collecting throughout the semester. Identify those that *all* of you judge to be excellent and those that *none* of you judges to be excellent. If you can't agree, identify the five best and five worst of the examples you have.
2. Pick two examples and have *each* member of the group analyze the style in each of the two examples. Compare results. If you are in essential agreement, proceed with the analysis of all of your examples. Do you find significant differences in various categories of analysis between your excellent examples and your adequate examples? Write up your results in a report for the class.

Writing about Writing

1. Analyze three of your earlier papers for aspects of writing style. Summarize your results in a table or chart. Does your analysis suggest ways in which you can improve your writing style? Write up a set of recommendations to yourself for improving your writing style.
2. To what extent is it possible to acquire habits of excellence in writing so that you can turn out excellent prose on your first draft? To what extent is excellence possible only through arduous revision? Write an essay exploring these questions.
3. Review the school writing that you have done for other college classes. Identify several situations where adequate writing is good enough. Identify several situations where excellent writing would have served you better. All in all, do you feel excellent writing is worth the time it takes? Gather your ideas on this topic into an essay.
4. Write an essay on the topic "There are algorithms for adequacy but only heuristics for excellence."
5. Identify what you feel to be the single most important aspect of style as it relates to excellence in technical writing. Write an essay justifying your selection, aiming your essay at your classmates.

□ Index □

Page references to tables and illustrations are in *italic* type.